How Deep is the Ocean?

THE UNIVERSITY OF WINNIPEG LIBRARY

THE ASHDOWN COLLECTION OF CANADIANA

THIS SPECIAL COLLECTION
OF
CANADIAN HISTORY AND LITERATURE
COMMEMORATES THE NAME
OF

HARRY C. ASHDOWN, 1886–1971

A DISTINGUISHED MEMBER OF
THE BOARD OF REGENTS FROM 1929 TO 1967
WHOSE GENEROUS BEQUEST
TO THE UNIVERSITY LIBRARY
PROVIDED THE INCENTIVE FOR
THE ESTABLISHMENT OF THE COLLECTION

How Deep
is the Ocean?

Historical Essays on Canada's Atlantic Fishery

Edited by James E. Candow and Carol Corbin

University College of Cape Breton Press

The University College of Cape Breton acknowledges the support received for its publishing program from the Canada Council's Block Grants program.

Cover design by Weldon Bona
Cover photos by William W. Warner
Map inside cover by Barry Gabriel
Book design by Gail MacEachern
Printed and bound in Canada by City Printers, Sydney, NS.

Canadian Cataloguing in Publication data
 How deep is the ocean?
ISBN 0-920336-86-8

Includes bibliographical references and index.

1. Fisheries - Atlantic Provinces - History
I. Candow, James E., 1954 -
II. Corbin, Carol, 1951 -

SH223.568 1996   639.2'09715   C96-950104-8

University College of Cape Breton Press
PO Box 5300
Sydney, NS
CANADA B1P 6L2

CONTENTS

Twentieth-Century Fishery

ACKNOWLEDGEMENTS

We acknowledge the Louisbourg Institute, an educational partnership between the Fortress of Louisbourg National Historic Site and the University College of Cape Breton, and the University College of Cape Breton Press. We wish to thank William D. Naftel, William A. O'Shea, Robert Morgan, and Silver Donald Cameron for their encouragement and support. In particular, we are grateful to Cathy Brousseau and Gail MacEachern for their invaluable help and commitment to the project.

FOREWORD

Leslie Harris

Although the "testimony of the spade" is weak and unsatisfactory for the very early pre-historic era in Atlantic Canada, we may assume that the earliest humans to come to our eastern shores took from the bountiful sea some part, at least, of their sustenance. While the proof to sustain such an hypothesis may well have been submerged in the process of coastal flooding, or otherwise lost, we can hardly imagine a culture that would have ignored the teeming anadromous species of estuaries and rivers, the rich molluscan and crustaceous resources of the intertidal zone, and the possibilities represented in the ubiquitous colonies of sea birds and the herds of marine mammals whose annual migrations brought them within range of the hunters' weapons.

Certainly, by the time the archaeological record becomes somewhat more clear, as in the case of the Maritime Archaic tradition and the subsequent Dorset culture, numerous clues point to an intimate association with, perhaps a profound reliance upon, the resources of the sea. Indeed, we might well ask what else but the abundance of marine species of fish, flesh, and fowl would have brought a people, some 8000 years ago, to settle at the rocky edge of the ocean amid the chill fogs and drifting floes that yearly choke that veritable funnel we now name the Strait of Belle Isle.

In any case, when the Vikings came, they too sought to establish their settlement on the shores of that same strait; although they were, after some few years, repulsed by ancestors of modern day Inuit into whose territories they had intruded. Five hundred years later, at the end of the 15th century, other Europeans came, seeking the fabled Northwest Passage that would lead them to the magical Indies where, they believed, untold wealth of spices, gold, and gems awaited the intrepid discoverer. Even after the dreams of instant wealth had faded, adventurous thirsters after knowledge and fame continued through the next three centuries and more to seek passage through the great northern ocean.

But for a great many others, their course had been stayed; not physically, despite the propaganda of the first Cabot voyage that described dense schools of fish that impeded the progress of ships; but by a promise more immediate than that of the elusive wealth of the Orient. For here, if I may quote myself, though there

> was neither the glitter of gold, nor the heady aroma of costly spices...[there] was a field of exploitation, sustenance and livelihood, a staple for commercial enterprise, a prospect of profit. Here was abundant life giving promise of life abundant. Here were mines that would not be exhausted; a treasure, if a prosaic one, that would bear time's stress and still retain its perennial newness contingent only upon relative geologic and climatic stability, and [as

subsequently became apparent] upon man's ability to tame his own rapacious greed. [1]

And, indeed, long after the gold-bearing quartz of Hispaniola had been mined to exhaustion, long after the accumulated wealth of cuzco had disappeared, long after the fabulous sugar loaf mountain of silver at Potosi had ceased to be anything but a splendid memory, the Labrador current still bore its burden of nutrients; the great whale and seal herds still passed and repassed the shores of vinland; the vast schools of caplin still came to shore and the cod still gave pursuit. And still, after 500 years, those newcomers, who had made fishing their avocation and the continental shelf their field of endeavour, clung to the edge of the sea, spurning the continent behind them and facing an untameable mistress who demanded constant devotion, even when her face of smiling benignity was turned in an instant to a mask of cruel ferocity.

Setting aside the Viking interlude, the first Europeans to come were French and Spanish Basques, Normans and Bretons, English from the southwestern counties, and Portuguese. Although David Quinn finds evidence for fishing expeditions across the western ocean from Bristol, perhaps to the territories soon to become the "New Found Lands," as early as 1481, it is most likely that Bretons and Basques were the more active entrepreneurs in the early years of the 16th century. In any event, by the middle of that century very substantial numbers of vessels, Spanish, French, English, and Portuguese, constituted a migratory fishery that tapped the virgin resources of the northwest Atlantic.

The magnets that attracted the essential venture capital and discounted the ever present perils of the stormy north Atlantic, of unknown lands and largely uncharted coasts, included the hitherto unexploited fish stocks of the great offshore banks that are such significant features of the Canadian continental shelf; the vast schools of fish that in season made their way to shallow coastal waters; the herds of mighty whales, and, in particular, the Right Whale, whose unsuspicious nature coupled with slow and ponderous movements made it the easiest of targets for Basque harpooners; the vast colonies of sea birds and, in particular, of great auks whose flightlessness ensured their butchery. In the 18th century, the exploitation of the herds of migrating harp seals provided an economic basis, in Newfoundland, at least, for an over-wintering population of crypto-colonists.

From the beginning of the 17th century, it had been clear that among the European powers with an interest in the northwest Atlantic, the struggle for dominance was essentially between England and France. The contest centred upon control of access to the rich fishing grounds, both coastal and offshore, and control of territorial bases for fishing operations. True, there was on occasion at play a strategic imperative related to larger imperial issues, and we may call to witness the great fortress at Louisbourg and the British answer to Louisbourg at Halifax. But it is not without significance either that the last battle of the Seven Years' War in North America was fought on Signal Hill in St. John's, or, that in negotiations leading to the Treaty of Paris, the irreducible French territorial demand was for St. Pierre et Miquelon or some other suitable base for an offshore fishery, together with fishing rights along considerable stretches of the Newfoundland coastline.

Even though, as Gillian Cell points out, the English had early decided that Newfoundland could be fully exploited without settlement, population nevertheless grew. The process was ribbon development run riot; as, seeking the closest possible access to fish, communities sprang up on 6,000 headlands and offshore islands along some 10,000 miles of coast. There had been attempts at formally planned colonization in the early-17th century, but all those had foundered on the hard realities of thin soils, exposed rock, harsh climate and a comparatively low value staple product whose worth in the market was a function of capricious natural variability and, as well, of fierce international competition. What emerged from the free-for-all that followed was an unplanned, unwanted colony, unrecognized as such until the 1820s, whose chaotic demography has been a nightmare for governments ever since.

But the people who comprised hundreds of isolated little anarchies, clung like proverbial limpets to the rocks they had chosen as home. Their attachments grew stronger with time and, in due course, the emergent lifestyle, to once more quote myself,

> became [itself] the object of passionate adherence . . . [informing] a political consciousness that . . . in recent decades, has witnessed a virtual renaissance. The personal and social values of hardihood, of endurance, of survival, of resilience, of neighbourliness, of sharing, of community and of independence have become the immutable elements of a mystique that might almost be described as the cult of the unique and distinctive Newfoundland.

In the more favoured lands to the south and west that now comprise the Maritime Provinces, there was an early recognition of a somewhat more diversified economy predicated upon greater possibilities for agriculture and exploitation of forest resources. The images that spring to mind when we speak of "the Garden of the Gulf," or of "the beautiful Annapolis Valley" are not those of rocky sea-smashed coasts. And when Bliss Carman, evoking images of his native New Brunswick, writes:

> Fair the land lies, full of August,
> Meadowy island, shingly bar,
> Open barns and breezy twilight,
> Peace and the mild evening star;

the mental picture we get is one of pastoral serenity.

But while timber, shipbuilding, and international trade became strong forces in the Maritime economy, and while agriculture played a much more significant role than in Newfoundland, the fisheries were dominant in the early days. And if the processes of settlement and development were somewhat less anarchic, the demographic pattern that emerged reflected, as it did in Newfoundland, an economic dependence upon fisheries and a relatively primitive technology that demanded close access to the resource. Look to Cape Breton, to the South Shore of Nova Scotia, to the shores of Fundy, to New Brunswick's North Shore, to the Gaspé, or to the North Shore of the Gulf of St. Lawrence and the pattern will be plain.

It is not by chance that the most widely recognized symbol of Nova Scotia is the beautiful Bluenose, exemplifying a class of schooners, superb creations of the shipbuilder's art, that sailed from Atlantic ports, skippered by "captains courageous," crewed by intrepid dorymen, who exercised their business in the great waters of the Scotian Shelf, and all of the banks from Georges to the Grand Bank of Newfoundland.

It is true that the stories of New Brunswick, of Prince Edward Island, and of the Gaspé are not redolent of the sea to the same extent as is the case for Nova Scotia and for Newfoundland. Nevertheless, fisheries had still their profound effects upon early settlement and development, as is made clear in the papers by Brière, Samson, and Andrew included in this volume. Indeed, they continue to this day to be of no inconsiderable consequence.

To make this point abundantly clear we may note the significance of fisheries to contemporary provincial economies. If we examine the role of fisheries among all commodity-producing industries in the Atlantic region, they represent 0.3% of the Quebec total, 5% for New Brunswick, 15.8% and 15.9% for Nova Scotia and Prince Edward Island respectively, and 20% for Newfoundland. More significantly still, the fisheries contribution to jobs in goods-producing industries is 1% in Quebec, 16.2% in New Brunswick, 23.3% in Nova Scotia, 33.3% in Prince Edward Island, and no less than 72% in Newfoundland. Nor should we fail to observe that while the numbers given for Quebec are quite small, when we look at them in the context of the regional economies of Gaspé and the Gulf of St. Lawrence North Shore, they become much more significant fractions.

The case is clear. For all of us who live beside the Atlantic, the fisheries were, in the vast majority of cases, the magnet that attracted our ancestors to these shores. In many cases communities had no other reason for being and, indeed, no other reason for continuing to exist. For many, fisheries represent the past, the present, and even the future. For without the economic base that a fishery represents, they will become moribund and eventually will die.

Nor should we fail to recognize that the magnet that drew our progenitors across the Atlantic still exerts its powerful attraction upon fishers, not only from European ports, but from the "four corners and the seven seas." They come, armed with the most advanced technology, to pursue steadily declining populations of fishes and other marine animals. Although their activities are often alarming to us, and sometimes appear to represent commercial greed, they are also a response to an apparently growing world appetite for protein from the oceans.

It will be apparent that the potential for conflict between coastal states, to whom the fish stocks are contiguous, and distant water fishers, from far away lands, is very real and very ponderable. And such is also the case as between major corporations deploying great factory freezer trawlers with their immense fish-killing capacity, on the one hand, and the coastal community small boat fishers and entrepreneurs on the other. This in turn implies significant policy issues involving the interests of a multiplicity of stakeholders and ranging from the local to the international in scope. It involves, as well, the much broader issue of the integrity of the ecosystem in which humans, as the ultimate predators, function; and it speaks to the moral awakening of human fishers to their responsibility for the conservation of species and the preservation of species diversity.

Certainly, our fisheries are not merely an artefact of historical scholarship, of interest only to dry-as-dust historians. Clearly, they are a major component of any conceivable future for a large number of eastern Canadians. But they have an intrinsic value far above the immediate and the economic. For it is the fisheries that, to a very large extent, formed us and made us what we are. That which we call our culture and our heritage derives in no small measure from our intimate association with the sea through centuries past. Our lifestyles, our community and family values, our society, our art, our music, our letters, all that we are, indeed, have been shaped to one degree or another, by that association. Like all humans, we carry the sea in our blood and in every cell of our bodies. But our affiliation goes further, for we have had the inestimable privilege of living by and drawing our sustenance from one of Canada's three great oceans.

From these discursive, sometimes trite, and altogether inadequate comments, it has been my intention to establish a perspective from which the essays in this volume may be viewed. These run the gamut from pre-European contact archaeology to policy issues of the 1990s. They throw the floodlight of sound research upon fascinating areas of the past;

they expose critical issues of the here and now; and they offer glimpses of the future. But, of course, they merely scratch the surface. Although they add an important volume to the steadily growing library of materials on the northwest Atlantic fisheries, they open more doors than they close. They suggest broad fields of knowledge yet untilled.

Perhaps they will invite the reader to reflect upon what it means to be people of the sea; perhaps to reflect upon the awesome responsibility of protecting our oceans and conserving the creatures living within it, in face of humankind's demonstrated capacity to pollute and to destroy. For we can no longer say, with Byron:

Roll on! thou dark and deep blue Ocean, roll;
Ten thousand fleets sweep over thee in vain,
Man marks thy shores with ruin,
His control stops with the shore....

Perhaps it is fair to say that our *control* stops with the shore, but assuredly the inference that our capacity to ruin stops there is a total misstatement of the facts.

Read these essays with pleasure, therefore. Read them to become more aware of who we Atlantic Canadians are. Read them to acquire insights into current social, economic, and environmental problems that confront us. Read them with a view to taking a stakeholder's position on policy issues that must be addressed. And, read them to acquire knowledge that is the foundation of wisdom.

Notes
1. Leslie Harris, "The Owl of Minerva," in *Early European Settlement and Exploitation in Atlantic Canada: Selected Papers,* ed. G. M. Story (St. John's: Memorial University of Newfoundland, 1982), p. 4.

INTRODUCTION

James E. Candow and Carol Corbin

When we ask "How Deep is the Ocean?" it conjures up an array of other questions. In song it has been the metaphorical answer to "How much do I love you?"[1] Oceans, like mountains or skies full of stars, represent an eternity that, to the poet, cannot be completely comprehended or measured and certainly cannot be depleted by mere human efforts. In more practical terms the question suggests that we may finally have reached the point where once-seemingly infinite marine resources have been exploited beyond their capacity—a dire predicament that many scholars are exploring from different perspectives. In the last half of the 20th century earlier notions of boundless resources have given way to concerns about limits. With a sense of loss we read frequent reports of another resource gone, another species extinct. The great cornucopia of life on earth is now measurable, and thus, finite.

The question also frames a field of study. The close social and economic ties that bind people to the ocean make it an immense arena of scholarship in the disciplines of history, sociology, folklore, biology, and so forth. Many of these disciplines are represented in this book. And finally, "How Deep is the Ocean?" is a rhetorical question, generating ideas about the dialectics of plenty and famine, science and art, culture and nature, and producing more questions than answers.

So we begin with what we know with some certainty. Fish stocks always have ebbed and flowed, existing like other natural systems in cycles of varying abundance. Early fishery records suggest that some years the fish nearly jumped into the boats, while other years they were scarce enough to create severe hardships in fishing economies. But the crisis afflicting Canada's east coast fishery today is different: never before have major fish stocks been as low as they were in the 1990s.

The loss of a fishery as a major world protein source has repercussions far beyond Canadian shores. As early as 1985, four of the world's 19 principal fisheries were completely depleted, and another nine were "fully exploited."[2] One of the most serious problems that the loss of a fishery creates is an increasing inability to improve nutrition in developing nations.[3] In addition to creating global shortages of protein-based foods, the closure of a fishery jeopardizes local maritime societies that for centuries have depended on the sea for livelihoods. How does one retrain a 50-year-old fisher who has fished for 35 years and wants only to continue in the industry? The fishery crisis has placed Canada in an economic and social quandary that few politicians have the answers to.

Finally, the fishery crisis poses a question of morality. Is it our prerogative to wield this god-like power to destroy hundreds of different species on earth through technology and human ingenuity? Do we have the right to exert complete control over the natural world? Philosophers and environmentalists, scholars and fishers debate these questions at length, and the answers dwell both in our souls and in our deeds.

The essays in this collection were gathered at a critical time in the history of Canada's east coast fishery. Canada's fishery problems became international front page news in 1995 when a Canadian Coast Guard ship fired at a Spanish trawler on the high seas in a dispute over the diminishing turbot stock. Fishers in extraordinary numbers were out of work, receiving government assistance, and staging protests about government regulations. The future of many eastern Canadian fishing communities looked particularly grim.

The essays paint a picture, albeit incomplete, of the history of the northwest Atlantic fishery, which, as a field of study, appears to be limitless. One fisheries scholar, the late Keith Matthews, used to joke during his Newfoundland history lectures that everybody quoted Harold Innis but nobody ever read him. The joke was a testament to the influence that Innis's *The Cod Fisheries: The History of an International Economy* (originally published in 1940) had on fisheries historiography in the 1970s. Innis's shadow is still discernible today, and no one has yet written a synthesis to rival his. But as the present collection shows, there is a new breadth to scholarship on the northwest Atlantic fishery.

Some of our authors evidently have read Innis, for they have taken issue with him. Peter Pope disputes his chronology for the development of the French bank fishery, and Darlene Abreu-Ferreira charges that he grossly exaggerated the scale of the early-Portuguese fishery in the northwest Atlantic. Innis's emphasis on Eurocentric politics and economics is yielding to more varied approaches, many of which are reflected in this volume. These include aboriginal participation in the fishery (Davis), fisheries science and ecology (Baker and Ryan, Neis), fishing societies (Andrew, Boyd, Macdonald, Samson, Sweeny), and modernization and technology (Balcom, Blake, Candow, Warner, Wright). Still, several articles (Blake, Brière, Candow, Pope, Warner, Tulloch) emphasize the fishery's international dimensions, and in so doing show that Innis's vision of the industry as an international economy has stood the test of time.

The majority of the essays in this volume approach the subject from a Newfoundland perspective. Far from being a case of editorial bias, this merely reflects the state of current scholarship. Readers would be hard-pressed to find much secondary literature on the history of Prince Edward Island's fishery other than Kennedy Wells's useful overview,[4] and in Nova Scotia, Balcom has largely had the field to himself for two decades.[5] New Brunswick and Quebec are better served,[6] but Newfoundland leads the way.[7] This is probably a function of the fishery's role in Newfoundland, at least since the commencement of European exploitation, as the employer of last resort. Elsewhere, other economic opportunities existed that, historically speaking, have come only recently to Newfoundland. Still, as the present collection shows, fishing has also played an important role in other regions of eastern Canada since the origins of human habitation. We modestly hope that this volume will stimulate broader geographic and intellectual approaches to the subject.

Notes

1. Irving Berlin, "How Deep is the Ocean?" 1932.
2. Lester R. Brown, "Maintaining World Fisheries." *State of the World*. World Watch Institute. (New York: W. W. Norton, 1985), p. 76.
3. Brown, p. 77.
4. Kennedy Wells, *The Fishery of Prince Edward Island* (Charlottetown: Ragweed Press, 1986).

5. B. A. Balcom, *History of the Lunenburg Fishing Industry* (Lunenburg: Lunenburg Marine Museum Society, 1977) and *The Cod Fishery of Isle Royale, 1713-58* (Ottawa: Parks Canada, 1984). But see also *Emptying Their Nets: Small Capital and Rural Industrialization in the Nova Scotia Fishing Industry*, ed. Richard Apostle and Gene Barrett (Toronto: University of Toronto Press, 1992).

6. For examples, see Maurice Beaudin et Donald J. Savoie, *Les Défis de l'Industrie des Pêches au Nouveau-Brunswick* (Moncton: Les Editions d'Acadie, 1992); Rosemary Ommer, *From Outpost to Outport: A Structural Analysis of the Jersey-Gaspé Cod Fishery, 1767-1886* (Montreal and Kingston: McGill-Queen's University Press, 1991); Nicolas Landry, *Les Pêches dans la Péninsule acadienne, 1850-1900* (Moncton: Editions d'Acadie, 1995); Roch Samson, *Fishermen and Merchants in 19th Century Gaspé: The Fishermen-Dealers of William Hyman and Sons* (Ottawa: Parks Canada, 1984); Laurier Turgeon, "Le temps des pêches lointaines: permanences et transformations (vers 1500-vers 1850)," *Histoire des Pêches Maritimes en France*, ed. Michel Mollat (Toulouse: Privat, 1987), pp. 130-86; Laurier Turgeon, "Basque-Amerindian Trade in the Saint-Lawrence during the Sixteenth Century: New Documents, New Problems," *Man in the Northeast*, Vol. 40 (Autumn 1990), pp. 81-89

7. The literature is voluminous, but some of the more important books include Sean T. Cadigan, *Hope and Deception in Conception Bay: Merchant-Settler Relations in Newfoundland 1785-1855* (Toronto: University of Toronto Press, 1995); Gerald L. Pocius, *A Place to Belong: Community Order and Everyday Space in Calvert, Newfoundland* (Montreal and Kingston: McGill-Queen's University Press, 1991); W. Gordon Handcock, *"Soe longe as there comes noe women": Origins of English Settlement in Newfoundland* (St. John's: Breakwater, 1989); Shannon Ryan, *Fish Out of Water: The Newfoundland Saltfish Trade 1814-1914* (St. John's: Breakwater, 1986); David Alexander, *The Decay of Trade: An Economic History of the Newfoundland Saltfish Trade, 1935-1965* (St. John's: Memorial University of Newfoundland, Institute of Social and Economic Research, 1977); and *The Peopling of Newfoundland: Essays in Historical Geography*, ed. John J. Mannion (St. John's: Memorial University of Newfoundland, Institute of Social and Economic Research, 1977). Although not a monograph, the *Dictionary of Newfoundland English*, ed. G. M. Story, W. J. Kirwin, and J. D. A. Widdowson (Toronto: University of Toronto Press, 1982) is the single most important work yet published on Newfoundland life, including the fishery.

Early Fishery

Credit: National Archives of Canada

chapter one

ARCHAEOLOGICAL EVIDENCE FOR PRE-CONTACT FISHING IN THE MARITIMES

Stephen A. Davis

Introduction

To date there have been few attempts to provide generalized frameworks for the pre-contact history of the Maritime provinces,[1] and the major difficulty has been a lack of stratified sites and an absence of concentrated regional investigation. Although sites have been excavated covering the entire time period from the retreat of the last glacier to European contact, they do not represent a continuum. The pre-European contact history is therefore comprised of single sites or groups of sites representing various time periods. These are interrupted by breaks in the sequence that have little or no data to allow for discussing the pre-European contact history as a continuous sequence of events. The following discussion incorporates the known data and utilizes a broad period concept (Table: 1). Data are drawn from northern New England and Labrador that are represented by detailed cultural chronologies. The primary focus of this discussion is the subsistence strategies as they changed over time, with a special emphasis on fishing as one component of subsistence activities.

The Palaeo-Indian Period: 11,000 - 9,000 B.P.

The peopling of the study area began some 11,000 years ago after the retreat of the last glacial ice sheet. This period, known throughout North America as the Palaeo-Indian Period, is the oldest generally recognized and accepted stage of human cultural development within the study area. The diagnostic elements form a complex of chipped stone tools with "fluted" projectile points of the Clovis tradition.

The Debert Palaeo-Indian site in Nova Scotia is perhaps one of the most important sites of the Clovis tradition in northeastern North America. The site was excavated by George MacDonald who uncovered living areas with hearth features. In his analysis, MacDonald was able to define eleven assemblage concentrations containing a wide range of Palaeo-Indian tools including fluted points, awls, spokeshaves, hammerstones, anvils, and abraders. Charcoal was collected from the individual hearth areas and a series of radiocarbon dates were obtained giving an average of $10,600 \pm 47$ B.P. for the site.[2]

Although Debert remains the only extensively excavated site for this period, the distinctive Clovis-type fluted points have been found in eight other locations throughout the three Maritime provinces. These isolated finds are not associated with any other evidence of cultural activity, and it is assumed that they are spent projectiles that either missed an animal and became lost, or hit an animal that carried the point and died without being retrieved by the hunter.

Table 1

Cultural Chronology of the Maritime Provinces

YEARS AGO	PERIOD	TRADITION	SUBSISTENCE FOCUS
Present-300	Historic	English	cities
		French	towns
		Acadian	farming/small
		Mi'kmaq	manufacturing
		Maliseet	farming/small
		Passamaquoddy	manufacturing
			farming year round
		Basque	fishing
			seasonal
400 - 300	Proto-Historic	Mi'kmaq/Malise et/ Passamaquoddy	hunting/gathering
400 - 3,000	Ceramic	Maritime Woodland	hunting/gathering coastal, seasonal interior activities
3,000 - 7,500	Late Pre-Ceramic	Broad Point Shield Archaic Maritime Archaic Laurentian Archaic	coastal, south interior, Boreal forest coastal, north Interior, river and lakes
5,000 - 8,000	The Great Hiatus	Prehistory Unknown	
8000 - 11,000	Palaeo-Indian	Plano Clovis	big game hunters

In 1989, tree clearing near Debert produced evidence for another habitation site now known as Belmont 1. A detailed survey and a testing program have shown that there are four additional sites in close proximity.[3] Thus, in a relatively restricted area of approximately 500 square acres, Nova Scotia has six Palaeo-Indian sites. To date, archaeological efforts have located their boundaries and confirmed their association with Debert.

A recurrent problem faced by archaeologists working in the Maritimes is the lack of direct evidence for how people fed themselves. The type of evidence most sought after is the bones of the various species that were hunted. The soils of the Maritimes are not conducive to the preservation of organics, so it is only in rare circumstances that actual bone is discovered in a cultural context. Thus Debert and its associated sites have not produced any direct evidence as to what their inhabitants were exploiting. However, given the location of the sites and evidence from outside the Maritimes, it is believed that Palaeo-

Indians were specialized big game hunters. Various natural sciences, notably palynology, have been used to reconstruct the environment in the Debert area at the time of its occupation. If we know what types of vegetation existed, then we can project the most likely kinds of species that would have been available for exploitation. In the Maritime provinces the earliest inhabitants entered at a time when the environment was not unlike the current Canadian Barrens, that is, a tundra setting. Given this and the stone technology found at Debert, the most probable species hunted was the now extinct caribou.[4]

It also seems reasonable to surmise that these early inhabitants would not have overlooked the rich maritime biome. A number of authors have suggested that the large wintering herds of harp seals may have attracted them.[5] Archaeological evidence exists along the coasts of southern Labrador to indicate that Palaeo-Indians in this area exploited marine resources. Tuck and McGhee concluded from a series of sites that a well-defined seasonal round of subsistence activities began in this period and continued until European contact.[6]

Unfortunately, the same conclusion cannot be substantiated for the study area. Whereas coastal Labrador has kept pace with rising sea levels, the Maritime Provinces have been submerging for the past 6,000 years.[7] This effectively means that if a coastal adaptive pattern was established, the evidence for it is now under water. The submerged coastline is probably the reason for the lack of evidence for people in the Maritimes after Debert. The environment changed dramatically from tundra to dense boreal forests that would not have supported large terrestrial mammals, thus confining subsistence activities to marine resources. The sites occupied at this time would have been along the coasts which, as suggested, have been drowned by rising sea levels. An alternative explanation is that the area was abandoned for more favourable environments to the north and south. There are, however, some indications of transition/late Palaeo-Indian occupation. These are in the form of Plano tradition sites on the Gaspé Peninsula, with a number of Plano-type projectile points from private collections in New Brunswick and southern Nova Scotia that should date to the period from about 10,000 to 8,000 years ago.

The Tuck and McGhee continuity model for southern Labrador begins with a late Palaeo-Indian occupation identified by a distinctive triangular projectile point. They see this type of point as a natural progression from the earlier fluted point type. David Keenlyside has recorded 22 similar specimens in private collections from Prince Edward Island and the north shore of New Brunswick. He supports the continuity hypothesis of Tuck and McGhee, who date their sites between 9,000 and 8,000 years ago.[8]

An interesting aspect of the above evidence, and one that is pertinent to the central theme of this chapter, is that all three archaeologists associate this point type with a shift in resource utilization. That is, the earlier Palaeo-Indians relied primarily on large terrestrial species, whereas the makers of the triangular points depended more upon marine and estuarine resources. At present, such a change in subsistence behaviour cannot be proven with direct evidence from the study area, because the Palaeo-Indian sites in the Maritimes have not produced any evidence to suggest that fishing played a role in the food procurement activities of these first inhabitants.

The Great Hiatus: 8,000 - 5,000 B.P.

The end of the Palaeo-Indian period is signalled by changes in technology; notably, the peoples of the Maritimes stopped producing chipped stone fluted points and the lanceolate, pressure-flaked plano points. These were replaced by a variety of forms that share a general trait of a contracting stem that allows them to be fixed to a wooden shaft. The most diagnostic trait that appears is the addition of ground stone technology. Although no sites have been excavated for this time period, evidence south and north of the study area suggests that sites must exist somewhere in the Maritimes. It is for this reason that the era between the late Palaeo-Indian occupation to approximately 5,000 years ago has been described as the 'Great Hiatus'[9.]

This lack of direct evidence of inhabitants after Debert and the scattered hints for later developments have also been used to support the abandonment hypothesis discussed earlier. The question remains as to where regional archaeologists should look to fill in this gap in our knowledge. In recent years some rather tantalizing clues have provided a possible direction. The clues, in the form of ground stone specimens, have been recovered by unorthodox means: they were scraped off the bottom of the Bay of Fundy and Northumberland Strait by scallop draggers. Three specimens—two ulus and a large ground stone point—were recovered by draggers working off Digby Neck. Two more ulus were recovered by the same means, one near the Bliss Islands, on the New Brunswick side of the Bay of Fundy, and the other off northeastern Prince Edward Island.

When I first reported these instances I was met with some scepticism after I suggested that people must have been living on islands that have long since vanished. Others thought that these specimens represented material that was lost when canoes in which people were traveling accidentally capsized. In either event, the specimens are representative of artifacts recovered from archaeological contexts in Labrador dated to 6,500 years ago. Additionally, ground stone ulus are used in the processing of soft-tissue species such as fish and sea mammals. Once again, although scant, the study area provides evidence for human habitation in the area, although unfortunately these specimens lie submerged and probably are unobtainable.

The Late Pre-Ceramic Period: 5,000 - 3,000 B.P.

After the 'Great Hiatus,' archaeological evidence of people in the Maritimes increases to a level where a more detailed account can be presented. This time-period is referred to as the Late Pre-Ceramic and equates with what is known in other areas as the Late Archaic period. The concept of the "Archaic" is used elsewhere to distinguish groups of hunters and gatherers from "Woodland Period" horticulturalists. This rather dramatic shift in subsistence and its consequences, seen elsewhere, does not occur within the Maritimes. Agriculture was never practised in northern Maine or the Atlantic Provinces. The only significant cultural change is found in the presence or absence of ceramic vessels, the implications of which are discussed below. The designations Pre-Ceramic and Ceramic are not meant to imply a break in the cultural history, but are used here as convenient markers to divide a rather long episode in the pre-European contact history of the area.

Although the evidence for human societies in the late Pre-Ceramic period is abundant, it is confusing. To date no less than four broad cultural traditions have been identified, of which a number could be further subdivided into chronological phases. Because the archaeological record is far from complete, it seems inappropriate to attempt a definite division of this period beyond the four major traditions set out below. The Laurentian Tradition (5,000 - 4,000 B.P.) was first defined from sites located in upstate New York, Vermont, and portions of western Quebec and southern Ontario. As a member of the Pre-Ceramic traditions it is defined partially by the trait of lacking pottery. It is also characterized by a collection of ground stone implements utilized in a wood-working context, including adzes, axes, and gouges. The ground stone technology is also represented in the production of slate 'bayonets,' semi-lunar knives (ulus), bannerstones, and pecked and ground plummets.

The ground stone industry shows slight variation throughout the period, but the chipped stone is more diagnostic in differentiating various phases within the tradition. The presence of a number of these diagnostic bifaces in collections throughout the study area would suggest that the interior-defined "Laurentian" may be represented in the region more fully than previously realized.

Most of the research conducted outside of the study area has concluded that this tradition was adapted to deciduous or mixed forest environments. Subsistence was based on species such as deer, moose, and smaller game. Fishing was also an important pursuit with anadromous species playing a dominant role in the diet during their migration runs. In all probability vegetable foods as well as nuts and berries were collected during specific seasons.

The question of exploitation of marine resources remains debatable, as some writers feel that the "Laurentian" was strictly an interior-based tradition.[10] They present the second major tradition, the "Maritime Archaic" as the most likely coastal-oriented manifestation. The counter hypothesis argues for a complete seasonal round for the "Laurentian" including exploitation of specific marine resources.[11]

The Maritime Archaic tradition (7,500 - 3,000 B.P.), was defined from sites located north of the St. Lawrence River. The ground stone and chipped stone industries of this tradition bear remarkable similarities to those that defined the "Laurentian" tradition. The distinctive character of the "Maritime" tradition derives from a prominent bone industry. It is in this aspect of the tradition that the primary evidence for the marine orientation is seen.

The type site for this tradition, at Port au Choix, Newfoundland, is a burial location situated on a raised beach that, owing to high calcium content, had excellent bone preservation. The assemblage includes a variety of barbed bone harpoons, including a number of toggle-types, bone foreshafts, and lance points, a technology strongly oriented towards marine resources.[12]

The economic interpretation suggested at Port au Choix is based on the bone tool assemblage rather than food bone as would be expected at a habitation site. Tuck reports that a variety of species were utilized in the manufacture of bone implements. These include terrestrial mammals such as caribou, beaver, and bear along with sea mammals. Bird remains were also recovered; however, not all of them may have been used for food. The Maritime orientation is also represented by a collection of fishing implements, notably harpoons and barbed bone points, perhaps utilized in taking the Atlantic salmon, a plentiful species in the area.[13]

The existence of a separate tradition with a strong adaption to marine resources is evident at other sites north of the St. Lawrence. Tuck also suggests that a number of closely related sites can be identified along the coast of Maine, most notably the Turner Farm site.[14] This location is a stratified habitation site with a number of discrete occupations. As with the Port au Choix site, there is excellent bone preservation at Turner Farm.[15] The economy is similarly varied with land mammals including deer, moose, bear, and beaver. The food refuse also contained swordfish, cod, sturgeon, grey and harbour seals, walrus, sea mink, and various sea birds.

The evidence suggests that the Maritime Archaic peoples spent most of the year on or near the coast. It further suggests that for brief periods, probably in the winter months, they travelled inland to hunt deer, moose, or caribou. It was during these activities that they came in contact with the "Laurentian" people of southern New England (somewhere around the Maine-New Hampshire border) and this accounts for "the presence of rare Laurentian traits in southerly Maritime Archaic assemblages."[16]

The Maritime area has numerous examples of the ground stone and chipped stone industries from the Late Pre-Ceramic. Unfortunately, with the exception of the Cow Point burial site in central New Brunswick, no other site from this period has been excavated.[17] A second hindrance to the Laurentian/Maritime debate is the virtual absence of bone specimens from this time at sites in the Maritime Provinces. Again, the rise in sea levels may have obliterated many of the early coastal sites. However, the collections housed in area museums and private collections suggest that other interior locations remain to be excavated.

The third tradition that may or may not be represented within the late Pre-Ceramic period is the Shield Archaic (4,000 B.P.), named after the region where it has been defined— the boreal forest of the Canadian Shield. In contrast to the others, this tradition is represented by a technologically inferior chipped stone industry. Typically the three main classes of artifacts recovered for this tradition—projectile points, non-stemmed bifaces, and end scrapers—are produced from massive silicious deposits such as quartzite and rhyolite.[18]

The principal reason for questioning the existence of the tradition within the Maritimes is that to date only one site has been excavated, with one private collection being assigned to the tradition.[19] The excavated assemblage is from the Dead Man's Pool site in the north-central highland region of New Brunswick.[20] Sanger concludes that enough traits exist at this location to differentiate it from the "Laurentian" tradition. The lack of a ground stone industry, along with the high percentage of formed endscrapers, indicate a "Shield" rather than a "Laurentian" occupation. He further concluded that the subsistence pattern was largely oriented towards hunting the woodland caribou and taking the spawning Atlantic salmon.[21] These subsistence pursuits are typical of Shield populations found in the Canadian boreal forests.

A small collection was recovered from a ploughed field in the highlands of Cape Breton Island. The McEvoy site material fits well with other Shield Archaic assemblages in both the artifact class frequencies and the specific artifact attributes.[22] The setting of the site is very similar to that of highland New Brunswick, and it probably supported similar resources.

The final, late Pre-Ceramic tradition, the Broad Point Tradition, has recently begun to attract attention within the region.[23] The parent tradition was first described from the Middle Atlantic States, especially along the Susquehanna River where it flows through Pennsylvania. In these southern areas the name applied to this tradition is the "Susquehanna." The use of the term Broad Point is preferred as it relates to the most common diagnostic trait found within the research area: a stemmed biface described in general terms by Ritchie as "half as broad as long, or less."[24] To date, there are no ceramics associated with the broad points nor have steatite vessels been recovered in a datable context. However, broad points and three-quarter grooved axes have been found in numerous locations.

The strongest representations of this tradition are found in Maine at the Turner Farm Site, the Hirundo Site, and the Young Site.[25] A distribution study shows that north of this area these diagnostic traits decline in numbers. It seems, from the evidence at hand, that this is clearly a southern, Middle Atlantic States-derived tradition that spread northward in response to climatic change.[26] With the apparent displacement of earlier Laurentian/Maritime Archaic cultural traits witnessed at Turner Farm, the mechanism of cultural spread was actual movement of peoples. Sanger's migration hypothesis contains some caution as to how far northward it occurred. The Maritime Provinces may not have been occupied by these people, but they did influence the indigenous population by introducing the broad point and the three-quarter grooved axe.

The subsistence activities of these people reflect a mixed hunting-fishing-gathering economy. With both interior and coastal sites represented along with a varied technology, they apparently were not oriented towards the marine specialization of the Maritime Archaic nor did they have the complete interior adaptation of their Laurentian predecessors.

Although more research is needed to clarify the Pre-Ceramic traditions, one fact remains: all of them pursued fishing as a subsistence activity. This is supported by technological evidence, including ulus and plummets, and, at least for Maritime Archaic sites, extensive collections of harpoons. Several sites, notably Turner farm, have produced food bone that includes various fish species.

Ceramic Period: 3,000 - 400 B.P.

As the name implies, the Ceramic Period began when the peoples of the Maritimes were introduced to the manufacture of clay vessels. At one time it was generally accepted that this technology had its origins to the south of the study area and diffused into the Maritimes. However, the work of Peterson and Sanger has established "that ceramics were manufactured in at least a portion of the far Northeast as early as or earlier than in most areas of northeastern North America."[27] The earliest type of vessel was decorated by impressing a cord-wrapped stick into the wet clay on both the interior and exterior walls of the vessel. This mode of decorating is representative of Vinette ceramics of which only a few examples are found in the Maritimes. In these early forms decorative techniques changed at various times throughout the period.

Although the presence of ceramics is a distinctive trait for identifying sites in this period, other technological changes also occurred. The ground stone industry that was prominent in major pre-ceramic traditions diminishes in importance. This is reflected in the archaeological record by the disappearance of a number of tool types, including gouges, adzes, bayonets, ulus, grooved axes, and plummets. The axe is the only significant tool manufactured by grinding, that continues in use. Although conjecture, the demise of the adze and gouge suggests that if dugouts existed in the pre-ceramic period they were replaced by bark canoes.

It is during this period that a unique site type appears within the study area. The sheltered bays and islands in many coastal areas of the Maritimes contain the remains of prehistoric shell middens. These are accumulations of discarded shells of molluscs: clams, quahogs, mussels, and occasionally oysters and scallops. The presence of molluscs has a positive effect on the archaeological record, as their high calcium content neutralizes the normally highly acidic soils. Thus, for the first time the opportunity exists to discuss substantive direct evidence for marine subsistence in the form of preserved food bones.

Ceramic Period shell middens show a high diversity in the species that were exploited.[28] The land provided large mammals including caribou, bear, and especially moose, while small game such as beaver, porcupine, and hares supplemented the diet. On a seasonal basis, ducks, geese, and shorebirds were added to the protein base. The intertidal zone provided the various molluscs that make up the midden deposits. The Atlantic Ocean and tidal rivers allowed for the exploitation of seals, small whales, cod, herring, salmon, shad, gaspereau, smelt, and eels.

The inhabitants of the shell midden sites also left behind evidence of their hunting and fishing technologies. The primary weapon used in hunting was the bow and arrow. The arrow was tipped with a small, chipped stone projectile point characterized by notches to aid in attaching it to the shaft. Other chipped stone tools include bifacially worked knives and small end scrapers. The technologies associated with fishing include various-sized harpoons and the three-pronged fish spear known as a leister. Other means of fishing included hooks, either as single or compound tools, and bipointed gouges. As yet, no direct evidence has been discovered for the use of nets. The interpretation of most of these fishing technologies can be enhanced by using the direct historical approach, because they appear as implements in use when the Europeans first began to arrive.

Proto-Historic Period: 400 - 300 B.P.

The following discussion reviews the ethnohistoric accounts of various fishing strategies to document the species that were exploited during the initial contact period. It should be noted that there are a number of difficulties in identifying particular genera and species. This stems from a lack of knowledge on the part of the original reporter or the inclusion of vague statements such as "...plenty of other fish and of waterfowl which serve as the usual food of the Indians."[29] It is also conceivable, albeit highly unlikely, that certain of the contemporary species were not exploited by the native peoples and therefore are not mentioned in 17th century documents.

The shell midden sites of the Ceramic Period have been identified as containing numerous resource types including salt-water fish and marine mammals. Although a list of contemporary offshore species available along the Maritime coasts would be extensive, the only detailed fishing strategy identified in the documents involves the capture of sturgeon (*Acipenser oxyrhyrchus*):

> It [sturgeon] is taken with a harpoon, which is made like a barbed rod, of eight to ten inches long, pointed at one end, and with a hole at the other in which is attached a line. Then it is fastened at the end of the pole, so that it may be used as a dart. The fishery is made in the night. Two Indians place themselves in a canoe; the one in front is upright, with a harpoon in his hand, the other is behind to steer, and he holds a torch of birch bark, and allows the canoe to float with the current of the tide. When the Sturgeon perceives the fire, he comes and circles all around, turning from one side to the other. So soon as the harpooner sees his belly, he spears it below the scales. The fish, feeling himself struck, swims with great fury. The line is attached to the bow of the canoe, which he drags along with the speed of an arrow. It is necessary that the one in the stern shall steer exactly as the Sturgeon goes, or otherwise it will overturn the canoe, as sometimes happens. It can swim well, but with all its strength it does not go with fury more than one hundred and fifty or two hundred paces. That being over, the line is drawn in, and it is brought dead against the side of the canoe. Then they pass a cord with a slip-knot over the tail, and they draw it thus to land, not being able to take it into the canoe because it is too heavy.[30]

Whereas the taking of sturgeon appears to have occurred in offshore waters, a number of other marine species were caught close to the intertidal zone. Denys reports that in the spring and autumn, bass (*Roccus saxatilis*) were speared in coastal coves from a canoe.[31] He gives a similar description for the capture of flounder (*Pseudopleuronectes americanus*), though he does not give the season, and he states that they were caught from canoes in sandy-bottomed coastal areas.

The economically important intertidal zone is well represented within ethnohistorical accounts. Both Denys and Lescarbot report that whales were not actively hunted but were exploited when found washed up on the shore.[32] Denys and Biard report on seal hunting on the rocky coastal islands and shorelines. Neither author provides specifics about the methodology employed, but Biard's descriptions indicate that the hunt took place in January.[33] If the season is accurate, then the most probable genus would be the grey seal (*Halichoerus grypus*) which is known to whelp during the mid-winter season along the shores of the Maritimes.[34]

Shellfish are the dominant resource species identified at archaeological sites along the coasts of the study area. The ethnohistorical accounts identify a number of strategies employed in gathering these species. Biard, although not specific to any particular species, stated that the women gathered shellfish beginning in the spring and continued to mid-September.[35] Denys identifies a technique that employed a fish spear to take lobster (*Homarus americanus*) among the rocky coastal areas. He also reports the use of fire as a technique for attracting squid (*Loligo paeleii*) during a rising tide.[36] The squid is found along the shores of the study area during the spring and summer months.[37] The final shellfish species referred to in the ethnohistorical accounts is the oyster (*Crassostrea virginica*). The oysters were taken during winter through a hole in the ice with the aid of a composite tool made from two lengths of pole joined to form a pincers device.[38]

Of the fish species that inhabit the intertidal zone, the tomcod (*Microgadus tomcod*), is of particular interest. The Mi'kmaq name for this fish is "*Bonodemequiche*," which is the same word used to identify the month of December. LeClercq, who provided this information, also indicated that they caught this fish using a hook and line through a hole in the ice.[39]

Table 2

Species	Seasons
Anadromous:	
Smelt *(Mallotus villosus)*	Mid-March/early April
Gaspereau *(Alosa pseudoharengus)*	End of April
Atlantic salmon *(Salmo salar)*	Spring/Autumn
Striped bass *(Roccuss saxatilis)*	Spring/Autumn
Shad *(Alosa sapidessima)*	Spring/Autumn
Catadromous:	
American eel *(Anguilla rostrata)*	

Numerous species of fish pass through the study area during annual spawning migrations (Table 2). The ethnohistorical accounts of fishing activities along the Maritime coasts are vague. For instance, Biard notes that only gaspereau was taken in coastal streams;[40] whereas Lescarbot describes a large community estimated at 300 people fishing near the mouth of the Restigouche River in New Brunswick, without indicating the season or the species being exploited.[41] In general, it would appear that these major fish runs were not exploited on the coast but rather on the inland rivers at tide head and in estuarial pools.

The resources found in the rivers, with one exception, are either the anadromous or catadromous species listed in Table 2. The single species reported in the ethnohistorical literature that is a year-round resident of Maritime rivers and streams is the brook trout (*Salvelinus fontinalis*). Denys reports a technique for the procurement of this species along with the Atlantic salmon (*Salmo salar*):

Where the pools are, there they carried their canoes through the woods, and launched them where the Salmon or the Trout were. These rarely are found together in the same pool. Being there, they lighted a torch. The Salmon or the Trout, seeing the fire which shines upon the water, come wheeling around the canoe. He who is standing up has in his hand a harpoon, which is the same as that used for Beaver, and likewise is fixed in the end of a long shaft. So as soon as he saw a fish passing he speared at it, and rarely missed. But sometimes the spear did not take hold, for want of catching on some bone; thus they lost their fish. This did not prevent them from taking a hundred and fifty to two hundred in a night.[42]

Another technique used to capture fish at river locations is one that is common throughout most hunting and gathering societies. In various forms, fish traps play an important role in most fisheries, especially in areas where large runs of fish occur seasonally. Again it is Denys who provides the most informative description of the trap used within the study area:

They make use also of another device. At the narrowest place of the rivers, where there is the least water, they make a fence of wood clear across the river to hinder the passage of the fish. In the middle of it they leave an opening in which they place a bag-net like those used in France, so arranged that it is inevitable the fish should run into them. These bag-nets, which are larger than ours, they raise two or three times a day, and they always find fish therein. It is in spring that the fish ascend, and in autumn they descend and return to the sea. At that time they placed the opening of their bag-net in the other direction.[43]

Summary

This chapter presents a general overview of cultural developments in the Maritime Provinces from the beginnings of habitation, approximately 11,000 years ago, to the first contact with Europeans. It should be readily apparent that owing to various factors, such as shoreline erosion and a lack of sites with organic preservation, we know very little about the earliest cultures. As with most archaeologically-defined cultures, the more recent events are better known, especially when it is possible to use historic documents.

The Palaeo-Indian Period represents a time when the Maritime Provinces were occupied by big game hunters. The primary species exploited was most likely caribou. Towards the end of this period the occupants of the far northeast probably began to exploit marine resources, and whether or not this included fish remains to be proven. The evidence for these people is represented by a technology based on various stone tools associated with the killing and processing of large animals. Until such time as a site is discovered that has other materials preserved, such as bone and wood, we may never know the complete story of these earliest inhabitants of the Maritimes.

The archaeological evidence for the early and middle phases of the Pre-Ceramic Period also is almost non-existent. There are tantalizing clues recovered by scallop draggers and the odd specimen housed in area museums or in private collections. However, the study area has yet to produce a site that could be excavated to fill this gap in our knowledge. The presence of ulus and a single large ground stone projectile point suggests an interest in marine species. Although minimal, this evidence can be taken as a precursor of the more substantive evidence revealed in the late Pre-Ceramic traditions.

The end of the Pre-Ceramic Period is represented by as many as four archaeologically-defined traditions. Their tool kits, site distribution, and in rare instances, the preserved remains of the species exploited, show that fishing played a role in their subsistence pursuits. During the Ceramic Period, shell middens at hundreds of sites situated along the lakes and rivers of the Maritimes show conclusively that fishing was a very important aspect of the lives of people just prior to the arrival of the Europeans. This is further supported by the 17th century chronicles of the Europeans themselves that show that the Mi'kmaq took full advantage of the resources found in a marine environment. They describe various strategies for the capture of a wide variety of species from the off-shore, near-shore, intertidal, and lake/river ecosystems.

Notes

1.   J. A. Tuck, *Maritime Provinces Prehistory* (Ottawa: National Museums of Canada, 1984). S. A. Davis, "Early Societies Sequences of Change" in *The Atlantic Region to Confederation*, eds. J. Reid and P. Buckner (Toronto: University of Toronto Press, 1994), pp. 3-21.
2.   G. F. MacDonald, *Debert: A Palaeo-Indian Site in Central Nova Scotia* (Ottawa: National Museums of Canada, Anthropology Papers 16, 1968).
3.   S. A. Davis, "Two Concentrations of Palaeo-Indian Occupation in the Far Northeast," *Rivista de Argueologia Americana*, 3, (1991), pp. 31-56.
4.   MacDonald, *Debert*, p. 120 and Davis, "Two Concentrations," p. 53.
5.   J. A. Tuck, "A Summary of Atlantic Canada Prehistory," *Bulletin: Canadian Archaeological Assocation*, 7, (1975), pp. 122-44. P. Renouf, "A Late Palaeo-Indian and Early Archaic Sequence in Southern Labrador," *Man in the Northeast*, 13, (1977), pp. 35-44. S. Loring, "Palaeo-Indian Hunters and the Champlain Sea: A Presumed Association," *Man in the Northeast*, 19, (1980), pp. 15-42.
6.   J. A. Tuck and R. McGhee, "Archaic Cultures in the Strait of Belle Isle Region, Labrador," *Arctic Anthropology*, 12, 2 (1975), pp. 76-91.
7.   D.R. Grant, "Recent Coastal Submergence of the Maritime Provinces, Canada," *Canadian Journal of Earth Sciences*, 7, 2 (1970), pp. 676-89.
8.   D. L. Keenlyside, "Late Palaeo-Indian Evidence from the Southern Gulf of St. Lawrence," *Archaeology of Eastern North America*, 13, (1985), pp. 79-92.
9.   Tuck, *Maritime Provinces*, pp. 14-16.
10. J. A. Tuck, "The Northeastern Maritime Continuum: 8,000 Years of Cultural Development in the Far Northeast," *Arctic Anthropology*, 12, 2 (1975), pp. 139-47. B. Bourque, "Comments on the Late Archaic Populations of Central Maine: The View from Turner Farm," *Arctic Anthropology*, 12, 2 (1975), pp. 35-45. W. W. Fitzhugh, "A Maritime Archaic Sequence from Hamilton Inlet, Labrador," *Arctic Anthropology*, 12, 2 (1975), pp. 117-38.
11. D. Sanger, "Culture Change as an Adaptive Process in the Maine-Maritimes Region," *Arctic Anthropology*, 12, 2 (1975), pp. 60-75.
12. J. A. Tuck, *Ancient Peoples of Port au Choix* (St John's: Memorial University of Newfoundland, Institute of Social and Economic Research, 1976), pp. 25-39.
13. Tuck, *Ancient Peoples*, pp. 36-37; 82-83.
14. Tuck, *Ancient Peoples*, p. 111.
15. Bourque, *Comments*, p. 40.
16. Tuck, *Ancient Peoples*, p. 118.
17. D. Sanger, *Cow Point: an Archaic Cemetery in New Brunswick* (Ottawa: National Museum of Man, Archaeological Survey of Canada, Mercury Series 12, 1973).
18. J. V. Wright, *The Shield Archaic* (Ottawa: National Museums of Canada Publications in Archaeology 3, 1972), p. 67.

19. J. A. Tuck, "The Archaic Period in the Maritime Provinces," *Prehistoric Archaeology in the Maritime Provinces: Past and Present Research*, eds. M. Deal and S. Blair (Fredericton: The Council of Maritime Premiers, Reports in Archaeology 8, 1991), pp. 2-65.

20. D. Sanger, "Deadman's Pool—A Tobique Complex Site in Northern New Brunswick," *Man in the Northeast*, 2 (1971), pp. 5-22.

21. Sanger, "Deadman's," p. 21.

22. Wright, *The Shield*, p. 67.

23. Tuck, *The Archaic*, pp. 2-65. D. Sanger and S. Davis, "Preliminary Report on the Bain Site and the Chegoggin Archaeological Project," *Prehistoric Archaeology in the Maritime Provinces,* pp 67-79.

24. W. Ritchie, *A Typology and Nomenclature of New York Projectile Points* (New York: New York State Museum and Science Service Bulletin 384 revised ed., 1971).

25. Bourque, *Comments*, pp. 35-45. Sanger, *Culture Change*, pp. 60-75. C. L. Borstel, *Archaeological Investigations at The Young Site Alton, Maine*, (Augusta: The Maine Historic Preservation Commission, Occasional Publications in Maine Archaeology, 2, 1982).

26. Sanger, *Culture Change*, p. 69.

27. J. B. Peterson and D. Sanger, "An Aboriginal Ceramic Sequence for Maine and the Maritime Provinces," *Prehistoric Archaeology in the Maritime Provinces*, pp. 113-52.

28. Unfortunately, due to space limitations it is not possible to review all of the reports on shell midden archaeology in the Maritime Provinces. This type of archaeological site has been the focus of attention beginning as early as the mid-19th century. The reader is directed to a couple of recent works that provide general overviews as well as extensive bibliographic references for this period: S. A. Davis, "The Ceramic Period of Nova Scotia," *Prehistoric Archaeology in the Maritime Provinces*, pp. 91-108. D. E. Rutherford, "The Ceramic Period of New Brunswick," *Prehistoric Archaeology in the Maritime Provinces,* pp. 107-19.

29. C. LeClerq, *The New Relation of Gaspesia* (Toronto: The Champlain Society, 1910), p. 110.

30. N. Denys, *The Description and Natural History of the Coasts of North America* (Toronto: The Champlain Society, 1908), pp. 353-54.

31. Denys, *The Description*, p. 173.

32. Denys, *The Description*, p. 403. M. Lescarbot, *History of New France* (Toronto: The Champlain Society, 1911), p. 185.

33. Denys, *The Description*, p. 349. P. Biard, *Relation of New France, and the Jesuit Father's Voyage to that Country* (New York: Pageant Book Company, 1959), p. 79.

34. D. Christianson, "Wabanaki Subsistence Strategies in the 16th and 17th Centuries," unpublished B. A. Thesis, Saint Mary's University, Halifax, (1976), p. 19.

35. Biard, *Relation*, p. 77.

36. Denys, *The Description*, p. 356.

37. Christianson, *Wabanaki*, p. 20.

38. Denys, *The Description*, p. 359.

39. LeClercq, *The New Relation*, p. 139.

40. Biard, *Relation*, p. 81.

41. Lescarbot, *History*, p. 46.

42. Denys, *The Description*, pp. 436-37.

43. Denys, *The Description*, p. 437.

chapter
two

THE 16th-CENTURY FISHING VOYAGE[1]

Peter E. Pope

Introduction

The fishermen of western Europe were remarkably prompt to exploit the waters explored by John Cabot and the Corte Reals at the dawn of the 16th-century. Although ships from Bristol made a number of early transatlantic fishing voyages, English participation in the industry was intermittent to about 1565.[2] It was the Bretons and the Portuguese who developed the industry. The former made voyages on a regular basis from at least 1508 (or perhaps as early as 1504) and guided other early participants to the northwest Atlantic fishery.[3] Norman voyages date from 1506 but in the early decades of the century the Normans were more often involved in the industry as financiers of Breton expeditions.[4] Breton pilots also played a significant role in the early development of a Basque transatlantic industry, which grew in the 1530s, although individual voyages were recorded as early as 1517.[5] Bordeaux did not usually crew fishing vessels, but acted instead as a provisioning port for Bretons, Normans, and especially the Basques; while La Rochelle assumed a similar role, particularly for early Breton ventures.[6] Bordeaux's well-preserved notarial records suggest that French Basque participation in the fishery was not very significant until about 1550.[7] From the 1540s on, Spanish vessels from the Basque provinces of Guipuzcoa and Viscaya participated in the transatlantic cod fishery in limited numbers, but participation weakened late in the 16th-century and they had largely withdrawn by 1610.[8] Portuguese participation in the transatlantic fishery declined about the same time, after significant activity throughout the 16th-century, particularly on the eastern shore of Newfoundland's Avalon Peninsula.[9]

It is rarely possible to determine where precisely in North America particular European vessels fished, since destinations were seldom documented except as "Terra Nova," "Terre Neuve," or revealingly "Terres Neufves," terms which included both Newfoundland and Nova Scotia in the 16th-century.[10] Intermittent references suggest that it was Breton fishermen who developed a migratory fishery at Cape Breton from the 1520s.[11] By the end of the century this was no longer solely their preserve: Cape Breton appears prominently in the northwest Atlantic rutter published by the Basque pilot Martin de Hoyarsabal in 1579, and there are other documentary indications of Basque exploitation of the area by 1565.[12] The archives of the Honfleur and le Havre notaries indicate that by 1560 the Normans also fished the waters of what they called the "Island of Cape Breton" and the "Coast of Florida" (i.e. Maine), and suggest intensified French exploitation of the area in the 1580s, culminating in Etienne Bélanger's 1587 expedition.[13]

An early 16th-century English ditty put the French transatlantic fleet at "above a hundred sail" and by 1580 Robert Hitchcock, in his *Politique Platt*, estimated a French fleet of 500 vessels. The latter figure may be somewhat exaggerated, but there were certainly several hundred French vessels in the transatlantic cod fishery by this time, indicating a remarkable growth in France's fishery in less than a century.[14] The Normans already had a fleet of over 60 vessels in the 1540s.[15] The French Basque fleet alone stood at 70-80 vessels in the later 16th-century.[16] The Spanish Basques' cod-fishing fleet was much smaller, apparently numbering less than 20 vessels.[17] The 16th-century Portuguese transatlantic fishing fleet has not been seriously investigated.[18] Both the large French fleets and the smaller Spanish and Portuguese fleets could trace their origins to limited beginnings in the first half of the century, followed by rapid expansion from about 1545 to 1565, a plateau between 1565 and 1585, and a marked decline after 1585, just when English participation accelerated from its first steady commitment in the 1560s.[19] How were these voyages organized?

The Voyage

The early modern fishing voyage was organized on a three-in-one basis. This formula is meant to emphasize that those involved were committed financially or otherwise only for a single voyage and that there were, in legal terms, three groups of participants in each of these adventures: the owners of the ship, the crew, and the merchants who hired or "freighted" the ship in the case of fishing voyages supplying provisions rather than actual freight, although of course they expected to see a cargo of fish return from North America. Normally, the agents for these three participants included: 1. the *bourgeois*, or predominant shareholder in the ship; 2. the master, who represented the crew; and 3. the *armateur* or *armador*, who organized victualling.[20] The persons who took on these various legal roles might sometimes overlap—that is, a fishing master might be a part owner or responsible in part for outfitting, the *bourgeois* might be one of the victualers as well as one of the owners—but for administrative, financial, and legal purposes, the responsibility for the voyage and the right to share in the catch (and hence any possible profit) were distributed among these three legal participants.[21]

In the early days of the industry, these three parties normally divided the catch and other miscellaneous receipts in thirds.[22] For ships provisioned at Bordeaux, it appears that the division of the spoils changed in favour of the victualers after about 1575. By the end of the century they were typically expecting half the returns from Atlantic voyages, with quarter portions allotted to the ship's owners and to the crew. This strengthened the victualers commercially, who then were in a financial position to purchase shares in the ships themselves. Their increased dominance in the industry may reflect a decline in the number of voyages (and therefore in demand for ships and crews) resulting from the military tensions of the late-16th century.[23] In this period, there were still no fishing or even fish-processing companies in the modern sense. Instead, specific ports regularly engaged in the supply of ships, provisions, or crews, for particular voyages in the transatlantic fishery. Although some of France's largest ports engaged in the early fishery, the 16th-century industry was remarkably dispersed, among some 50 different ports.[24]

There is a paradox inherent in the early-modern transatlantic fishery, a paradox typical of early resource industries. While the industry was in its organization extremely localized or even "atomistic," it was at the same time "an international economy," to use Harold Innis' term.[25] Each of the essential elements of the fishing voyage—the ship, its provisions, and crew—was normally vernacular, that is, the indigenous product of a circumscribed locale.[26] Even the crew itself was a traditional product, for fishermen and pilots were customarily trained through apprenticeships, formal or informal.[27] On the other hand, the products of these locally-sponsored voyages were sold in national or even international markets.[28] Furthermore, the three key participating groups involved in a particular voyage were not necessarily from the same locale. A ship and crew from one region might be victualled by merchants of another region, before setting sail across the Atlantic for fish.[29] From an early date, Breton vessels victualled for the fishery at southwestern French ports like La Rochelle and Bordeaux.[30] A Guipuzcoan ship might be chartered to a French Basque merchant or a Norman ship might even be chartered to English interests.[31] The early transatlantic fishery was, as Jacques Bernard has emphasized, a collective effort.[32] The voyage itself was often a national or even international collectivity, despite the fact that each of its three components were normally organized within regional bounds.

The organizational success of the vernacular components of this industry surely rested, in part, on the strongly seasonal character of the fishery. The various parts of the collective that made up the fishery of any region could predict with some certainty when their respective inputs would be required. A common parameter of most early modern industries, seasonality was very strongly marked in the fishery. Crews set sail for North America in mid-March or April, although they might have set out from home for a provisioning port weeks earlier.[33] The northern fleets would set out in convoy from Brittany or France's middle Atlantic coast on a northern route; the Basque fleets preferred a southern route via the Azores.[34] Either would bring fishermen to their intended station in late May or early June, when they would anchor or careen their ships and turn to the task of reproducing the infrastructure of their industry: the wooden shore stages, drying flakes, cookrooms, and cabins in which they passed their limited leisure time.

Crews had only a few months between June and August to catch and preserve a profitable cargo, at which point they would set out once more into the cold Atlantic to their intended market, a voyage which could easily take another month, so that crews would not be back in Europe until late September or October.[35] It has been pointed out that the 18th-century French migratory fishery operated under close time constraints; in other words, the practical window of opportunity for making a safe voyage to the northwest Atlantic fishery, catching and preserving a profitable cargo, and making a safe return required virtually uninterrupted labour during the brief months ashore—a circumstance which gave the early-modern fishing station something of the character of industrialization *avant la lettre*. These temporal constraints operated also in preceding centuries. In its division of labour, work discipline, and size of work force, the early-modern fishing station was a kind of pre-industrial factory.[36]

Return to Europe was not necessarily a return home, for fish was brought to an international market. Cod in the 16th century was not a regional commodity but one distributed throughout France and indeed western Europe.[37] In the 1530s Norman venturers exported fish to English markets, and after 1570 France opened markets in Spain and Italy, although not as successfully as the English, who greatly expanded their trade to Iberia and the Mediterranean at this time.[38] When ships eventually returned to Europe, their cargoes of fish were often carried to markets far from their home ports and exchanged for a second cargo. Before a ship reached home, before it was open to charter again, and before its crew was paid off, over a year would usually have passed.[39] The financing of vessels intended for such extended voyages was a serious matter.

The Ship

The 16th-century ship was a sophisticated sailing machine. The product of several centuries of evolution, the efficiency achieved served for at least two centuries, until the spread of fore and aft rigging in the later-18th century. The three-masted square-rigged ship could be built in a range of tonnages. Ships of only 40-50 tons made early voyages, but the ships used in the transatlantic, migratory, inshore, dry fishery later in the 16th century ranged in size to over 200 tons.[40] Ships of 70 or 80 tons were common and about 100-120 tons came to be considered best for the industry.[41] Vessels of about 100 tons would embark a crew of 20-30.[42] The bank fishery, which began to develop in the late-16th century, employed smaller vessels. Many similar ships, of 30-50 tons, were later employed in the English shore-based industry, particularly as "sack" or market export ships.[43] The French dry fishery of later centuries saw many ships of greater capacity, upwards of 200 tons, but only a few 16th-century cod-fishing vessels were this large, even among the Guipuzcoan ships, which seem to have been larger on average than their Breton and Norman competitors, though only half the size of the vessels used in the Spanish Basque whaling industry.[44] In the context of the dry fishery of the period, a "fishing ship" was something of a misnomer. First, the vessels used in the migratory fishery were not a specialized type.[45] Second, these vessels carried fishermen and their equipment to North American stations but there they were used, if they were used at all, for storage or as something against which to erect a tent or lean-to.[46] Shore-based fishermen caught fish from relatively small inshore boats, not from the decks of the ships that had brought them across the Atlantic.

A vessel of approximately 100 tons represented a considerable investment in the 16th century. It is, in fact, difficult to think of larger investments in this period, except perhaps in the mining industry. An 80-ton ship built in the Norman port of Honfleur in 1576 cost over 1300 livres. Bernard puts the value of a ship suitable for the fishery at about 1500 to 1900 livres, which would represent about $ 75,000 to $ 95,000 Canadian in today's funds.[47] These are substantial amounts and raise the obvious question of how such capital was raised.

Property in ships had been organized, since medieval times at least, in shares.[48] This partition of ownership was usually expressed in quarters, eighths or subdivisions thereof. Such shared ownership was customary throughout western Europe and had been recognized in the 12th century Laws of Oleron, a kind of international law of the sea, widely observed from the Bay of Biscay to the Baltic.[49] The distinguished French Maritime historian Michel Mollat concludes that property in ships was more subdivided in 16th-century France than it was in England. Within the English fishing industry, for example, outright or joint ownership among two or three shareholders was common in the East Anglian-Iceland fishery.[50] In France, on the other hand, ownership was commonly shared by up to six or eight investors, one of whom might be recognized as the principal investor and therefore as effective manager of the vessel as an investment.[51] This was what the Bordelais called the *aunier*, an adaption of the English term "owner."

This form of ownership constituted a simple hedge against loss, for the part-owner of several ships was much less likely to suffer a total loss than the outright owner of a single vessel. Commercial vessel and freight insurance became available in France about 1530, but it was expensive and full coverage was unobtainable.[52] The French Basque ports of Bayonne, St. Jean de Luz, and Biarritz had their own mutual insurance society for vessels venturing to the new lands.[53] The inland town of Burgos was an important and very competitive insurance market not only for Spanish Basque ventures but also the French, until the military and economic woes which beset Spain after 1588.[54] But even at Burgos, freight insurance cost 10-12% of insured value, while insurance on a vessel might be 15% of insured value. Insurance rates at Bordeaux, where Norman and Breton ships often sought financial services, were closer to 20% in this period.[55] It is not surprising that investors in the fishery looked for alternatives to formal insurance and found these in shared risks and in loans on bottomry, a form of quasi-insurance.

The legal instrument that brought owners and freighters together for the purposes of a voyage was the charter party, a signed and notarized document specifying the terms and conditions for the lease of the vessel by the owners to the freighters.[56] The charter party also specified the interest of the master and crew, for the crew normally came with the ship to the freighter.[57] Custom in the fishing trade required ships' owners to hire crews expert in the fishery and this would be the responsibility of the master.[58] Investors in the 16th-century French transatlantic fishery normally chartered ships for 15 months, from January of one year to March of the following.[59] Charter parties are among the most common type of document surviving in continental records of Maritime business and are often vivid snapshots of a ship, its crew, and provisioning at a moment in time, although these documents must be read with caution, since they represent intentions for a voyage rather than events as they actually unfolded.[60]

The Crew

In order to appreciate the role of the crew in the 16th-century transatlantic fishing voyage, and to make sense of systems for subdivision of the crew's share, it is necessary to understand clearly the tasks actually undertaken by migratory fishermen in North America. Despite occasional evidence for the shipment of green or wet-salted fish, it has become clear that Harold Innis was mistaken in his belief that the wet fishery preceded the dry fishery.[61] In fact the early transatlantic 16th-century fishery appears to have been predominantly an inshore dry fishery, like the one the Bretons had carried out in European waters in the Middle Ages.[62] No mention of voyages to the banks is made in the Bordeaux records until about 1585.[63] Most of the fishing voyages to Cape Breton waters in the 16th-century would have been adventures in a shore-based fishery, combined later in the century with trade for furs.

The best surviving description of the dry fishery is that published by Nicholas Denys in 1672.[64] This is a detailed and indeed almost scientific picture of the French migratory fishery in the Maritimes in the mid-17th century. Since no description of the 16th-century fishery survives, this is, *faute de mieux*, the best source available. It is not likely that practices in the dry fishery had changed in a major way in the century preceding Denys' publication, but we must allow the possibility of variations in custom and practice which have not come to light. As noted, a typical vessel in the 16th-century transatlantic fishery would have been of about 70-100 tons with a crew of 20-30. Such a crew was, as Denys remarked, about twice the size of that carried by a banking vessel of similar tonnage.[65] This was a very large production unit by early-modern standards, characterized too by a definite, if simple, division of labour. What tasks faced the crews of these ships as they arrived in North America in May or early June?

The actual process of catching cod fish and making them into a lightly-salted, dried, preserved food suitable for ocean transport to distant markets is at least as old as the European fishery in North America. The Basques preferred this to a wet or green cure, and many Breton and Norman crews used the same techniques.[66] Denys thought that the Basques were the most skilled in the dry fishery, followed by fishermen of the Isle de Ré, Isle d'Oléron, La Rochelle, Bordeaux, and Brittany.[67] The work was not unskilled but required small boat handling skills and familiarity with the habits of *Gadus morhua* (cod), preferably on a particular stretch of shore. Each crew member baited two hooks on two or three lines. A crew of three handled each boat: a master, his boatswain, and a relatively unskilled "stower."[68] In other words, although every fisherman was not necessarily fully skilled, each crew was. Sundays excepted, boat crews worked long hours from dawn to late afternoon, when they returned to port. With pews or gaffs they off-loaded the catch, which then became the responsibility of the shore crew.[69]

The shore crew began the task of making fish right on the stage, the combination wharf and processing plant where the fish was unloaded. The header gutted and decapitated the fish, setting aside the cod livers in a train vat, where the oil rendered out in the sun. The splitter opened the gutted fish and removed the spine. Two shore workers could handle the catch of a three-person boat, so that when numbers of workers and boats are reported for the migratory dry fishery they usually occur in a ratio of about five to one.[70] Young apprentices moved the split fish in hand barrows and piled it up for salting. The beach-master who supervised the salting was a skilled hand. After the fish spent a few days in salt, the youngsters rinsed it in seawater and piled it up for a day or two before spreading it out to dry on a cobble beach or on wooden flakes.[71] When the wind and gentle northern sun had fully dried the fish, the young shore workers would stack it in large round piles, protected from the weather by covers and from the ground by circular stone pavements, some of which are still visible along the coasts of the Gulf of St. Lawrence.[72]

There were other tasks. Through the fishing season crews deployed herring seines or mackerel lines for bait.[73] At the end of the season the ship would have to be loaded with the processed catch, including a few quintals of late-season wet-cured fish and barrels of train oil rendered from the fish livers. Crews had a more onerous task at the beginning of each season, when they might spend a month reproducing the infrastructure of their industry: the boats, train vats, stages, flakes, cookrooms, and cabins. Parallel tasks awaited the crew at season's end, when they recycled their shore structures as firewood for the return journey.[74] Of these structures, the stage was the largest and most costly. This was, essentially, a rough wooden quay projecting up to 60 metres from shore, with a partially-closed structure at its seaward end.[75] Cabins, cookrooms, and the stage head work space were of wattle construction: fir posts, woven with boughs, sealed on the inside with fir rinds and roofed with rinds and turf or a sail. Sometimes the crews' lodging was simply a tent of fir poles and a canvas sail. The "fishing" ship itself might be careened and used as the centre-piece of an extended lean-to.[76] If it remained anchored, no one slept on board.[77]

The making of dry fish required skills, although the production of competitively priced fish—and, in fact, the reproduction of the whole system—required the employment of young, untrained apprentices too. Each boat crew included a novice; and such youngsters, some as young as eight or ten years old, worked on shore as well, assisting the skilled headers, splitters, and salters.[78] In typical early modern fashion, the passing-on of skills was built into the employment structure of the industry. In effect the "society" of the fishing ship consisted of three classes: 1. Officers (captain, pilot, beach-master, surgeon and steward); 2. Skilled workers (boatmasters, fishermen, shoremen and carpenters); and 3. Unskilled apprentices (*grommets*).[79] These workers were, as far as we know, invariably male.

The roles of the pilot, the beach-master, steward, and carpenters in the dry fishery are unproblematic. The pilot guided the vessel to and from its fishing station; the beach-master supervised processing; and the steward or purser supervised the use of provisions. The carpenter would have boats, stages, and vats to attend to, as well as the ship itself. The surgeon's role was more complex, at least in the 17th-century French fishery. According to Denys, he carried a chest of medicines and instruments to care for the crew. This included acting as their barber, for which he received a small fee per head. He also could expect a share in the catch, but in return was expected to work on shore filleting fish and carrying hand barrows, "like the least of the sailors," as Denys remarks. The surgeon was also in charge of the cookroom, the construction of the flakes, and he might tend a garden or hunt for the captain's table, where he himself ate with the pilot and beach-master.[80]

Terms of employment varied from region to region, but had at least one thing in common. The marked division of labour within the dry fishery and the importance of the skills passed on in informal apprenticeship are recognized, in each case, by marked differentials in remuneration.[81] In Denys' time, boatmasters could expect to earn at least four times what would be apportioned to the *grommets*.[82] If the evolution of pay differentials in the French fishery resembles that in the English industry, then there would have been an even larger gap in pay levels in the 16th century.[83]

The Basques operated on a straight share system. In the 17th century a third of the prospective catch was divided into 200 or 300 shares, portioned out by contract in varying amounts to the master, beach-master, pilot, boatmasters, filleters, etc.[84] Some later Basque crews made more complex agreements, e.g. for a two-fifths share of the catch, less two barrels of the train oil.[85] Bordelais masters contracted with the owners and victualers for a share of the cargo (often one-third, but with some variation).[86] The master in turn contracted with the crew for their fractions of this share, the payment of youngsters being at his discretion. His remuneration consisted of the remainder left once the crew had been paid off. Rochellais crews used a similar system, but shared only one-quarter of the catch, making up for this by receipt of a small fixed wage, determined by rank.[87] Many French systems of remuneration included an advance paid to fishermen before they embarked, with the balance, variously calculated, paid on their return.[88] In the 18th century, Breton fishermen of St. Malo or St. Brieuc received such an advance, called a *pot de vin*, and the custom may well have been of long duration.[89] These crews then divided one-fifth of the catch *à la mode du nord*, on their return home. Norman fishermen shared a third of the cargo, less advances, or followed other more complex systems of remuneration.[90]

Such differences persisted in the French migratory fishery well into the 19th century and it is very likely that they were of long standing. Two enduring aspects of this variation in patterns of remuneration are worth comment. First, the system of remuneration depended on the regional origin of the crew, not of the ship or where it was outfitted.[91] Second, patterns of remuneration were interdependent on customs of victualling. For example, Bretons, from the earliest days of the transatlantic fishery, appear to have been less well paid than Normans and to have expected less ample provisions of food, because they expected to be provisioned with drink in the form of wine or cider, while the Normans did not.[92] Given the vernacular organization of both crewing and provisioning arrangements, it should not be surprising that the latter also varied somewhat from region to region.

Provisions

"Victualer," the English equivalent of the French term "avitailleur," underestimates the function of the merchants who outfitted ships for the fishery. Hitchcock came nearer to the mark in 1580, when he described French fishermen conferring with a "Money man" who underwrote provisions on the expectation of repayment "either in fish or in money."[93] In the case of this industry the merchant did not supply freight but supplied capital for the materials used in an extended transatlantic voyage. Hitchcock speaks of "victuals, salt, lines and hooks," the requirements for catching and curing fish, but he might also have mentioned the tools and iron work, particularly nails, necessary to prepare or repair boats, stages, and the other components of the shore-based infrastructure, and he apparently took for granted the arms that mariners carried with them on almost all deep sea voyages in this period (for reasons discussed below).

These provisions would represent about 300 livres in victuals alone and, all in all, a considerable investment of perhaps 1000 livres, that is something in the order of $50,000 in today's currency.[94] Provisioning merchants shared out this investment, just as ship ownership was partitioned. These investments were normally in the form of loans *à la grosse aventure*, or "on bottomry" to use the English terminology.[95] Such loans incorporated a form of insurance. On completion of a voyage the provisioning merchants expected to be repaid their original capital plus a significant interest payment, in the order of 20-40%.[96] On the other hand, if the ship failed to return safely, the provisioning investors accepted their losses, just as the ship's owners and indeed the crew had to accept theirs. The merchants of La Rochelle and Bordeaux made similar loans for coastal voyages, but investments in transatlantic voyages were larger. Significantly, there was no interest differential between coastal voyages and ventures in the transatlantic fishery—confirming that the migratory fishery in North America was a safe voyage, relative even to the *dangers civils* of western Europe.[97] What was it that investments in the North American fishery bought?

Salt: Salt preservation, developed in western Europe in the Middle Ages, made the transatlantic migratory fishery economically feasible. In this sense salt was the most important provision in the fishery, certainly in terms of bulk. The shore-based dry fishery required proportionately only half the salt used in the emerging bank fishery, but even its requirements were substantial.[98] France produced a gray "bay salt" on the coasts of Brittany, Aunis (particularly the Isle de Ré), and Saintonge. Nevertheless French fishermen also obtained salt in Portugal, where the price was lower and the salt purer. There was much discussion of the merits of salt from different sources, but competition in this developing international market seems to have been essentially price-based.[99]

Boats: It was the *chaloupe* that was used "in the fishery in the new land and elsewhere," according to a Bordeaux notarial record of 1549.[100] These boats were about nine metres in length but the exact form of 16th-century fishing shallops is unknown and, indeed, they almost certainly varied regionally.[101] Denys tells us that *chaloupes* were built in France and shipped knocked-down as kits to be re-assembled upon arrival in North America. He describes in meticulous detail a vessel for sailing or rowing, with a fine high bow and a transom stern, pinched at the waterline, heavily-framed, carvel-planked, and finished with a mixture of turpentine and cod-liver oil. Inside, the *chaloupe* was divided by five or six thwarts, each with vertical staves, creating a series of rooms (*rums*). The amidships room of about 1.5 m. in length, was larger than the others and was used as a fish hold. Fully laden, a *chaloupe* could hold 500 to 600 cod, according to Denys. So burdened, or in rough weather, the fishermen erected tarred canvas sideboards along the gunwales to increase their freeboard.[102] The fishermen of Denys' time rigged their *chaloupes* "like those of La Rochelle."[103] Denys describes a single-masted asymmetrical square rig, employing a sprit: in other words, a square-rig evolving into a fore and aft sprit or gaff rig. Each boat carried 30-40 m. of mooring rope of about three centimetres diameter and a grapnel of about 25 kg. At the end of the season these boats were hidden in the woods, or even sunk with a few stones in a pond, where they might remain for several years before being retrieved.[104]

Hooks and Lines: Each of the three fishermen aboard a boat required two lines, a dozen hooks and enough lead to make six sinkers of about 1.5 kg. apiece. Fishermen kept their lines wound on simple square wooden frames, of a type still familiar. They had to tie hooks to lines, for the hooks of the early modern period did not have eyes, as modern hooks do, but rather flattened tips at the upper end. Since a crew of 25-30 would keep five or six boats busy, it is clear that each fishing voyage required significant provision of equipment of this type, as well as processing equipment like vats, hand-barrows, barrels, gaffs, pews, knives, and even grindstones to keep these edge tools sharp.[105] Provisioning of this sort would not vary greatly by region, although we should note that in the mid-16th century Spanish Basques sometimes equipped a vessel for the cod fishery, whaling, and privateering in a single voyage.[106] The intention in such cases, the charter parties suggest, was to concentrate on whatever business turned out to be most productive.

Arms: Given the risk of privateers, in an age when European states commonly authorized this kind of legally-mandated private warfare, almost all deep-sea vessels of the period carried arms. In fact French law required this.[107] Everyone on transatlantic voyages of the period would normally be armed with sword, arquebus, or crossbow.[108] The charter party for the 1552 voyage of the Saint-Esprit of St. Jean-de-Luz specified that it would be provisioned at Bordeaux with an arquebus or crossbow for each of its men as well as with 20 cannon, shot, powder, and 50 pikes, in two sizes. Even a small ship carried at least four heavy guns and eight swivel guns or *versos* of the type recovered from the 1565 wreck of the *San Juan* in Red Bay Labrador.[109] A heavy gun of this period was recovered many years ago in Louisbourg harbour and is more likely to have been carried on an early fishing vessel than to have arrived through the more fanciful circumstances proposed when it was dredged up.

Victuals: Extant provisioning lists give us a good indication of the diet of 16th-century mariners. The origin of the crew and the region in which the ship was provisioned determined diet. Outfitters made provisions for bread, the major staple, in three forms: wheat, flour, and biscuit. Although supplies might be purchased in the form of wheat or flour, this was processed by the victualers and ship's biscuit is what was packed into casks and what crews eventually consumed in the amount of about 800 g. per person per day. Limited quantities of flour might be shipped for those on board of higher status. The other major staple shipped was pulses: dried peas and broad beans, in quantities of about 60 g. per person per day. With wheaten biscuit this would provide the carbohydrate base for the diet and some of the protein as well as vitamins A and B.[110]

Ships' provisions often included small quantities of salt meat, sardines, herrings, and dried cod. Such stocks of protein would have lasted for the voyage out. Fresh fish, for the crew, and game for the captain's table at least, were available in North America. Spanish Basque victualers supplied fat for crews in the form of bacon and olive oil, in roughly equal quantities totalling about 30 g. per person per day. When the *Saint-Esprit* was provisioned at Bordeaux in 1552, the fat laded was mostly lard, with a smaller quantity of olive oil. Denys mentions (salt) pork and butter, which would be characteristic northern provisions. To dress their meals crews had salt, vinegar, garlic, and mustard. Vessels sometimes stocked the latter in truly surprising quantities of over 4 g. dry weight per person per day, which may indicate that the seeds were used for sprouting mustard greens as well as being ground as a condiment.

Despite such condiments, the 16th-century seafarer's diet sounds dull, by modern standards. It should be emphasized that this diet did not differ greatly from that expected on land at the time, with one significant difference for an ordinary worker: on board ship, the crew was guaranteed a ration in some form, a significant attraction in a period that still saw famine during economic crises and harvest failures. Much of the fat and protein was salt-preserved, giving the mariner's diet the salty character it has retained to this day. A number of 16th-century victualling contracts for the French migratory fishery acknowledge crews' preference for biscuit and salt meat over pulses, although the latter were almost certainly healthier provisions. This preference may in part have functioned as a symbol of identity.[111] All in all the diet was reasonably healthy, especially if vitamin C could be obtained from mustard greens, cider, or even spruce beer.[112] A decline in standards appears in the French fishery after 1575, towards a regime solely of biscuit and wine.[113] This seems to have reflected a temporary economic circumstance, for standards improved again by the second half of the 17th century.[114]

Drink: Guipuzcoan provisioning lists suggest that Spanish Basque mariners consumed between 2 and 2.4 l. of cider per day on voyages to the northwest Atlantic. This was by far the most important beverage shipped, at least from Spain. French Basque voyages shipped considerably more wine and less cider. The apple-based beverage may have been a good source of vitamin C. Fortified wines like sherry, when shipped by Spanish Basques, were consumed at the rate of about 120 ml. per person per day; unfortified wines at a rate of about 700 ml.[115] Bretons also drank cider and, as already noted, expected generous rations of drink. Denys tells us that in the 17th-century fishery, crews drank their wine diluted with three or four parts water to one of wine, except on their Sunday rest.[116] Stores of this dilute wine and the other standard victuals were open to all. The 17th-century French custom was for crews to eat in messes of seven or eight at a time, the youngsters eating what was left over, the captain dining apart with the other officers.[117] The evening meal would be the only hot dinner served each day; lunch would consist only of a little watered wine with bread and perhaps some left-over fish.

Personal Stores: Crew members also shipped their own personal supplies, or *regalos*, as the Spanish called them.[118] In the case of one Spanish Basque mariner these included about 25 kg. of white biscuit, 24 l. of wine, a shoulder of bacon, some cheeses, eggs, raisins, and almonds. Denys notes that French crew members who wanted to drink brandy brought their own private stores, which they might sample on the job. He also observes that masters used a round of brandy as a reward for a particularly good day's catch.[119]

Dress: Clothing is another form of consumption in which origins, social and regional, inevitably play a significant role. It is quite possible that crew members of highest status (i.e. masters) would frequently wear the ornate and impractical clothes favoured by merchants of similar status in their own regions. The dress of ordinary fishermen and shore crews would be more practical, but even in their case there seems to have been some regional variation. Denys stresses that Basque boat crews were generally the best outfitted, with large leather sea boots, sheepskin aprons, well-oiled on the leather side, and jackets of the same material with a hood. Some even had two sets of this wet-weather gear. Mariners of Gironde, Saintonge, and Aunis had sea boots, but often made do with hooded cloth coats with leather or tarred linen sleeves, under a sheepskin apron.[120]

Daily Life: How fishermen passed their limited leisure time will probably remain a matter for speculation, pending the archaeological excavation of a 16th-century fishing station. The excavation of a contemporary whaling station at Red Bay uncovered numerous rosary beads as well as a wooden cross, suggesting that Spanish Basque mariners, at least, brought Christian practices with them on their annual migrations to North America. Finds of coins at this site do not prove that these crews also brought with them the practice of gambling, but this would seem to be one of the most likely uses for cash in an isolated location.[121]

Conclusion

The development of the European migratory fishery along the coasts of 16th-century Atlantic Canada required the refinement of a number of late-medieval inventions. These include not simply the new and more manageable three-masted ships of the period but also the wide-spread legal acceptance of the tripartite economic structure of the voyage, in which the interests of owners, victualers, and crew were reconciled by a charter party, for the purposes of a kind of vernacular capitalism: "capitalism" because it was designed to maximize profits in a market context; "vernacular" because the organization of the inputs was rooted in local custom, synchronized less by administrative directive than by long-standing seasonal rhythms. By 1600 this vernacular capitalism had succeeded, paradoxically, in creating an international economy in fish. This early capitalism rested, in turn, on the evolution of economic culture not simply among merchants who might invest in the ownership or victualling of ships, but also among the *gens de mer* who actually caught and made salt fish.

The economic culture of early transatlantic fishermen may shed some light on the development, within the matrix of the fishery, of a fur trade with native peoples. During the second half of the 16th century, fishermen coming to North America, particularly to stations in Cape Breton and the north shore of the Gulf of St. Lawrence, began to count on bartering with native Amerindian peoples.[122] The early modern mariner's perquisite of "portage," the right to venture private cargo on his own account, has a direct bearing on such non-fishing activities.[123] One Spanish Basque brought the usual moderate store of wine, bacon, and raisins with him, but also 720 l. of cider![124] Such a store must have been intended for barter. What began informally, probably on the initiative of individual mariners exploiting their right to portage, developed into merchant ventures, designed to obtain furs in commercial quantities. The fishery would be the staple economy of the Atlantic region for many centuries, but by 1580 the European newcomers had already begun to diversify on a commercial scale.[125]

Notes

1.   The author thanks the Heritage Section, Nova Scotia Department of Municipal Affairs, for sponsoring the initial version of this paper.

2.   See Peter E. Pope, "The South Avalon Planters, 1630 to 1700: Residence, Labour, Demand and Exchange in Seventeenth-century Newfoundland," unpub. Ph.D. diss., Memorial University of Newfoundland, 1992, pp. 27-29.

3.   [Pierre Crignon] (1539), in David Beers Quinn, *New American World*, vol. 4, "Newfoundland from Fishery to Colony. Northwest Passage Searches" (New York: Arno Press, 1979), pp. 156-57; Laurier Turgeon, "Le temps des pêches lointaines: permanences et transformations (vers 1500 - vers 1850)," in *Histoire des pêches maritimes en France,* ed. Michel Mollat (Toulouse: Privat, 1987), pp. 134-81, see p. 136; J.A. Dickinson,"Les précursurs de Jacques Cartier," in *Le monde de Jacques Cartier: l'aventure au XVIe siècle,* eds. Fernand Braudel et Michel Mollat du Jourdin (Montreal: Libre Expression, 1984), pp. 127-48; Jacques Bernard, *Navires et gens de mer à Bordeaux (vers 1400 - vers 1550),* vol. 2 (Paris: S.E.V.P.E.N., 1968), p. 809; Michel Mollat, *La vie quotidienne des gens de mer en Atlantique (IXe-XVIe siècle)* (Paris: Hachette, 1983), p. 189.

4.   Michel Mollat, *Le commerce maritime normand à la fin du moyen age, étude d'histoire économique et sociale* (Paris: Plon, 1952), p. 262.

5.   Bernard, *Navires et gens de mer,* p. 811; H.P. Biggar, "The Precursors of Jacques Cartier 1497-1534, a Collection of Documents Relating to the Early History of the Dominion of Canada," *Publications of the Canadian Archives no. 5* (Ottawa, 1911), pp. 124-25; Selma Huxley Barkham,"Aperçu

de l'évolution de la pêche sur les côtes de l'Est canadien," in *L'aventure maritime, du golfe de Gascogne à Terre-Neuve, Comité des Travaux Historiques et Scientifiques,* eds. Jean Bourgoin and Jacqueline Carpine-Lancre (Paris, 1995), pp. 173-80.

6.   Michel Mollat, Review of Bernard, *Navires et gens de mer,* in *Etudes d'histoire maritime* (1938-1975) (Turin: Botega d'Erasmo, 1977), pp. 119-35, see p. 134; Dickinson,"Les précursurs," p. 138; Bernard, *Navires et gens de mer,* p. 811.

7.   Turgeon, "Temps des pêches lointaines," p. 136ff.

8.   S.H. Barkham, "Guipuzcoan Shipping in 1571 with Particular Reference to the Decline of the Transatlantic Fishing Industry," in *Anglo-American Contributions to Basque Studies: Essays in Honor of Jon Bilbao,* ed. W.A. Douglas, Desert Research Institute Publications on the Social Sciences no. 13 (Reno, Nevada, 1977), pp. 73-81; Harold A. Innis, "The Rise and Fall of the Spanish Fishery in Newfoundland," in *Essays in Canadian Economic History,* ed. M. Innis (Toronto: University of Toronto Press, 1956), pp. 43-61.

9.   A.R. Mitchell, "The European Fisheries in Early Modern History," in *Economic Organization of Early Modern Europe,* vol. 5 of *Cambridge Economic History of Europe,* eds. E.E. Rich and C.H. Wilson (London: Cambridge University Press, 1977), pp. 132-84; David Beers Quinn, *North America from Earliest Discovery to First Settlements: the Norse Voyages to 1612* (New York: Harper and Row, 1977), pp. 347-68.

10.   Charles de la Morandière, *Histoire de la pêche francaise de la morue dans l'Amérique septentrionale (des origines à 1789)* (Paris: G-P. Maisonneuve et Larose, 1962), vol. 1, p. 256 (All subsequent references to La Morandière are from this volume.) For "Terres Neufves" see Y. Raymond (1533) in Biggar, *Precursors,* pp. 181-82.   The French adventurer Marc Lescarbot used "terre-neuvier" to refer to the Nova Scotian-based Basque captain Savalete, in 1618. On dispersion of the industry see J. Mannion and S. Barkham, "The 16th-century Fishery," in *From the Beginning to 1800,* vol. 1 of *Historical Atlas of Canada,* eds. R.C. Harris and G.J. Matthews (Toronto: University of Toronto Press, 1987), Plate 22.

11.   A map in the Kunstmann Atlas of c. 1514 to 1520 calls Cape Breton "*terra a foy descubierto por bertomas*" (land discovered by the Bretons); see S.E. Dawson, "The Voyages of the Cabots in 1497 and 1498; with an Attempt to Determine their Landfall and to Identify their Island of St. John," *Transactions of the Royal Society of Canada* (1st series) 12, Section II, pp. 51-112 (p. 75); Dickinson, "Les précursurs," p. 137.   Pierre Crignon claimed, in 1539, that Bretons and Normans had discovered Cape Breton in 1504; P. Crignon (1539), *loc. cit.* The talented Breton pilot Jean Alphonse was evidently familiar with the area in the 1520s; S.E. Morison, *The European Discovery of America: The Northern Voyages* (New York: Oxford University Press, 1971), p. 231. It may also be significant that John Rut reported only one Breton vessel at St. John's in 1527, with 11 Norman and two Portuguese vessels, suggesting that the Breton fleet fished elsewhere, perhaps at Cape Breton. See Quinn, *New American World,* vol. 1, p. 190 and vol. 4, p. 62; La Morandière, *Histoire,* p. 259.

12.   M. Hoyarsabal, *Les voyages avantureux du Capitaine Martin de Hoyarsabal, habitant de Cubiburu, contentant les regles & enseignments necessaires à la bonne & seure navigation* (Bordeaux: Jean Chouin, 1579); La Morandière, *Histoire,* p. 253. Fransisco de Souza, 1570, in Biggar, *Precursors,* pp. 196-97; Morison, *European Discovery,* p. 231; Samuel de Champlain, *Voyages of the Sieur de Champlain* [1613], ed. H.P. Biggar, Champlain Society Publications (Toronto, 1922), vol. 1, pp. 462-64.

13.   La Morandière, *Histoire,* p. 238; L. Turgeon, "Basque-Amerindian Trade in the Saint Lawrence during the Sixteenth century: New Documents, New Perspectives," *Man in the Northeast* 40 (1990), pp. 81-87; Quinn, *New American World,* vol. 4, p. 306.

14.   Turgeon, "Temps des pêches lointaines," pp. 136-37 and "Basque-Amerindian Trade," p. 82. For Hitchcock's estimate see Quinn, *New American World,* vol. 4, p. 105.

15.   La Morandière, *Histoire,* p. 237.

16.   L. Turgeon, "Pêches Basques en Atlantique Nord (XVIIe-XVIIIe siècle)," thèse doc. 3e cycle, Université de Bordeaux, 1982, pp. 13-14.

17. Barkham, "Guipuzcoan Shipping," pp. 73-77; Turgeon, "Pêches Basques," p. 14; W. D. Phillips, "Spain's Northern Shipping Industry in the Sixteenth century," *Journal of European Economic History* 17(2) (1988), pp. 267-301.

18. Quinn, *New American World*, vol. 4, pp. 81, 85.

19. L. Turgeon, "Pour redécouvrir notre 16e siècle: les pêches à Terre-Neuve d'après les archives notariales de Bordeaux," *Revue d'Histoire de l'Amérique Francaise* 39(4) (1986), pp. 523-49; K.M. Matthews, "History of the West of England-Newfoundland Fisheries," unpub. D.Phil. dissertation, Oxford, 1968, pp. 34-46.

20. Quinn, *New American World*, vol. 4, p. 82, notes that the *armateur* in France often owned a substantial share in the ship, while in Spain the *armador* was more likely to restrict his investment to provisioning.

21. Bernard, *Navires et gens de mer*, p. 539 ff.; Mollat, Review of Bernard; *Le commerce maritime normand*, p. 265; La Morandière, *Histoire*, p. 50ff. For further discussion see John Gilchrist. "Exploration and Enterprise in the Newfoundland Fishery c. 1407-1677," in *Canadian Business History: Selected Studies, 1497-1971*, ed. D.S. Macmillan (Toronto: McClelland and Stewart, 1972), pp. 7-26.

22. J. Tredian (1523), in Biggar, *Precursors*, p. 162; La Morandière, *Histoire*, p. 51; Bernard, *Navires et gens de mer*, p. 813; Quinn, *New American World*, vol. 4, p. 82. Cases of individuals fulfilling more than one function, for example a master taking some responsibility for provisioning, complicate assessment of whether this continued to be the norm in the fishing fleets of the various regions of France, as it did in England's West Country into the 17th century. *Cf.* J. de Mendicabal (1541) in Quinn, *New American World*, vol. 4, p. 91; on West Country customs see Gillian T. Cell, *English Enterprise in Newfoundland, 1577-1660* (Toronto: University of Toronto Press, 1969), pp. 15-17.

23. Turgeon, "Notre 16e siècle," p. 541.

24. These included the Norman ports of Fécamp, Havre de Grâce, Honfleur, Rouen, and Granville. The more important Breton ports participating in the early industry were St. Malo and the many small ports in the admiralties of St. Brieuc and Brest, among the latter, Le Croisic and Pornic. The mid-Atlantic port of La Rochelle was very important in provisioning as was Bordeaux; La Morandière, *Histoire*, pp. 231-53. St. Jean de Luz, Bayonne, and Labourd were the main French Basque ports, while Basque crews from the Guipuzcoan ports of Fuenterrabia, St. Sebastian, Pasajes, Orio, and Lequeito sailed under the flag of Spain; see Turgeon, "Pêches Basques"; Barkham, "Guipuzcoan Shipping."

25. H.A. Innis, *The Cod Fisheries: the History of an International Economy* (rev. ed. 1954, rep. Toronto: University of Toronto Press, 1978); *cf.* Quinn, *New American World*, vol. 4, p. 82.

26. *Cf.* Mitchell, "European Fisheries." On the term "vernacular," see Pope, "South Avalon Planters," pp. 472-77.

27. Aingeru Zabala Uriarte, "L'apprentissage de la pêche pour les jeunes 'terrestres' aux XVIe-XVIIe siécles," in *L'aventure maritime*, pp. 389-402.

28. Quinn, *New American World*, vol. 4, p. 84.

29. Charter party of the *Saint-Esprit* of Saint-Jean-de-Luz (1552), in Quinn, *New American World*, vol. 4, p. 94.

30. P. Jourdain (1523) in Biggar, *Precursors*, pp. 160-62; Bernard, *Navires et gens de mer*, p. 807.

31. Charter party of the *Salvador* of Fuenterrabia (1541); Charter party of the *Clemence* of Cherbourg (1603); both in Quinn, *New American World*, vol. 4, pp. 91-93, 125. For further examples see G.V. Scammell, "Shipowning in England, c. 1450-1550," *Transactions of the Royal Historical Society*, 5th series, 12 (1962), pp. 105-22; see p. 107.

32. Bernard, *Navires et gens de mer*, p. 807.

33. Bernard, *Navires et gens de mer*, p. 817; C. de Barros (1574) in Quinn, *New American World*, vol. 4, pp. 103-04.

34. Mollat, *Les gens de mer en Atlantique*, p. 190; Turgeon, "Pêches Basques," p. 35.

35. On stored boats see J. Cordier (c. 1544) in Mollat, *La vie quotidienne*, p. 191. On return dates see de Barros (1574).

36. J.-F. Brière, *La pêche francaise en Amérique du Nord au XVIIIe siècle* (Quebec: Editions Fides, 1990), pp. 58-60; Turgeon, "Temps des pêches lointaines," p. 134.

37. On French markets see Turgeon, "Notre 16e siècle," pp. 546-47; "Temps des pêches lointaines," pp. 140-42.

38. Turgeon, "Temps des pêches lointaines," p. 138, *cf.* Cell, *English Enterprise*, pp. 22-33.

39. See the discussion below of charter parties.

40. La Morandière, *Histoire*, pp. 232-33.

41. Mollat, *Le commerce maritime normand*, p. 265; Bernard, *Navires et gens de mer*, p. 816ff.

42. Bernard, *Navires et gens de mer*, p. 817; Mollat, *Les gens de mer en Atlantique*, p. 190.

43. La Morandière, *Histoire*, p. 41; Pope, "South Avalon Planters," pp. 119-25.

44. Turgeon, "Temps des pêches lointaines," p. 138; Barkham, "Guipuzcoan Shipping;" A.P. Usher, "Spanish Ships and Shipping in the Sixteenth and Seventeenth Centuries," in *Facts and Factors in Economic History* (New York: Augustus Kelly, 1967), pp. 189-213.

45. Bernard, *Navires et gens de mer*, p. 816.

46. Bernard, *Navires et gens de mer*, p. 819ff.

47. Bernard, *Navires et gens de mer*, p. 814. One estimate values English ships of c. 1500 to 1550, in their first decade afloat, at roughly £1 10s per ton, or about $750 today, making a 100 ton ship worth $75,000; see Scammell, "Shipowning in England," p. 111. On currency equivalents see Innis, *Cod Fisheries*, pp. 27-29; on inflation see E.H. Phelps Brown and S.V. Hopkins, "Seven Centuries of the Prices of Consumables, compared with Builders' Wage-rates," *Economica* 23 (1956), pp. 296-314.

48. R. Davis, *The Rise of the English Shipping Industry in the 17th and 18th Centuries* (1962, rep. London: National Maritime Museum, 1972), p. 82ff.

49. Mollat, *Le commerce maritime normand*, pp. 345, 408-09; review of Bernard, *Navires et gens de mer*, p. 466; Scammell, "Shipowning in England," p. 114ff. On the Laws of Oleron, see Bernard, *Navires et gens de mer*, II, p. 678ff; Mollat, *La vie quotidienne des gens de mer*, p. 85ff; Elisa Ferreira, "L'incorporation du Nord-Ouest ibérique à la vie èconomique du golfe de Gascogne," in *L'aventure maritime*, pp. 331-40.

50. R. Tittler, "The English fishing industry in the sixteenth century: the case of Great Yarmouth," *Albion* 9(1) (1977), pp. 40-60.

51. Mollat, Review of Bernard, p. 130; Jean-Philippe Priotti, "Des financiers de la mer: les marchands de Bilbao au XVIe et au début du XVIIe siècles," in *L'aventure maritime*, pp. 181-96.

52. Mollat, "Review of Bernard," p. 131.

53. Bernard, *Navires et gens de mer* II, p. 818.

54. S. Barkham, "Merchantile Community in Inland Burgos," *Geographical Magazine*, November 1973, pp. 106-13 and "Burgos Insurance for Basque Ships: Maritime Policies from Spain, 1547-1592," *Archivaria* 11 (1981), pp. 87-99; Priotti, "Des financiers de la mer," p. 183.

55. Barkham, "Burgos Insurance"; J. Lopez de Soto (1570) in Quinn, *New American World*, vol. 4, p. 101.

56. See Bernard, *Navires et gens de mer* II, p. 682ff.

57. *Cf.* Quinn, *New American World*, vol. 4, p. 99.

58. Mollat, "Le commerce maritime normand," p. 265.

59. Mollat, *Les gens de mer en Atlantique*, p. 191.

60. See the charter party for the *Saint-Esprit* of Saint-Jean-de-Luz in the French Basque country, provisioned at Bordeaux for a transatlantic fishing voyage in 1552; Quinn, *New American World*, vol. 4, p. 94.

61. Compare La Morandière, *Histoire*, pp. 233, 255 or Bernard, *Navires et gens de mer*, p. 819 and Innis, *The Cod Fisheries*, p. 25. For an example of green fish in 1517 see Bernard, *op. cit.*, p. 806; but it was normal for vessels engaged in a dry fishery to return with a small quantity of wet fish, caught too late in the season to dry ashore. Dickinson, "Les précursurs," pp. 137-38, insists on an early bank fishery, but offers no evidence.

62. M. Mollat, "The French Maritime Community: a Slow Progress up the Social Scale from the Middle Ages to the Sixteenth Century," *Mariner's Mirror* 69(2) (1983), pp. 115-28; Turgeon, "Notre 16e siècle," p. 534.

63. Turgeon, "Notre 16e siècle," p. 534.

64. Nicholas Denys, *Histoire naturelle des peuples, des animaux, des arbres & plantes de l'Amérique septentrionale, & de ses divers climats*, vol. 2 [1672], pp. 1-252, in W.F. Ganong (ed.), *The Description and Natural History of the Coasts of North America (Acadia) by Nicholas Denys*, Champlain Society, (1908, rep. New York: Greenwood, 1968). References here are to the original pagination.

65. Denys, *Histoire Naturelle*, p. 57.

66. La Morandière, *Histoire*, pp. 244, 252-58, 308-11.

67. Denys, *Histoire Naturelle*, pp. 56-57. Diffusion of the word "bacalao" suggests early Basque expertise in the industry; see Miren Egaña Goya and Brad Loewen, "Dans le sillage des morutiers basques du Moyen Age: une perspective sur l'origine et la diffusion du mot bacallao," in *L'aventure maritime*, pp. 235-50.

68. Denys, *Histoire Naturelle*, p. 65.

69. Denys, *Histoire Naturelle*, pp. 151-53, 173.

70. Denys, *Histoire Naturelle*, pp. 197-98.

71. Denys, *Histoire Naturelle*, pp. 162-66.

72. Denys, *Histoire Naturelle*, p. 222.

73. Denys, *Histoire Naturelle*, p. 177ff.

74. See Denys' *Histoire Naturelle* for illustrations of some of this infrastructure.

75. Denys, *Histoire Naturelle*, pp. 89-101.

76. Bernard, *Navires et gens de mer*, p. 819; Denys, *Histoire Naturelle*, pp. 82-86.

77. Denys, *Histoire Naturelle*, p. 150.

78. La Morandière, *Histoire*, p. 95.

79. *Cf.* Denys, *Histoire Naturelle*, p. 171 and R.C. Smith, "Vanguard of Empire: 15th- and 16th-century Iberian Ship Technology in the Age of Discovery," unpub. Ph.D. dissertation, Texas A&M University, 1989, pp. 173-77.

80. Denys, *Histoire Naturelle*, pp. 62-64, 111, 117ff., 171, 217, 229.

81. These marked differentials in pay were typical of maritime employment in the period; see Mollat, "French Maritime Community", p. 123; Tittler, "English Fishing Industry"; K.R. Andrews, "The Elizabethan Seaman," *Mariner's Mirror* 68(3) (1982), pp. 245-62.

82. Denys, *Histoire Naturelle*, p. 62; Ganong (ed.), *Description and Natural History*, p. 272n.

83. *Cf.* Pope, "South Avalon Planters," p. 334ff.

84. Denys, *Histoire Naturelle*, p. 59.

85. La Morandière, *Histoire*, p. 135ff.

86. Turgeon, "Notre 16e siècle," p. 541, notes a decline from one-third to one-quarter in crew's shares out of Bordeaux in the late-16th-century; Denys speaks of a one third share in the late-17th century.

87. Denys, *Histoire Naturelle*, pp. 59-62.

88. La Morandière, *Histoire*, p. 53.

89. La Morandière, *Histoire*, p. 128ff.

90. La Morandière, *Histoire*, p. 122ff.

91. Brière, *La pêche francaise*, pp. 109-32.

92. Mollat, "The French Maritime Community," p. 122 and Review of Bernard, p. 131.

93. Hitchcock, *Politique Platt* (1580).

94. Bernard, *Navires et gens de mer*, p. 814.

95. La Morandière, *Histoire*, p. 53.

96. Bernard, *Navires et gens de mer*, p. 815. Mollat, *Le commerce maritime normand*, p. 265, suggests rates of 30 to 40% in the early-16th century. Hitchcock, *Politique Platt* (1580) says 35%.

97. Bernard, *Navires et gens de mer*, p. 815.

98. Denys, *Histoire Naturelle*, pp. 57, 58.

99. Mollat, *Les gens de mer en Atlantique*, p. 190; La Morandière, *Histoire*, p. 59ff; Turgeon, "Pêches Basques," Figure 6.6, p. 244; Ferreira, "L'incorporation du Nord-Ouest ibérique," p. 339.
100. Cited in Bernard, *Navires et gens de mer*, p. 821.
101. La Morandière, *Histoire*, reproduces an illustration of c. 1679, which shows six distinct types of boats of the port of Bayonne alone and another, from the same source, with Rochellais boats "used on the coasts of America and Canada and also for the fishery."
102. Denys, *Histoire Naturelle*, p. 117ff.
103. Denys, *Histoire Naturelle*, p. 65.
104. Denys, *Histoire Naturelle*, pp. 230-31.
105. *Cf.* Denys, *Histoire Naturelle*, pp. 155, 156.
106. Barkham, "Documentary Evidence," p. 55.
107. La Morandière, *Histoire*, p. 91.
108. *E.g.*, S. Barkham, "A Note on the Strait of Belle Isle during the Period of Basque Contact with Indians and Inuit," *Etudes Inuit Studies* 4 (1980), pp. 51-58.
109. Barkham, "Documentary Evidence," p. 57; J.A. Tuck and R. Grenier, "A 16th-century Basque Whaling Station in Southern Labrador," *Scientific American* 245(5) (1981), pp. 180-90.
110. Anon., Provision list for the *Guadalupe* (1566); Anon., Provision list for the *Concecio* (1566); both in Barkham, "Documentary Evidence," 1982, pp. 90-92, 93-95; Charter Party for the *Saint-Esprit* (1552), *loc. cit*; and Denys, *Histoire Naturelle*, p. 58.
111. D. Dickner, "L'alimentation des marins pêcheurs au XVIe siècle-début XVIIe siècle: de la fonction nutritive à la fonction culturelle," unpublished paper, Institut d'Histoire de l'Amérique Francaise conference, Trois-Rivières, Quebec, 1988.
112. On the latter see La Morandière, *Histoire*, p. 79.
113. Dickner, "L'alimentation des marins-pêcheurs."
114. Judging by Denys' endorsement of fishermen's diet at that time.
115. Provision lists, *Guadalupe* and *Concecion* (1566).
116. Denys, *Histoire Naturelle*, pp. 143, 172.
117. Denys, *Histoire Naturelle*, p. 171ff.
118. M. Barkham, "Aspects of Life aboard Spanish Basque Ships during the 16th century," Microfiche Report no. 75 (Ottawa: Parks Canada, 1981), p. 10.
119. Denys, *Histoire Naturelle*, pp. 177, 186-87.
120. Denys, *Histoire Naturelle*, pp. 74-76.
121. On the finds see J.A. Tuck, "A 16th-century Whaling Station at Red Bay, Labrador," in *Early European Settlement and Exploitation in Atlantic Canada,* ed. George M. Story (St. John's: Memorial University of Newfoundland, 1982), pp. 41-52.
122. Turgeon, "Notre 16e siècle" and "Basque-Amerindian Trade."
123. Peter Pope, "The Perquisite of Portage in the Early Modern North Atlantic: Introduction to an Issue in Maritime Historical Anthropology," unpublished paper presented to Canadian Historical Association, Montreal, 1995.
124. M. Barkham, "Aspects of Life."
125. Marcel Delafosse, "Les premiers développements du commerce des forrures à La Rochelle," in *La traite de la fourrure, les Français et la découverte de l'Amérique du Nord*, Musée du Nouveau Monde (La Rochelle, 1990), pp. 12-15; Raymonde Litalien, "Marins basques, ports normands et Nouvelle-France" and Laurier Turgeon, "Pêcheurs basques du Labourd dans le golfe et l'estuaire du Saint-Laurent au XVIe siècle," in *L'aventure maritime*, pp. 417-27 and pp. 213-34.

chapter
three

PORTUGAL'S COD FISHERY IN THE 16th CENTURY: MYTHS AND MISCONCEPTIONS

Darlene Abreu-Ferreira

Most books on Portugal's 16th-century overseas exploits mention Newfoundland, but it is difficult to locate a study in which the Portugal-Newfoundland connection is accurately and adequately developed. Opinions abound on Portuguese pre-eminence in the early cod fishery, but these affirmations are usually shrouded in nationalistic garb, and most fail to provide adequate evidence for their claims. The reason for this lack of evidence is quite simple: none exists. Despite this, or perhaps because of it, study after study has elevated the Portuguese-Newfoundland connection to mythic proportions.

The secondary literature on the early Portuguese connection with Newfoundland is weak, albeit colourful. Although hardly any research has been done on the early modern Portuguese cod fishery off Newfoundland,[1] fantastic statements about the importance of Portugal's involvement in the region abound. Samuel Eliot Morison, for example, alleged that Newfoundland was practically a transatlantic province of Portugal in the first quarter of the 16th-century. David Quinn, moreover, suggested that "unofficially" Newfoundland was considered "almost part of the Iberian peninsula itself," in the 16th and 17th centuries.[2] Neither historian provided any evidence for these claims.[3]

Some historians, including a few Canadian scholars, have gone to great lengths to maintain the legend that the Portuguese were major players in the 16th-century cod fishery. George Patterson contended that contemporary sources, such as reports from Richard Hakluyt, underestimated the Portuguese presence in Newfoundland waters.[4] D.W. Prowse argued that the lack of evidence should not impede generous conclusions about the Portuguese role in the early cod fishery.[5] Finally, Harold A. Innis claimed that much evidence exists on the early Portuguese cod fishery, but that little of it was ever divulged.[6]

Because of the scanty evidence available for a strong Portuguese presence in Newfoundland waters in the early-16th century, historians have traditionally followed one of three paths: omit any reference to sources, cite one another as sources, or rely on the policy of secrecy allegedly practised by Portuguese monarchs. This secrecy theory has been especially popular among those wishing to prove that the Portuguese were in North America by the mid-15th century, if not earlier. Portugal kept little information on these explorations, the argument has gone, in order to safeguard its findings from other competing nations. Since early Portuguese chronicles contain much on Portugal's overseas expansion but nothing on Terra Nova, it has been argued that they were tampered with in order to protect state secrets.[7] R.A. Skelton believed that some of the most revealing original documents on 15th- and 16th-century Portugal were destroyed during the Lisbon earthquake of 1755,[8] a possibility that will remain forever safe from scrutiny. As for the element of secrecy, Skelton suggested that since the coasts of Newfoundland provided rich fishing grounds, fishers of any nationality would want to keep the location secret;[9] apparently Bristol fishers maintained this policy as well.[10]

Not surprisingly, the "policy of secrecy" has been expounded especially by Portuguese historians, notably Eduardo Brazão, Jaime Cortesão, and Vitorino de Magalhães Godinho. Brazão, who referred to "nos secrets si bien gardés sur la navigation," contended that numerous and contradictory errors on earlier maps and charts were sometimes made purposely by Portuguese officials in order to confuse their rivals.[11] Cortesão claimed that Portugal knew of America by 1452 at least, when it began its serious work in the northwestern Atlantic. These early expeditions were undertaken clandestinely, he stated, particularly during the time of Henry the Navigator (1394-1460), who wished to build a monopoly, a view also shared by Magalhães Godinho.[12] It should be noted, however, that a number of historians have flatly rejected all claims of secrecy. Morison contended that the "phony voyages are the alleged secret ones," and that secrecy arguments have been put forth by all western countries, but especially by Portugal. Malyn Newitt pointed out, perhaps more succinctly than most, that when nothing can be shown to have happened during a particular period, Portuguese historians were fond of stressing the policy of secrecy as proof of a "great national plan." He concluded outright that these historians were simply "schizophrenic."[13]

Although the "policy of secrecy" debate adds little value to the question of Portuguese involvement in the North Atlantic region, it is worth noting because it has often been used by historical apologists to explain the lack of concrete material on the Portuguese in 16th-century North America. Yet, a careful examination of the few surviving records throws serious doubts on some of the most blatant assumptions found in the secondary literature about the early Portuguese cod fishery. Notwithstanding the probability that some records have been lost, a void is still a void. While historical interpretation should not be solely document-driven, the fact remains that *nothing* has been uncovered showing a systematic, large-scale, and long-term Portuguese presence in and around Newfoundland in the early modern period. Admittedly, few historians have made such a bold claim outright, but many have intimated its possibility if only more were known, and especially if Spain had left Portugal alone.

Conventional wisdom maintains that Spain's annexation of Portugal, 1580-1640, made the latter vulnerable to Spain's traditional enemies—the French, English, and Dutch—and that this led to an onslaught of attacks on Portugal's overseas stations and oceanic voyages. Iberian historians continue to debate the merits of this assessment, but there is little evidence to connect the ascendancy of Phillip II (r.1556/1580-98)[14] to Portugal's disappearance from the north Atlantic. Insufficient documentation impedes a proper reconstruction of Portuguese involvement in Newfoundland's cod fishery prior to and during the annexation period; even less is known about campaigns to assault only Portuguese fishing ships. Though some records show that a few Portuguese cod-carrying vessels were attacked in the early 17th century, an equal number of references exist on the English, French, and Dutch attacking one another in and around Newfoundland waters. There were even cases of English pirates attacking English ships. Sir Walter Raleigh, in his expedition of 1618, "taxed" fishers all along Newfoundland's harbours, and kidnapped 130 men.[15] Clearly piracy and privateering posed a real threat to everyone engaged in overseas trade.

Despite the contradictory evidence, historians have had difficulty resisting the notion that Spain's annexation was a source of Portugal's ills. The Portuguese fishing fleet was especially targeted, the argument has gone, because after 1580 Portugal was considered part of Spain's burgeoning empire, a source of resentment in northern Europe. Gillian Cell, for example, concluded that the "unfortunate Portuguese bore the brunt of Drake's fury" during his 1585 naval attack in Newfoundland, and that his raid in fact killed the Portuguese fishery. If that were true, then the Portuguese fishery could not have been very significant at that time, since Drake apparently raided 20 Portuguese ships only, some of which were not fishing vessels.[16] Portuguese historians, too, have long maintained that the English and Dutch war against Phillip II eliminated the Spanish and Portuguese from the Newfoundland fishery.[17] The Spanish Armada of 1588 in particular ruined the Portuguese cod fishery, some have argued, because many Portuguese vessels involved in the fishery were pressed into the Armada and were destroyed during the campaign. Indeed, the effect of the Spanish annexation must have been profound, for although it ended in 1640, the Portuguese fishery did not recover until the 19th century.[18]

Not all historians have promulgated the annexation theory. As early as the late-19th century, J.P. Oliveira Martins put the full blame for Portugal's decline on the Portuguese themselves. In his opinion, the Portuguese marine was lost long before the loss of independence. He blamed incompetent administration for the decay of Portugal's navigation, stating that by the time of King D. Sebastian (r.1557-78), the country's prospects were dim. A navigational law of 3 November 1571 aimed to promote and protect Portuguese shipping. It prohibited the use of foreign bottoms to transport goods to any of Portugal's overseas posts, including the Atlantic islands, offered financial incentives to the shipbuilding industry, and opened overseas trade to Portuguese merchants. But it was too little too late.[19] As Fernand Braudel put it, the union of Spain and Portugal in 1580 was a union of two monumental weaklings.[20]

Whether the decline of Portugal's cod fishery was caused by piracy, the Spanish Armada, or internal factors is ultimately irrelevant because there was probably little Portuguese activity in Newfoundland waters. Prior to and during the Spanish annexation, Portugal's involvement in the Newfoundland cod fishery was sporadic at best. Though incomplete data reinforces this intermittency, it is substantially correct. Existing sources suggest that cod was only occasionally available at local markets in northern Portugal. Even Ernesto Canto, a flagrant nationalist, minimized Portugal's role in the early Newfoundland cod fishery.[21] In fact, he contended that only in the third quarter of the 16th century did Newfoundland begin to be called "Terra do Bacalhao," following the increase of European activity in the area.[22] Indeed, European interest in Newfoundland was haphazard for the first quarter of the 16th century, and a systematic cod fishery did not emerge until mid-century or later.

A fair amount of documentation has survived from the early modern period showing Portuguese dealings with large volumes of cod. Admittedly, much less is available for the 16th than for the 17th century, but enough exists to construct a reasonable hypothesis about Portuguese involvement in the early cod fishery.[23] What all these records fail to do, however, is prove that the Portuguese were first among western Europeans in Newfoundland waters in the modern period, or that they were at any time truly successful in the 16th-century cod trade. The evidence suggests that cod was not a prominent fixture in Portuguese kitchens in the 16th century, and that Portugal's interest in the trade was ill-defined and intermittent. Though success is a relative term, the documentation shows no sign of Portuguese control of the cod fishery, even if such a goal were ever pursued. Periodic participation in an economic activity does not amount to domination.

With the exception of Portugal's contribution, the history of the 16th-century cod fishery in its international dimension is fairly well understood, but a brief overview of the major participants will be considered in order to place the Portuguese in historical perspective. Overall, better documentation exists for the French, Spanish, and English presence at Newfoundland than that which has survived for the Portuguese, suggesting further that Portugal's role in the early cod fishery was minimal. For example, a French record from 1510 describes a conflict on board a vessel involved in the trade of "poysson...de la Terre Neusfve,"[24] and an early French reference to the "dixme du poesson," dated 1514, stipulated that the tax applied to all fish from everywhere, including "morues" from "la Terre-Neuffve."[25] A total of 104 voyages to Newfoundland were registered in Bordeaux from 1517 to 1549,[26] but a large-scale fishery did not materialize until the 1540s and 1550s. One of the first French Atlantic port towns to show a real interest in Newfoundland,[27] Bordeaux sent out 20 vessels per year in the 1550s, and 40 in the 1560s, while Rouen accounted for 73 in 1549 and 94 in 1555.[28]

The French fished off the Avalon Peninsula in the 16th century, but they also showed an interest in the Gulf and bank fisheries.[29] The bank or green fishery was preferred by some because two and even three voyages could be made annually, and the bank fish was considered "larger, thicker, and sweeter" than inshore cod.[30] French green cod was especially appreciated in the Paris region, a market supplied by the French Channel ports as well as Biscayan vessels, particularly those from les Sables d'Olonne.[31] The French Basques, on the other hand, fished off the south and west coasts of Newfoundland, as well as along the Gaspé Peninsula and Acadia.[32] They concentrated on the dry fishery in part because of different market demands. Basque catches went to southern France and Spain, where wet fish spoiled easily in the hot climate.[33]

The Spanish Basques arrived a little later in Newfoundland waters than their French Basque neighbours, probably because the former were profitably engaged in the West Indian trade during the early-16th century. As the century progressed, however, the Spanish Basques took up the Newfoundland fishery. In the first half of the 16th century, Basque expeditions often combined cod fishing and whaling, but specialization became the norm later on, with whaling in the Strait of Belle Isle, and cod fishing on the west coast of Newfoundland.[34] Both Selma Barkham and Laurier Turgeon have concluded that the Basque fishery was relatively successful until the end of the 16th century, when Phillip II's costly Armadas severely undermined Basque shipping.[35]

Phillip II's foreign policies have often been blamed for the collapse of the Spanish and Portuguese fishery, as well as the rise of the English presence in Newfoundland, but this argument needs further analysis. Embargoes and the impressment of ships and mariners for royal campaigns had their effect on maritime commerce in Spain, but the correlation between them has not been properly investigated. Reasons for the decline of the Spanish cod fishery were undoubtedly complex and multifaceted; increased English involvement in Newfoundland was certainly a factor, but hardly the only one.[36] Keith Matthews, for one, has contended that the English cod fishery expanded not because of the defeat of the Armada, but because of the Elizabethan law of 1563 that established Wednesdays and Saturdays as fish days, and the exemption from import taxes on fish from Newfoundland. The object was to encourage fish consumption and thus promote the fishery to serve as a training ground for mariners. Furthermore, religious wars in France (1562-98) meant that the French could not always meet their own needs; interestingly, English cod exports were first sent to France, not to the Iberian Peninsula.[37]

The English cod fishery off Newfoundland became significant only toward the end of the 16th century.[38] According to Charles de la Morandière, the English were at first less interested in the Newfoundland fishery because they were adequately supplied with cod from Iceland,[39] though this could not always have been true, given that foreigners were still supplying the English with fish in 1584.[40] By 1597 the English were selling cod to the Dutch, Irish, French, and Spanish.[41] It was not until the first two decades of the 17th century, however, that the English cod fishery off Newfoundland, New England, and the Gulf of Maine was truly developed.[42]

Harold Innis concluded that the English cod fishery in Newfoundland was successful owing to a concentration of the trade in the West Country, and to the employment of large carrying vessels, or sack ships, to transport cod directly from Newfoundland to European and Mediterranean markets.[43] Matthews, too, noted that the English cod fishery grew simultaneously with general commercial growth in the West Country; still, Newfoundland did not become the "stay of the West-Countries" until the 17th century.[44] According to Innis, the English did not have adequate salt deposits in their midst and thus were forced to establish a sedentary fishery in the Avalon Peninsula. This arrangement allowed them to trade with foreign vessels for supplies and to concentrate on producing dry cod, which required less salt.[45]

The Portuguese were among those who provided the English with salt right in Newfoundland,[46] and according to Anthony Parkhurst's report of 1578, the Portuguese numbered 50 sail that year.[47] Parkhurst also stated that the Spanish had approximately 130 ships in the region, including 20 or 30 whalers; but Selma Barkham's archival research has shown that there could not have been nearly as many Spanish vessels in Newfoundland as Parkhurst claimed.[48] If this were true for the Spanish, could Parkhurst have been wrong about the Portuguese as well? The Portuguese, French, Spanish, and English began fishing for cod off the coasts of Newfoundland at different times in the 16th century. Of the four major national groups, the Portuguese are the least understood because their contribution has been more difficult to pinpoint—possibly because theirs was the least significant.

Historians have suggested that in the early modern period Portugal was an "agrarian monarchy" as well as a "maritime monarchy," meaning both that the king owned much of the land in the countryside, and that the crown dominated the maritime economy through its monopoly of trade.[49] Yet even if royal intervention in Portugal's economy led to "monarchical capitalism,"[50] there is little evidence that the crown committed itself to anything beyond the large-scale operations of its overseas empire. While the monarchy had a well-developed system of collecting royal revenues throughout the country, it did very little to promote the national economy. Towns were empowered to regulate local markets, but without consistent support from above, the Portuguese effectively had to settle for a piecemeal economy.

Nowhere is this more apparent than in Portugal's fishery. Though fishers and mariners were protected through certain royal privileges,[51] and many belonged to *confrarias*, or confraternities, that enjoyed some advantages,[52] there is no evidence that the crown sought to foster a national fishery.[53] The monarchy was quick to collect fish tithes, but did little to nourish the trade. The fishery actually declined over the 16th century as more and more fishers and mariners sought their fortunes in the Asian and Brazilian trades.[54] Equally ruinous to the national fishery was the pressing of fishing vessels for overseas trade, a trend that began in the early 1500s but became more pronounced towards the end of the century.[55] Since the crown would have wished to protect its own intake, this suggests that the fishery was not a major source of revenue.

In his work on medieval Portugal, A.H. Oliveira Marques cautioned that, in the Middle Ages at least, the fishery was a minor industry in Portugal compared with agriculture. Despite its long coast, Portugal did not provide as much fish as has been believed, primarily because of its coastal winds and storms.[56] As a result, from early times the Portuguese were forced to seek fish in other regions. Those in the south fished off the coasts of Africa, while northerners went to French and English waters. In the Treaty of 1353, Edward III of England granted the Portuguese the right to fish off the coast of England for the next 50 years. Paradoxically, the Portuguese fishery was open to foreigners as well. The Florentines were involved in the large-scale sardine fishery off Portugal from as early as 1399, and in whaling and the tuna fishery in the Algarve in the mid-15th century.[57]

It would be helpful to know whether or not the Portuguese were motivated to participate in the early cod fishery off Newfoundland because of insufficient fish supplies at home, but evidence of Portuguese fish-eating habits is lacking. Little work has been done on consumption patterns in Portugal,[58] and while 16th-century municipal records for Ponte de Lima and Braga[59] reveal something about their respective fish markets, the details are contradictory and inconclusive. Both sets of records deal with municipal regulation of food prices; Braga's documents referred to cod in 1561 and 1582, while Ponte de Lima's dealt with cod in 1581 and 1583. Though in both cases the records are incomplete, and references to cod are infrequent, the available data suggest that in the 16th century cod was often more expensive than beef, and significantly more costly than most other local fish.[60] Furthermore, a greater amount of other types of fish was found in Braga's records, suggesting that cod only occasionally reached a market in Portugal.

The scanty references to cod in 16th-century Portuguese markets coincide with the overall meagre evidence of Portugal's involvement in the cod fishery. Early Portuguese documents dealing with Newfoundland are conspicuously silent about cod, unlike corresponding English records for the same period. Following John Cabot's 1497 voyage, the English crown was informed of land surrounded by water "covered with fish which are caught not merely with nets but with baskets," and "that they will fetch so many fish that this kingdom will have no more need of Iceland, from which country there comes a very great store of fish which are called stock-fish."[61] Though there has been some debate over the actual location that Cabot visited during his search for a northwestern passage to Asia, there is little doubt that the fish about which he and his companions boasted was the North Atlantic cod, likely sighted in or near Newfoundland waters.

The Portuguese also organized reconnaissance voyages to the region during this period, but their reports emphasized different things. The original reports have not survived, but Albert Cantino, who was residing in Lisbon when news of the Corte Real expeditions arrived,[62] wrote a letter to a contact in Italy in which he highlighted the Corte Real findings. His description included sightings of land covered with "trees and pines of such measureless height and girth, that they would be too big as a mast for the largest ship that sails the sea."[63] Though Portuguese explorers noted the abundant fish in the area, they appear to have been more interested in what they saw ashore.

Reasons for the dissimilarity between the Cabot and Corte Real reports may never be fully understood. Part of it could be that the two expeditions took different routes. H.P. Biggar has concluded that the Corte Real expedition explored the southeastern coast of Labrador, not Newfoundland.[64] But another possible explanation for the differences between the Cabot and Corte Real reports is that the returning adventurers knew how best to impress their respective monarchs. The English may have had a greater need for fish for the populace, though historians have not reached a consensus on this issue.[65] The Portuguese, with their eagerness to conquer the East, needed timber to build ships. Indeed, another second hand report by Pietro Pasqualigo, dated 18 October 1501, stated that the Corte Real findings:

will be most useful to [the king's] plans in several respects, but chiefly because being very near to his kingdom, he will be able to secure without difficulty and in a short time a very large quantity of timber for making masts and ships' yards, and plenty of men-slaves, fit for every kind of labour, inasmuch as they say that this land is very well populated and full of pines and other excellent woods. And said news has so pleased his majesty that it has made him desirous of sending ships again to said region, and of increasing his fleet to India....[66]

It appears, therefore, that the Portuguese were more impressed with the resources on land than with those in the sea. This would explain Portugal's early interest in establishing a settlement in the area, notably the Fagundes colonization of 1521,[67] and the 1567 venture by Manoel Corte Real, nephew of Gaspar and Miguel Corte Real. Having inherited the Terra Nova captaincy,[68] he received royal approval to send three ships with passengers and supplies to begin a settlement in Newfoundland.[69] For reasons unknown, Portuguese attempts to colonize the area were unsuccessful, but it is significant to note that, unlike England in the 17th century, Portugal's interest in Newfoundland in the 16th-century was probably not motivated by a desire to control the cod fishery.

Even the numerous royal letters confirming the Corte Real family's patent to Newfoundland, dated 1500, 1506, 1522, 1538, 1574, and 1579, never mentioned fish.[70] A document dated 2 May 1537 dealing specifically with fish tax revenues in the island of Terceira, Azores, did not refer to cod either.[71]

This absence of cod in Azorean records is noteworthy, since many historians have claimed that Portuguese vessels returning from the banks of Newfoundland stopped off in the Azores before reaching Portugal.[72] The earliest available port records for Ponta Delgada, São Miguel, show no such connection, and a collection of copied documents, dated 1568-1603, makes only general comments about fish taxation by the local municipality without identifying any fish categories.[73] The earliest originals, dated 1603-38, do not mention cod, or any fish for that matter; most references are to woad, wheat, and sugar.[74] It is possible that ships coming into port for shelter or for provisions would not have their merchandise listed, yet one would expect that some of the cod would stay in the Azores.[75]

The Azorean historian Julião Soares de Azevedo found only two possible references to cod passing through the Azores in the 17th century, both documents from French archives. First, notarial records from La Rochelle showed four vessels were contracted in 1633 to take dry fish and train oil[76] to Terceira. Second, the French Consul in Faial, Azores, reported on 26 November 1686 that the English had great commercial dealings in the Azores. The account stipulated that an average of 70 to 80 vessels passed through the Azores on their way to Barbados, Jamaica, and the Carolinas, loaded with cod, train oil, herring, and other salted fish, as well as wood products and manufactured goods from England. Unfortunately it is not clear whether the English left any of these goods in the Azores.[77]

Nothing has been found to support the notion of Portuguese pre-eminence in the early cod fishery. One would think that such a long-standing assumption would be based on some solid foundation, but that is not the case. This view has survived for so long because no one has bothered to put the premise to a rigorous test. For instance, the most-cited document used to prove Portugal's early dominance of the Newfoundland cod fishery is the *alvará* from King D. Manuel I, dated 14 October 1506. The royal letter refers to fish from "Terra Nova" arriving in a few northern ports located between the Douro and Minho Rivers, and royal rights to revenues from this fishery.[78] The document has been construed to mean that the Portuguese cod fishery at Newfoundland was already well established by 1506, and extremely profitable, given the interest of the crown to regulate its tithes.[79] This interpretation is rather optimistic, and not very true to the spirit of the letter. First, the document is a copy, not the original letter of 1506, and it is not known when the copy was made. The importance of this point is that the scrivener titled the document "*Trelado de hua Carta del Rey, nosso Senhor, acerqua da Dízima dos bacalhaos,*" or "Copy of a Letter from our Lord the King in regard to the Tithe of Codfish." But while the title refers to cod, it must be stressed that the letter itself does not. It is possible that the fish in question was cod, but even that cannot be ascertained. The reference to Terra Nova, too, should be viewed with caution, for in the early modern period several "newly-discovered" lands were referred to as "Terra Nova."[80]

Another crucial point about this royal letter is that the crown's desire to control the fish tithes in northern Portugal has been interpreted as an indication that the cod fishery must have been well-developed. Yet the document gives no such hint. What appears to have been a concern to the king, and what motivated the order in the letter, is that rights to fish tithes in some northern Portuguese ports were being granted to individuals. The crown's interest, therefore, was to affirm its own jurisdiction over the collection of fish tithes.

Ironically, the few records of a Portuguese presence in Newfoundland waters in the 16th century are not of Portuguese origin. One of the earliest is a report dated November 1527, made by Spanish officials in the West Indies after they apprehended an English ship. The English informed their captors that they had been on a discovery mission, and had explored Newfoundland where they saw approximately 50 fishing vessels, including Spanish, French, and Portuguese.[81] An English account of the 1527 voyage indicated that "eleven saile of Normans, and one Brittaine, and two Portugall Barkes" were spotted, "all a-fishing."[82] English court records also offer some testimony on the Portuguese in Newfoundland waters, including complaints by merchants/mariners from Viana and Aveiro of English privateering attacks on Portuguese ships between 1579 and 1591.[83]

Portuguese documents, on the other hand, seldom mention anything about an actual Portuguese fishery in Newfoundland. Some scattered records allude to a cod trade in the early-16th century,[84] but there is little evidence that the Portuguese themselves procured their own cod supplies. Because most early port records did not survive, there is no way of verifying the identity of ships and shipmasters entering Portugal's northern ports in this period. In Aveiro, the best customs records are available from the mid-18th century only.[85]

Enormous voids in the original sources hinder any attempts at definitive conclusions, but the hypothesis that Portugal possessed a significant Newfoundland fishery or even a special interest in imported cod during the 16th century is not supported by the available evidence. The lack of documentation on Portugal's 16th-century maritime endeavours is real, extensive, and regrettable, and any discussion of the early Portuguese-Newfoundland connection must take this into account. But nothing has survived to indicate that the Portuguese had a *bona fide*, established, and regulated cod fishery in the 16th century. Portugal's active participation in Newfoundland was generally sporadic. All indications point to a lack of commitment to the Newfoundland cod fishery, and to Portual's subsequent dependence on cod imports in the 17th century.

Notes

1.   A noteworthy exception is found in Viana do Castelo where serious archival work has been done on the city's early cod trade. See Manuel António Fernandes Moreira, "O Porto de Viana do Castelo e as Navegações para o Noroeste Atlântico," in *Viana—o Mar e o Porto* (Viana do Castelo: Junta Autónoma dos Portos do Norte, 1987).

2.   Samuel Eliot Morison, *Portuguese Voyages to America in the Fifteenth Century* (Cambridge: Harvard University Press, 1940), p. 72; David B. Quinn, "Newfoundland in the Consciousness of Europe in the Sixteenth and Early Seventeenth Centuries," in *Early European Settlement and Exploitation in Atlantic Canada*, edited by G.M. Story (St. John's: Memorial University of Newfoundland, 1982), p. 17.

3.   Unsubstantiated and exaggerated claims about Portugal's early connection to Newfoundland have been made by many more historians, on both sides of the Atlantic. This is true of older works as well as more recent studies. See, for example: Franco Maria Ricci et al, eds., *Portugal-Brazil: The Age of Atlantic Discoveries* (Lisbon: Bertrand Editoria/Brazilian Cultural Foundation, 1990), p. 99; G.V. Scammell, *The First Imperial Age: European Overseas Expansion c.1400-1715* (London: Unwyn Hyman, 1989), p. 277; A.H. de Oliveira Marques, *Portugal Quinhentista (Ensaios)* (Lisbon: Quetzal Editores, 1987), p. 175; John A. Dickinson, "Old Routes and New Wares: The Advent of European Goods in the St. Lawrence Valley," in *Le Castor Fait Tout: Selected Papers of the Fifth North American Fur Trade Conference, 1985*, (Montréal: Lake St. Louis Historical Society of Montréal, 1987), p. 33; Mário Moutinho, *História da Pesca do Bacalhau: por uma antropologia do "fiel amigo"* (Lisbon: Editorial Estampa, 1985), pp. 19-23; Fernand Braudel, *Civilization and Capitalism, 15th-18th Century: The Perspective of the World*, Vol. 3, translated from the French by Siân Reynolds (New York: Harper & Row, 1979), p. 143; Samuel Eliot Morison, *The European Discovery of America: The Northern Voyages, A.D. 500-1600* (New York: Oxford University Press, 1971), p. 225; René Bélanger, *Les Basques dans l'Estuaire du Saint-Laurent, 1535-1635* (Montréal: Les Presses de l'Université du Québec, 1971), p. 9; Carl O. Sauer, *Northern Mists* (Berkeley: University of California Press, 1968), p. 57; J.H. Parry, *The Establishment of the European Hegemony, 1415-1715: Trade and Exploration in the Age of the Renaissance* (New York: Harper & Bros, 1961), pp. 76-78; Jaime Cortesão, *Teoria Geral dos Descobrimentos Portugueses* (Lisbon: Seara Nova, 1940), p. 36.

4.   George Patterson, "The Portuguese on the North-East coast of America, and the first European attempt at Colonization there. A lost chapter in American History," *Transactions of the Royal Society of Canada*, Sec, 2, Vol. 8 (1890), p. 146.

5.   D.W. Prowse, *A History of Newfoundland from the English, Colonial, and Foreign Records* (London: Eyre and Spottiswoode, 1896), pp. 35, 44-45, 59.

6.   Harold A. Innis, *The Cod Fisheries: The History of an International Economy* (Toronto: University of Toronto Press, r.1954 [1940]), pp. 13-14; and "The Rise and Fall of the Spanish Fishery in Newfoundland," *Transactions of the Royal Society of Canada*, Section II, Third Series, Vol. 25 (1931), p. 51.

7.   See, for example, Eduardo Brazão, *La Découverte de Terre-Neuve* (Montréal: Les Presses de l'Université de Montréal, 1964), p. 20; Edgar Prestage, *The Portuguese Pioneers* (London: Adam & Charles Black, r.1966 [1933]), pp. 168, 171, 211; David B. Quinn, *North America from Earliest Discovery to First Settlements* (New York: Harper and Row, 1977), p. 110; Dan O'Sullivan, *The Age of Discovery, 1400-1550* (London: Longman Press, 1984), pp. 11-19; David Arnold, *The Age of Discovery, 1400-1600* (London: Methuen, 1983), pp. 11-12.

8.   R.A. Skelton, *Explorers' Maps: Chapters in the Cartographic Record of Geographical Discovery* (New York: Frederick A. Praeger, 1958), p. 47.

9.   R.A. Skelton, *The European Image and Mapping of America, A.D. 1000-1600*, The James Ford Bell Lectures, No. 1, (1964), p. 15.

10.   James A. Williamson, *The Cabot Voyages and Bristol Discovery Under Henry VII* (Cambridge: Hakluyt Society, 1962), p. 43.

11.   Brazão, *La Découverte de Terre-Neuve*, pp. 20, 28.

12. Cortesão, *Teoria Geral dos Descobrimentos*, pp. 38-39; Vitorino de Magalhães Godinho, *Dúvidas e Problemas à cerca de algumas teses da história da expansão* (Lisbon: Edições Gazeta de Filosofia, 1943), p. 5. All three historians failed to provide concrete proof for their assertions.

13. Morison, *Portuguese Voyages*, pp. 73-81, and *The European Discovery of America*, 82; Malyn Newitt, *The First Portuguese Empire*, (Exeter: University of Exeter, 1986), pp. 13-14. Also see Henri Harrisse, *The Discovery of North America* (Paris: H. Welter, 1892), p. 272, and Duarte Leite, *História dos Descobrimentos* (Lisbon: Edições Cosmos, 1958), p. 360. More recent work in Portugal has also questioned the entire theory of secrecy. See Francisco Contente Domingues, "A Política de Sigilo e as Navegações Portuguesas no Atlântico," *Boletim do Instituto Histórico da Ilha Terceira*, Volume XLV, Tomo 1 (1987), pp. 189-220.

14. Phillip II's reign in Spain began in 1556 and in Portugal in 1580. He invaded and captured Portugal following the deaths of Portuguese rulers D. Sebastian in 1578 and Cardinal Henry in 1580, both of whom left no heirs.

15. Innis, *The Cod Fisheries*, p. 62.

16. Gillian Cell, *English Enterprise in Newfoundland, 1577-1660* (Toronto: University of Toronto Press, 1969), pp. 47-48.

17. The most commonly-cited expert on this question in Portuguese historiography is still A.A. Baldaque da Silva, *Estado Actual das Pescas em Portugal* (Lisbon: Imprensa Nacional, 1892), p. 166 (see his footnote).

18. Moutinho, *História da Pesca do Bacalhau*, p. 22.

19. J.P. Oliveira Martins, *Portugal nos Mares*, Vol. 1 (Lisbon: Ulmeiro, r.1988 [1881]), pp. 36-39.

20. Fernand Braudel, *La Méditerranée et le Monde Méditerranéen à l'Epoque de Philippe II*, Deux Volumes, 9e édition (Paris: Armand Colin, 1990 [1949]), Vol. 1, p. 207.

21. For reasons of efficiency and clarity, all cod references in this study will be associated with "Terra Nova" —that is, present day Newfoundland— even though this was not always stipulated in the records. It is recognized that some of the cod reaching Portugal's ports in the early modern era may have come from other regions, notably from waters near Greenland and especially from Norway. See Moses Bensabat Amzalak, *A Pesca do Bacalhau* (Lisbon: Museu Comercial, 1923), p. 8; Braudel, *La Méditerranée*, Vol. I, p. 177; O.A. Johnsen, "Les relations commerciales entre la Norvège et l'Espagne dans les temps modernes," *Revue Historique*, T.CLXV (Paris, 1930), pp. 77-78, as cited in Virginia Rau, *Estudos sobre a História do Sal Português* (Lisbon: Editorial Presença, 1984), p. 171; and "Adventencias para la yntroducción del Trato y Navegación de la Provincia de Terranova y Norvega en los Puertos de Cantabria," *Colección Vargas Ponce III*, No. 68, as cited in Innis, "The Rise and Fall of the Spanish Fishery in Newfoundland,"p. 65.

22. Ernesto do Canto, ed., *Archivo dos Açores,* 14 vols. (Ponta Delgada: Universidade dos Açores, r.1983 [1882]), Vol. 4, p. 415. A letter written 13 September 1512, however, in which King Ferdinand of Aragon requested a visit with Sebastian Cabot, mentioned the "Isla de los Bacalhaos," or "Island of the Codfish." See "Sebastian Cabot Consulted about Newfoundland," cited in H.P. Biggar, ed., *Precursors of Jacques Cartier, 1497-1534: A Collection of Documents Relating to the Early History of the Dominion of Canada* (Ottawa: Government Printing Bureau, 1911), pp. 115-16. Furthermore, the Ruysch map of 1508 includes the name "In. Baccalauras" (present day Baccalieu Island), and the Jorge Reinel planisphere c.1519 has "Bacalnaos" on the eastern coast of Newfoundland. See H.P. Biggar, "The voyages of the Cabots and of the Corte-Reals to North America and Greenland, 1497-1503," *Revue Hispanique*, Nos. 35-36 (r.1961 [1903]), p. 556; Armando Cortesão and Avelino Teixeira da Mota, eds., *Portvgaliae Monvmenta Cartographica*, 6 vols. (Lisbon: Comemorações do V Centenário da Morte do Infante D. Henrique, 1960), Vol. 5, unpaginated.

23. The best archival documentation on the Portuguese early cod trade was found for the 17th century, and especially from 1639 to 1679. See my Ph.D. Dissertation, "The Cod Trade in Early-Modern Portugal: Deregulation, English Domination, and the Decline of Female Cod Merchants," (Memorial University of Newfoundland, History Department, 1995), chapters 3 and 4.

24. "Pardon to the Mate of a Newfoundland fishing-vessel," cited in Biggar, *Precursors of Jacques Cartier*, pp. 116-18. For more references to the French early fishery, see Charles de la Morandière,

Histoire de la Pêche Française de la Morue dans l'Amérique Septentrionale (des origines à 1789), Tome I (Paris: G.-P. Maisonneuve et Larose, 1962), pp. 38, 77, 161, 164, for examples.

25. "Agreement between the monks of the Abbey of Beauport and the inhabitants of the Island of Brehat, wherein mention is made of Newfoundland cod," cited in Biggar, *Precursors of Jacques Cartier*, pp. 118-23; see also pp. 124-27, 132-33, 159, and 177-82.

26. Jacques Bernard, *Navires et Gens de Mer à Bordeaux (vers 1400-vers 1550)*, Trois Volumes (Paris: S.E.V.P.E.N., 1968), Vol. 2, p. 807.

27. La Morandière, *Histoire de la Pêche Française*, p. 250.

28. Laurier Turgeon, "Le temps des pêches lointaines: permanences et transformations (vers 1500 - vers 1850)," in *Histoire des Pêches Maritimes en France*, sous la direction de Michel Mollat (Toulouse: Bibliothèque Historique Privat, 1987), p. 137.

29. Innis, *The Cod Fisheries*, pp. 23-26, 46.

30. Innis, *The Cod Fisheries*, p. 47.

31. La Morandière, *Histoire de la Pêche Française*, 185; Innis, *The Cod Fisheries*, pp. 48-49.

32. The Basques were also involved in trade with the local inhabitants in the 16th-century. See Laurier Turgeon, "Basque-Amerindian Trade in the Saint Lawrence during the Sixteenth Century: New Documents, New Perspectives," *Man in the Northeast*, No. 40 (1990), pp 81-87.

33. La Morandière, *Histoire de la Pêche Française*, pp. 308, 344. Owing to the more complex preparation process, the dry fishery employed a greater number of personnel than the wet fishery, but La Morandière suggested that because the Basque region was relatively poor in agriculture, the sedentary fishery was more feasible because a pool of cheap labour was available from the hinterland.

34. Selma Barkham, "Documentary Evidence for 16th-century Basque Whaling Ships in the Strait of Belle Isle," in *Early European Settlement and Exploitation in Atlantic Canada, Selected Papers*, edited by G.M. Story (St. John's: Memorial University of Newfoundland, 1982), pp. 53-56.

35. Barkham, "Documentary Evidence," 62; Laurier Turgeon, "Sur la Piste des Basques: La redécouverte de notre XVIe siècle," *Interface* (Sept-Oct, 1991), pp. 14-15.

36. Gillian Cell, especially, expounded the view that English expansion of the Newfoundland fishery was intricately connected with the demise of the Spanish fishery, though she also noted that sources for English activities in Newfoundland prior to 1610 are scattered in "bits and pieces." See Cell, *English Enterprise in Newfoundland*, pp. 22, 36.

37. Keith Matthews, "A History of the West of England-Newfoundland Fishery," Ph.D. Dissertation (Oxford University, 1968), pp. 45, 57.

38. Despite the excited tone of Cabot's reports about the abundant fishing grounds he "discovered" in 1497, English participation in the Newfoundland cod fishery was limited in the first half of the 16th century, though some evidence survives showing an English presence already in the early 1520s. See "A projected expedition to Newfoundland," and "Notices of the return of the English fishing fleet from Newfoundland," cited in Biggar, *Precursors of Jacques Cartier*, pp. 134-43. The English government also passed an Act on 25 February 1548/9 that the Newfoundland fishery was open to everyone in the kingdom. Leo Francis Stock, ed., *Proceedings and Debates of the British Parliaments respecting North America, 1542-1688* (Washington: Carnegie Institution of Washington, 1924), Vol. 1, 2.

39. La Morandière, *Histoire de la Pêche Française*, p. 221.

40. Innis, *The Cod Fisheries*, 31. Unfortunately it is not clear who exactly supplied this fish, and how much of it was cod from Newfoundland.

41. Innis, *The Cod Fisheries*, p. 32.

42. Innis, *The Cod Fisheries*, pp. 70-72.

43. Innis, *The Cod Fisheries*, p. 54.

44. Matthews, "History of West of England," pp. 57-61. Quote from Sir Walter Raleigh's speech to British Parliament on 23 March 1592/3, in Stock, *Proceedings and Debates*, Vol. 1, 7. The number of English ships engaged in the Newfoundland fishery for the first three decades of the 17th century ranges from 150 to 600 annually, depending on the source. Innis, *The Cod Fisheries*, pp. 69-70.

45. Innis, *The Cod Fisheries*, p. 36. Not everyone agrees with this view, however. C. Grant Head, for example, has argued that the English specialized in dry cod not because of a lack of domestic salt

but because of a lack of a domestic market for cod. The English had to export their cod to warmer regions in southern Europe, and this cod had to be well dried otherwise it would deteriorate. See Head, *Eighteenth Century Newfoundland: A Geographer's Perspective* (Toronto: McClelland and Stewart/The Carleton Library, 1976), p. 72. This argument might be valid if the English concentrated on Mediterranean markets, for the French had no problem selling large volumes of green cod, or *pasta*, to the Portuguese in the first half of the 17th century, while the English imported dry cod, or *vento*, primarily. See my Ph.D. Dissertation, "The Cod Trade in Early-Modern Portugal," Chapter 4.

46.   Innis, *The Cod Fisheries*, p. 35. Innis suggested that by 1578 the English had established some form of "overlordship" in the region, especially in connection to the Portuguese in return for protection from the French, but the nature of this arrangement is not very clear. It also seems strange that the English would be in a position to "protect" the Portuguese when Parkhurst stated that the number of English and Portuguese vessels engaged in the fishery at that time was approximately the same.

47.   "A letter written to M. Richard Hakluyt...containing a report of the true state and commodities of Newfoundland, by M. Anthonie Parkhurst," cited in H.A. Innis, ed., *Select Documents in Canadian Economic History, 1497-1783* (Toronto: University of Toronto Press, 1929), p. 10.

48.   Barkham, "Documentary Evidence," p. 58. She concluded that there were probably an average of ten Spanish Basque whaling ships involved annually. Though she was less specific about the number of cod fishing vessels, Barkham rejects the greatly exaggerated figure of 200 Spanish Basque ships in Newfoundland from 1570 to 1586. See Selma Huxley Barkham, "Guipuzcoan Shipping in 1571 with Particular Reference to the Decline of the Transatlantic Fishing Industry," in *Anglo-American Contributions to Basque Studies: Essays in Honor of Jon Bilbao*, edited by William A. Douglass, Richard W. Etulain, and William H. Jacobsen (Desert Research Institute Publications on the Social Sciences, No. 13, 1977), p. 79 (her note 5).

49.   Frédéric Mauro, "Political and Economic Structures of Empire, 1580-1750," in *Colonial Brazil*, edited by Leslie Bethel (Cambridge: Cambridge University Press, 1991 [1987]), p. 39.

50.   Nuñez Diaz, cited in Fernand Braudel, *Civilization & Capitalism, 15th-18th Century*, Vol. 2, p. 444.

51.   Fishers in Porto, for example, were exempted from forced military service, except in cases where the monarch participated in the expedition as well. Arquivo Histórico Municipal do Porto (A.H.M.P.), *Livro das Vereações*, No. 37 (1606), f.199-f.200v, as cited in Francisco Ribeiro da Silva, *O Porto e o seu Termo (1580-1640): Os Homens, as Instituições e o Poder*, 2 vols. (Porto: Câmara Municipal do Porto/Arquivo Histórico, 1988), Vol. 1, pp. 186-87. In 1503 Aveiro fishers were granted the same liberties and privileges enjoyed by fishers elsewhere in the kingdom, though it is not clear what all those privileges and liberties were. See António Gomes da Rocha Madahil, ed., *Colectânea de Documentos Históricos, 959-1516*, (Aveiro: Edição da Câmara Municipal de Aveiro, 1959), pp. 252-53; the "original" is in *Tombo da Confraria de Santa Maria de Sá*, f.86v, a duplicate of which is in the Arquivo Distrital de Aveiro.

52.   For instance, members of the "Confraria da Casa do Corpo Santo de Setúbal" were granted tax exemptions in the purchases of materials essential for their line of work. See Rau, *Estudos sobre a História do Sal*, p. 172.

53.   There is evidence that the crown offered Aveiro financial support in the early-16th century to stimulate the local shipbuilding industry, but it is not clear whether the ships were destined for the fishery or for the Asian trade. See Rocha Madahil, *Colectânea*, Vol. 1, p. 262; original is in the Arquivo Nacional da Torre do Tombo (A.N.T.T.), *Corpo Cronológico*, Part I, M. II, Doc. 6.

54.   Samuel Eliot Morison agreed that after the discovery of Brazil, "who cared for codfish, mast trees, and icebergs?" In his *European Discovery of America*, p. 210; see also Bruce G. Trigger, *Natives and Newcomers: Canada's "Heroic Age" Reconsidered* (Kingston/Montréal: McGill-Queen's University Press, 1985), p. 120.

55.   Oliveira Martins, *Portugal nos Mares*, pp. 207-13.

56.   Frédéric Mauro, too, has contended that Portugal had a dangerous coast which made maritime endeavours quite difficult: "Des caps battus par les vents, des estuaires de torrents souvent ensablés au XVIIe siècle; il n'y a pas de quoi attirer spécialement les flottes de commerce." See Mauro, *Le Portugal et l'Atlantique au XVIIe Siècle (1570-1670)* (Paris: S.E.V.P.E.N., 1960), pp. 89-90.

57. It is not clear whether the Florentines in question were naturalized Portuguese citizens or actual Italian merchants involved in the fish trade in the Mediterranean. A.H. Oliveira Marques, *Portugal na Crise dos Séculos XIV e XV* (Lisbon: Editorial Presença, 1987), pp. 109-12.

58. The most comprehensive work on the early period was done on fruit and fish consumption at the court of King D. Afonso V (r.1438-1481), whose records for 1474 reveal that meat was almost always preferable to fish, though fresh sole and flounder were also appreciated. Maria José Azevedo Santos, "O Peixe e a Fruta na Alimentação da Corte de D. Afonso V—Breves Notas," *Brigantia: Revista de Cultura*, Vol. 3, No. 3 (July-Sept. 1983), p. 309.

59. Both inland towns in northern Portugal. Unfortunately, no similar documents on cod prices have been located for Portuguese coastal towns where fish prices were probably different from prices found in the interior.

60. João Gomes d'Abreu Lima and Ovidio de Sousa Vieira, "Ponte de Lima nas Vereações Antigas," *Arquivo de Ponte de Lima*, Vol. 2 (1981), pp. 16-17, 20; Vol. 4 (1983), pp. 12-13. Câmara Municipal de Braga "Acordos e Vreações da Câmara de Braga...1561/1582," *Bracara Augusta*, Vol. 30 (1976), p. 766; Vol. 28 (1974), pp. 534-536; Vol. 27 (1973), pp. 620-621; Vols. 25-26 (1971/1972), p. 455; Vol. 24 (1970), p. 433.

61. "Second Dispatch of Raimondo di Soncino to the Duke of Milan [London, 18 December 1497]," cited in Biggar, *Precursors of Jacques Cartier*, pp. 19-21.

62. Gaspar and Miguel Corte Real were brothers from the Azores who led expeditions to Newfoundland or Labrador (the exact geographic location is still debated) in 1501 and 1502, respectively.

63. "Dispatch of Albert Cantino from Lisbon to the Duke of Ferrara, Hercules D'Este [Lisbon, 17 October 1501]," cited in Biggar, *Precursors of Jacques Cartier*, pp. 63-65.

64. Biggar, "The voyages of the Cabots," p. 579.

65. Some historians have argued that the English demand for fish in the early period was small because their needs were met from their own inshore fishery and from Icelandic waters. See Matthews, "History of the West of England," p. 35, and Innis, *The Cod Fisheries*, pp. 13-14. Head, too, has contended that there was no English market for cod, in *Eighteenth Century Newfoundland*, p. 72. Prowse, however, claimed that "codfish was gold in these old days" for the English nation in the first half of the 16th century, because only the very rich had access to fresh meat during the winter months. He contradicted himself, however, when he stated that Elizabethan attempts to create a Protestant Lent were unsuccessful because "the proud stomachs of our ancestors have always resented interference with their beef and beer." Prowse, *History of Newfoundland*, pp. 18, 56.

66. "Letter from Pietro Pasqualigo to the Signiory of Venice," cited in Biggar, *Precursors of Jacques Cartier*, pp. 66-67.

67. "Confirmation of the Letters Patent to Fagundes," cited in Biggar, *Precursors of Jacques Cartier*, pp. 127-31.

68. This document suggests that in 1567 the Corte Real family, descendants of Gaspar and Miguel Corte Real, still claimed rights to Terra Nova. What exactly this meant in real terms is not clear.

69. A.N.T.T., "Carta régia de 4 de Maio de 1567, de que consta mandar Manoel Corte Real uma expedição de 3 navios com colonos da ilha Terceira para povoarem a Terra Nova," *Livro 6 dos Privilégios de D. Sebastião*, f.237, cited in Canto, *Archivo dos Açores,* Vol. 4, p. 537. See also the "Tratado das Ilhas Novas e Descobrimento dellas...," cited in Biggar, *Precursors of Jacques Cartier*, pp. 195-97.

70. Canto, *Archivo dos Açores*, Vol. 4, pp. 497-502.

71. "Alvara de merce do disimo do pescado da ilha Terceira a Pedro Annes do Canto, 2 maio de 1537," cited in Canto, *Archivo dos Açores*, Vol. 12, pp. 410-11.

72. See endnote 3.

73. Alfândega de Ponta Delgada (A.P.D.), *Livro 1o dos Registos (1568-1603),* ff. 43-45.

74. A.P.D., *Livro 3 de Registos, 1603-1638.* Book No. 2 has not survived, and No. 1 is not a port registry but a collection of regulations dealing with customs and tariffs imposed on merchandise entering the port. Woad, or *pastel*, is a plant that yielded a blue dye, and in which the Azores were prolific.

75. Another possibility is that Terceira, and not São Miguel, was the port of call for Newfoundland ships in the early modern period, but unfortunately this could not be confirmed. After a visit to the

local archives, followed by two years of correspondence, the public library and archives at Angra do Heroismo has had to concede that its Terra Nova collection has been misplaced.

76. Train oil was oil produced from the fat of whale, seal, or cod.

77. Julião Soares de Azevedo, "Nota e documentos sobre o comércio de La Rochelle com a Terceira no século XVII," *Boletim do Instituto Histórico da Ilha Terceira*, Vol. 6 (1948), p. 20. The author concluded that the dry fish was cod but it is not clear whether the French document stipulated that or not.

78. A.N.T.T., "Trelado de hua Carta del Rey nosso Senhor acerqua da Dízima dos bacalhaos," in *Livro dos Registos del Rei noso snor, das cartas & alvaras, mandados & outras cartas que o dito Snor manda a esta Alfândega*, Nucleo Antigo, No. 110, f. 46. Portuguese text and English translation also found in Biggar, *Precursors of Jacques Cartier*, pp. 96-98.

79. The tendency has been to quote one another's interpretation of this document. See Matthews, "History of the West of England," p. 38; Innis, *The Cod Fisheries*, p. 14; Prowse, *History of Newfoundland*, p. 15; Patterson, "The Portuguese on the North-East coast," p. 145.

80. Referring to a document from 1570, George Patterson explained that "Terra Nova was at this time used generically for all the newly-discovered lands to the north-west, and not specifically for Newfoundland, though it was the best known." See, Patterson, "The Portuguese on the North-East Coast," 163 (his note 2). The Cantino planisphere of 1502 refers to the region as "Terra del Rey de portuguall"; the "Kunstmann III" chart of c.1506 shows "Terra de Cortte Riall"; and the Lopo Homem-Reinal atlas of 1519 mentions "Terra Corte-Regalis." See, Cortesão and Teixeira da Mota, *Portvgaliae Monvmenta*, Vol. 5. The Ruysch map of 1508, with map names in Latin, indicates "Terra Nova" for part of Newfoundland. See Biggar, "The voyages of the Cabots," pp. 502-03.

81. "La rrelacion que se ovo de la nao ynglesa quando estovo en la ysla de la Mona que venia de camino para la ysla espanola," cited in Biggar, *Precursors of Jacques Cartier*, pp. 165-77.

82. Letter written by John Rut of the "Mary of Guilford," from St. John's harbour on 3 August 1527, cited in Innis, *The Cod Fisheries*, pp. 12-13.

83. Maritime History Archive - Memorial University of Newfoundland, Frances Fernando, "To the Queenes maties most honorable pryvie Councell [1579]," 16-A-1-013; Thomas Pyres, "Examination [1583]," 16-A-1-005; John Heimers, "Examination [1583]," 16-A-1-004; William Durston, "Examination [1591]," 16-A-1-006. See also Cell, *English Enterprise in Newfoundland*, p. 47.

84. I have explored this theme in more detail in a paper entitled, "The Portuguese at Newfoundland: Documents Examined," *Portuguese Studies Review* (University of New Hampshire), Volume 4, Number 2 (Fall-Winter, 1995-96), pp. 11-33.

85. Early port records have not survived for Lisbon, Porto, Viana, and most other major maritime communities in Portugal.

Eighteenth-Century Fishery

Credit: National Archives of Canada

chapter
four

THE FRENCH FISHERY IN NORTH AMERICA
IN THE 18TH CENTURY

Jean-François Brière

Although they were not the first Europeans to reach Newfoundland, the French fol-lowed close behind the English and Portuguese in exploiting the extraordinary cod resources of the northwestern Atlantic. The first ships set out from Brittany for the Newfoundland fishery in 1508 (one ship from Bréhat), from the Basque ports in 1512, and from Nor-mandy in 1524. Western Europeans' exploitation of the northwest Atlantic cod seems to have been a response to a latent need of the western European economy in the early-16th century. The wealth of North America, unlike that of South America, lay initially in the sea, and exploitation of this wealth did not lead directly to colonization. The fishery, indeed, had a repetitive pattern; it involved returning regularly to the same known locations along the coasts, and provided no incentive for exploration.

Two types of fishing were practised as early as the 16th century. The green-cod fishery was carried on offshore, on the banks; the fish was simply salted on board and immediately taken back to Europe for sale. The dry-cod fishery was practised from chaloupes based on shore; the fish was salted and dried on the spot, then transported to the Atlantic or Mediter-ranean regions of southern Europe. Dry cod withstood the heat better and kept better; for that reason, it very early became a major article of export from the cold countries to Medi-terranean, subtropical ones: Italy, Provence, the Iberian Peninsula and, later, the West Indies. The French practised both types of fishing until the 19th century: green cod was sold north of the Loire, dry cod south of the Loire. In contrast, the English, whose domestic market remained small, in the second half of the 16th century became great exporters of dry cod to the richest and most heavily populated countries of southern Europe: France, Portugal, Spain, and Italy.

Notable changes appeared in the Newfoundland fishery in the course of the 17th cen-tury. After the rapid decline of the Spanish and Portuguese fisheries, a de facto Franco-English monopoly was established; it was given concrete expression by the permanent settlement of small colonies of fishermen on the eastern and southern coasts of Newfound-land, around St John's (English) and Plaisance (French). Numerically and economically weak, these colonies were not a threat to the fishing vessels arriving each year from France and England. But the fishermen of New England, a colony that became populated much more quickly than Newfoundland and had more varied resources, soon acquired a range of action that brought them into conflict with French fishermen and colonists frequenting the banks and coasts of Acadia. Commercially, they were already a threat to the English cod trade, particularly in the West Indies, where the growing masses of slaves and the crews of European ships offered a growing market for the sale of dry cod from New England.

The extension of colonization and the development of trade relations between Atlantic Europe and the rest of the world in the second half of the 17th century gave the Newfoundland fishery a political role. This fishery trained large crews who were well seasoned and easy to enrol in the royal navies, since they returned to the mother country each winter. French government officials became aware of this role during the reign of Louis XIV and began to regulate and support the Newfoundland fishery. The 18th century marks the geographic, economic, and political peak of the French fishery in North America: at no other time did it mobilize so many ships and men or hold so large a place in the minds of rulers.

The world of the transatlantic Newfoundland fishery was subject to technical and geopolitical constraints very different from those of conventional maritime trade and colonization. It seems clear today that the most serious threat to the French migratory fishery in North America did not come so much from England as from French and English fishermen who settled on the shores of North America.

Financial Organization

The financial organization of the Newfoundland fishing fleets did not differ from that of the conventional commercial ones. The outfitters were either individual or family companies, or else—which was much more frequently the case—partnerships of two to 15 investors, nearly always linked by close personal relationships. Each associate (*consort*) would purchase capital shares (from 1/64th to 1/2) in the ownership of one or more ships and would be assigned equivalent shares of the profit or loss at the end of the voyage. The outfitter—generally the chief investor, but not the only one—received a commission. Captains often took shares in the vessels they commanded. Such partnerships remained loosely structured, forming and dissolving with each expedition. The origin of the investments was local or regional: traders, people of means, members of the *parlements,* and other officeholders would invest their money in maritime undertakings. Companies with capital from outside the region investing in the Newfoundland fishery—companies such as that of the Baron d'Huart in the 1750s—remained exceptional cases.[1] Versatility was the rule in the maritime trade in the 18th century, and no trader was, strictly speaking, specialized in the Newfoundland fishery. Similarly, ships were fitted out for all sorts of destinations at the same time, depending on the profit outlook. If in certain ports, such as Granville, traders fitted out only for the cod fishery, it was because no better option was available to them.

The Green-cod or Migratory Fishery

The French ships fitted out for fishing green cod on the banks—the bankers (*banquais*)—were, in the 18th century, nearly always small or very small vessels: 50 to 100 *tonneaux*, occasionally 150 *tonneaux*; some did not exceed 25 or 30 *tonneaux*. Most of them had two masts, but variety was the rule in the types of hull and rigging: snows (*senaus*), doggers (*dogres*), sloops (*corvettes*), schooners (*goélettes*), and brigs (*bricks*). Ships with one or two decks, but no quarterdeck, were more practical on the banks, since the absence of quarterdecks facilitated fishing operations.

The supplies for a Newfoundland fishing vessel fell into three essential categories: provisions, salt, and fishing gear. The bankers did not call on shore—a unique situation for voyages that could last more than six months—and thus had to take with them all the food and fresh water necessary for their crews until they returned to France. The crews were therefore very vulnerable to scurvy. To lack provisions meant risking an early return halfway through the fishing season. The royal authority gradually imposed standards to be followed in this area.

Salting and drying were the main ways of preserving fish in the 18th century.[2] Without salt, there could be no cod fishery. The salt supply was therefore a question of vital importance. But salt was a problem commodity primarily because the production sites were geographically limited. In France, the great salt-producing regions extended from southern Brittany (Le Croisic) to the shores of the Aunis and Saintonge (Brouage, Isle de Ré, Seudre River). It also could be obtained in Spain (Cadiz) and in Portugal (Setubal, Averio). Salt also was a commodity unlike others because of the taxes placed on it. The differences in status between provinces with regard to the salt tax (*gabelle*) singularly complicated the situation, for they resulted in enormous variations in the price of salt from one province to another. However, salt used for the cod fishery had been exempted since 1680.[3] The counterpart to this privilege was that Newfoundland vessels whose home ports were in non-exempt regions were required by law to pick up their salt directly at the production sites after they had departed for the fishery, because the introduction of tax-free salt into the province of their home port was still prohibited. In the course of the 18th century, most of the ports involved in the Newfoundland fishery that were located in non-exempt regions obtained authorization to keep stores of tax-exempt fishing salt in their territory, under close administrative surveillance.[4] Ships could thus obtain supplies of salt in their home ports. Nevertheless, ships commonly travelled to the Ile de Ré at the start of their voyages, in order to take advantage of the lower prices there.

Fishing gear—lines, hooks, leads, knives, gaffs (*piquoirs*), nets, etc.—was an important part of the supplies of a banker. It took up little room and was of fairly small total value. All the products necessary for supplying Newfoundland fishing vessels had been exempted, since 1717, from payment of numerous duties, particularly export duties.[5] To prevent smuggling, the quantity of products that could be embarked duty-free on each ship was strictly limited by law to the presumed needs of the crew. The law, then, put consumption on the Newfoundland fishing vessels on the same basis as domestic consumption.

Banker crews were very small: 10-20 men in general, including officers. The largest vessels might have up to 30 men, the smallest only six or seven. The cramped conditions aboard vessels, the impossibility of fishing over both sides of a craft at once (one had always to be "to windward") and the absence of stopovers in the course of the voyage account for such extremely small crews. The ship's function as an instrument of work narrowly limited the flexibility of the physical ratio of men to ships. No other transoceanic navigation was practised with such small crews.

The composition of these crews followed very precise rules related to fishing techniques or imposed by the administration of the Marine department. The officers were frequently limited to a captain and a second in command. One or two lieutenants would be taken aboard larger vessels. The bankers never took on a chaplain, as their crews were fewer than 40, the threshold above which embarkation of a chaplain was required by law.[6] Most also set out without a surgeon: "We are looking for a seaman who knows how to perform a bleeding," notes Duhamel du Monceau.[7] But when the crew exceeded 19, a surgeon was required.[8] This legal requirement seems to have been well respected. It certainly explains the very high proportion of crews of 19 men among the bankers (particularly at Granville) and it no doubt contributed to limiting the number taken aboard. In reality, there was less fraud concerning the presence of the surgeon (which was checked by the officers of the Admiralty) than concerning the surgeon's qualifications.

The crew of a banker included two to four petty officers—former seamen with solid technical experience in a given trade—as splitter, salter, carpenter, or header. Ordinary "rated" seamen were at least 18 years old and had two years of navigation experience behind them (17 years and one year of navigation starting in 1745).[9] These were the real fishermen, the *lignotiers*. One to three ship's boys, ranging in age from 12 to 18, rounded out the crew.[10] In 1739, the government created the echelon of apprentice (*noviciat*), intermediate between those of boy (*mousse*) and rating (*matelot classé*) for men from 16 to 25 years of age without previous navigation experience. In 1745, the government required all outfitters to take one apprentice aboard for every four crew.[11] Thus on the average, the bankers had to take on from two to six apprentices—hence the extreme youth of the Newfoundland crews, whose average age was always under 30 years.

The green-cod fishery at the end of the century seems to have been centred in Normandy (109 ships out of 192 in 1786, or 56.5% of departures) and northern Brittany (63 ships out of 192, or 32.7% of departures). The Atlantic ports, despite their nearness to salt supplies, lost the leadership role in this fishery that they had held at the beginning of the century. The Channel ports gradually monopolized the fitting out of vessels for the banks.

The bankers would leave France in February or March, in advance of the ships bound for inshore fishing along the coasts of Newfoundland and Canada. The banks were always accessible, and cod was present there practically year-round; but storms and icebergs made fishing difficult and very dangerous in the winter, and few people ventured to try it. Taking on salt on the Isle de Ré would occasion a delay of two weeks to a month on the outward voyage to the banks, which explains the very early departures of the cod vessels from Normandy. Fishing in spring and summer was most common; this was the best time of year, because of the prevailing south-southwest winds, which brought fine weather. Some ships would leave between the end of April and the end of June for a fishing voyage with a lateness premium (*prime de retard*). The late fishery (*pêche de tard*), with a departure in July or August, was exceedingly rare; the only bankers that practised it were those that made two voyages per year, which was exceptional in the 18th century, except at Les Sables d'Olonne.

The ships, packed with salt, went to take up their positions between the 43rd and 45th parallels, in the same latitude as the southern edge of the Grand Bank. In case of a departure very late in the season, they would position themselves between the 45th and 47th parallels, as the fishery then shifted toward the north.[12] They would then run straight toward the west. The Atlantic crossing lasted about a month, often a bit more, but never less than three weeks. During the crossing, the crew would begin preparations for fishing by installing the *bel*, a movable wooden gallery a meter in width that ran along the port side, overhanging the rail. The approach to the banks was indicated by a change in water color, fogs, an abrupt drop in air temperature, and flights of birds. Once these signs had been registered, the sounding lead was cast; when it touched bottom, the ship was "banked."

Once at its selected position, the ship furled its sails. The fishing was done with the vessel adrift and the wind on the port beam. Only a small mizzen sail, *culetin* or ring-tail (*tape-cul*), was lowered, so that one could beat up slightly to windward and maintain the drift. Since the cod were on the bottom, too strong a drift had to be avoided.

The fishermen lined themselves up on the staging or *bel*, facing to windward so that the lines did not pass under the vessel. Precautions were taken to ensure that the fishermen were not overly bothered by bad weather and were not at risk of falling overboard. In order to fish without getting soaked and to maintain their balance while keeping both hands free, the *lignotiers* stood in a kind of barrel; these barrels, widening out toward the bottom, were arranged along the *bel* and firmly secured to it. Over the upper part of their bodies, they wore a coat of tarred cloth, the lower part covering the outside of the barrels. A *cuirier*, a large leather apron treated with grease and oil, and *manigots*, leather mitts, completed their outfit.

Each fisherman had two lines from 75 to 90 fathoms in length, at the end of which was a large lead weighing 5-6 pounds. Attached to this main line was a very short secondary line, the snood (*avançon*), to the end of which the hook was fastened, baited with a *boette* (bait) consisting of a piece of salt herring or cod entrails. As soon as the cod was lifted aboard, the *lignotier* would hook it onto the *élangueur* ("tonguer"), an iron stake placed near by him. The fisherman would cut out the tongue and keep it to count the catch. He would then remove the stomach (*meulette*) to see whether it contained any shellfish that could be used as bait.[13] Behind the *lignotiers*, on the deck, was a sort of table, the *étal*, around which three fishermen were installed, firmly wedged in their barrels: the header (*étesteur*), the splitter (*habilleur*), and the *nautier*. The detongued cod slipped along to the header through a wooden conduit or else was handed to him at the end of a *piquoir*, a kind of long pointed gaff. The header would cut the head off the cod and gut the fish, carefully preserving the liver, which was placed in a special barrel. The splitter, frequently the captain, would then open up the cod with a cleaver and remove the bones and part of the spine (*raquette*). The splitter then passed the cod to the *nautier*, a boy, who would detach the air bladder (*noue* or *naut*), in order to salt it separately. The cod were then thrown into an opening (*éclaire*) in the middle of the table, from which they slid into the *carniau*, a kind of pipe which took them directly to the hold. There the salter would rub them in salt, then let them drain for 24 or 48 hours, after which they were salted *à demeure*, i.e. permanently. In the salt which filled the hold, the salters would dig spaces called *rums*, in which they placed the cod. The fish were arranged head to tail in *arrimes* (piles), each layer being separated from the others by a sort of floor of branches covered with a mat and a layer of salt.

The Newfoundland fishing vessels never left the banks before the end of July. The first bankers thus arrived in France in late August. There was lively competition among crews to be the first to arrive with green cod, because of the assured high price of *morue nouvelle*. The bulk of the fleet headed for home between August and October, after five or six months of fishing. They would reach the French ports between September and November.

The geographical distribution of green-cod landings was roughly the same as that of the ports where vessels were fitted out. In 1787, 75.7% of the green cod from Newfoundland arriving in France was landed north of Brest: 4.6% in northern Brittany, 55.3% in Normandy, 15.8% in Dunkirk. The following year, the ports north of Brest received 71.7% of the green cod: 4.1% in northern Brittany, 63.1% in Normandy, 4.5% in Dunkirk.[14] The small Norman ports that dominated the supply end of the business thus took the lion's share of landings as well. But the hierarchy of landing ports was appreciably different: at the top were the three ports of Dieppe, Honfleur, and Le Havre (47.5% of arrivals in 1787 and 54.2% in 1788), whose importance was a result of their closeness to the Paris market. Granville and Saint-Malo, although major supplying ports, received little green cod. Outside the Channel, there were four notable points of landing: Nantes, La Rochelle, Bordeaux, and Dunkirk. Here again, we note an increased concentration of returns on the Channel at the expense of the Atlantic coast. The bankers would return to their home ports either in ballast, with some goods for the outfitter, or with freight, goods from the islands picked up at Le Havre, tar, pitch, and timber from the north.

One of the most remarkable aspects of the green-cod fishery, at the technical level, resides in its industrial character, with its role of ensuring mass output of a finished product. The elaborate division of labour and the existence of a chain of workers repeating the same actions remind us of the fragmented labour of 20th-century factories; but it was Taylorism in appearance only, for the system arose chiefly out of the necessity of treating the cod very quickly to prevent its decomposition. The bankers, veritable floating factories on a small scale, seemed to prefigure the era of industrial mechanization.

The Dry-cod or Sedentary Fishery

The second major type of fishing practised by the French—the dry-cod or sedentary fishery—was carried out along the coasts of Newfoundland and, until 1763, the Atlantic coast of Canada. The cod could only be dried on shore. Although the extent of the North American coastline frequented by cod is immense, only a small portion of it was usable in the 18th century because of the need for a place where the ship could be moored in complete safety and where the fish could be dried. This meant a sheltered site, with a fairly wide beach, preferably covered with pebbles or gravel, well exposed to the wind but not too exposed to the sun.[15] These requirements created a qualitative hierarchy among the various sectors of the coastline, and thus resulted in competition among users for these areas.

Practised from shore—hence "sedentary" fishery—the dry-cod fishery escaped the requirement limiting the number of fishermen who applied to the banker fleets. The number of fishing chaloupes could be increased at will, and on shore one had all the room necessary for setting up operations and processing cod. The Newfoundland vessels fitted out for the sedentary fishery were therefore not fishing vessels, but rather carriers of workers and cod; for that reason, their tonnage was often greater than that of the bankers—between 100 and 350 *tonneaux*.

Supplying the ships in the sedentary fishery raised fewer problems than was the case with bankers, since water and fresh provisions could be obtained on shore. And preparing dry cod required far less salt than green cod. To prevent decomposition of the fish, cod was lightly salted before the process of dehydration by drying—slower than salting, but much more effective for preservation—began to produce its effect. The problem of salt supply was the same as for the bankers. However, when the ports situated in regions not exempt from the salt tax obtained the right to procure supplies in Brittany (an exempt region), and subsequently the right to store their fishing salt, it was mostly the sedentary fishermen who took advantage of these opportunities, while the bankers continued going to the Ile de Ré for their supplies. The fishing gear was little different from that employed on the bankers, except for the fishing chaloupes. These boats, of 4 or 5 *tonneaux*, and with one small square sail, were taken apart for the trip across the Atlantic. To avoid cluttering up the ships, fishing crews never brought them back to France. They would be left on the coast at the end of the voyage, to be used again the next year.

Crews aboard the ships in the sedentary fishery were much larger than banker crews: between 30 and 160 men, in most cases. The record for the century was no doubt set by the *Baron de Binder* (Saint-Malo, 1787) with 190 men.[16] Sedentary Newfoundland vessels carried crews three times as large as trading vessels of equivalent tonnage. Two-thirds of the men aboard the largest sedentary vessels were in fact destined for fishing work after making land. The regulations imposing quotas of youth and apprentices applied easily to sedentary-fishery vessels, since part of the work of processing the cod on shore could be done by individuals with no experience in maritime navigation or fishing.

A high proportion of Saint-Malo sedentary-fishery vessels set sail with stowaways on board, nearly always children or adolescents. Out of the 62 such vessels that left Saint-Malo in 1742, 37 carried stowaways—between one and 16 per vessel. The work of the sedentary fishery adapted very well to a surplus of labour. Captains willingly kept on these adolescents, whom they could force to work without having to pay them. They were granted absurdly low pay, very much less than what they would have obtained by joining the vessel in the regular way. The certainty of being brought back to their points of departure, of being fed, of acquiring some experience and of collecting a tiny sum of money seems to have been enough to drive these wretched boys to sneak aboard "over the side." Previous complicity by the crew and the outfitter appears to have been highly probable in many cases. Everyone benefitted from these illegal embarkations, including the royal administration, which as a matter of principle looked favourably on increasing the number of seafarers.

The fishery's reputation for "salubrity," commonly mentioned in the documents of the time, deserves some analysis: one is struck by the considerable proportion of accidental deaths, especially drownings. Accidents were a far greater threat than disease to the safety of the workers in Newfoundland. The small number of Newfoundland vessels wrecked is fairly easy to explain. To escape the risks of a storm, two solutions were available: either take shelter in a roadstead or port, or move offshore. This was the quasi-permanent status of the Newfoundland vessels, since the sedentary vessels remained at anchor in their harbor while the bankers were always at sea. With average death rates of about 1 or 2% per cruise, the Newfoundland fishery in the 18th century appears to have been much less hazardous than transoceanic voyages to the West Indies, Africa, and Asia.

Outfitting for the sedentary fleet was geographically even more concentrated than for the bank fleet. This concentration increased over the course of the century and its axis moved toward the Channel. The dry-cod fleets disappeared from the ports situated south of Brest (Nantes, Les Sables d'Olonne) or collapsed (Basque country), while Granville, a banker port, turned to the sedentary fishery and, with Saint-Malo and Saint-Brieuc, monopolized the bulk of this activity. In 1786, these three ports alone sent out 92.5% of French vessels for the Newfoundland coast![17]

Ships going to the sedentary fishery on the coasts of Newfoundland and Canada would leave France between March 15 and May 15, four weeks later than bankers. French sedentary fishermen were active in four main regions:

◆ the coasts of Newfoundland to which access was permitted by the Treaty of Utrecht; from 1713 to 1783, these coasts were situated between Cape Bonavista in the north and Point Riche in the west (French Shore). After 1783, the corresponding limits were Cape St. John and Cape Ray. The north coast, or Petit Nord, the fiefdom of the Saint-Malo and Granville fishermen, was by far the most heavily frequented. The Basques, who were less numerous, were concentrated on the west coast, in the regions of Portachoix, Ferol, and Ingornachoix.

◆ the coasts of Labrador known as Grande Baye, located to the west of the northern tip of Newfoundland.

◆ the coasts of the Gaspé, north of Baie des Chaleurs.

◆ the coasts of Isle Royale (Cape Breton Island).

Until the late-18th century, two intangible principles, on both the French side and the English side, regulated the competition for occupying harbours: 1) any portion of coast belonged to the first ship to arrive from Europe; and 2) such occupation was limited to the current fishing season. The first arrivals therefore took the best places, which explains why sedentary fishermen tried to set sail as early as possible. They—and especially those who set sail first—were engaged in a veritable transatlantic race that could prove very dangerous when the coasts were blocked by ice. To prevent clashes between crews, the usage of giving the title of admiral (*amiral*) to the captain of the first ship to arrive in a harbour became established. This captain would select his own place and then designate the places for the captains of the vessels that followed. The captian judged disputes concerning the occupation of the shoreline. These general rules were applied until the end of the 18th century wherever French sedentary fishermen were active, except in the Petit Nord of New-foundland, where special regulations, originating in Saint-Malo, were in force: all captains had to go to the harbor of Petit Maître, roughly halfway between the two extremities of the Petit Nord, to declare in writing their date of arrival there, the number of crew under their command, and the name of the harbour and place they were choosing on the coast. As they arrived, the captains thus knew what places were taken and would choose among those still available. A man belonging to the first ship kept the register and turned it over to the captain at the end of the period for occupying harbours. In this way, there was theoretically only one admiral for the whole Petit Nord coast, whose authority people could turn to in exceptional cases. This system was abolished in 1764. In 1766, the government limited the extent of coast which could be used by each ship, but apparently without great success.[18] It required captains, starting in 1770, to pick up from the *commissaire des classes* in their outfitting port a printed form on which, upon arriving in Newfoundland, they would enter the name of the harbour and the place chosen. The form corresponded to a single place, and each captain had only one form (two for large crews) that had to be given to the *commissaire* upon return.[19] This system, applied until the Revolution, did not put an end to the disputes.

Once the harbour had been chosen, everything was unloaded and the unrigged ship remained at anchor for several months. If the place had never been occupied before, the crew had to build huts for themselves and their stores, as well as the fish stage (*échafaud*), a sort of jetty extending into the water, on which a primitive shed was erected. In this shed one or more tables (*établis*) and salt tubs (*salines*) were placed. If the beach was of pebbles, it was cleaned; if it was sand, it had to be covered with *rances*, i.e. a layer of branches, or with flakes (*vignaux*), a similar surface, but suspended two feet above the ground on racks (*claies*). When everything was ready, the chaloupes went out fishing at daybreak, each with three men aboard: the chaloupe master (*maître de chaloupe*), the *avant* and the *ussat*. If the fishing looked poor, the captain could decide to send some of the chaloupes out (*en dégrat*) to other harbours. In the *grand dégrat*, practised after 1730, schooners were sent far away for several weeks; the cod were then split and salted on board.

The chaloupes would return in the late afternoon, and the fishermen's day would end as soon as they touched land. The "shore workers" (*garçons de grave* or *graviers*) would then go aboard the chaloupes, pick up the cod by the head with a gaff (*siste*) and throw it into a tub situated at the extremity of the stage platform (*galerie*). Shore workers known as throaters (*piqueurs*), standing up to their waist in the cod tub, would remove the tongues and throw the fish to the headers (*décolleurs*). These in turn would pass it along to the splitters (*habilleurs*), who with three strokes of the knife would open up the cod and re-move its backbone. The split cod would then be thrown through a hole into the *esclipot*, a kind of case whose bottom was sloped toward one side, which was closed by a sliding board; when the board was drawn to one side, the cod slipped directly onto a kind of stretcher called a *boyard*; on such stretchers, they would be carried to the salters, who were positioned at the other end of the stage. The salters would arrange the fish in superimposed layers, separated by a layer of salt. The cod were allowed to sweat for two to eight days, after which shore workers would wash them in a kind of large wooden cage (*lavoir* or *timbre*) inserted between stakes driven into the water close to shore. Then piles of cod would be built up in the form of haystacks, known as *javelles*. After eight days, the *javelles* would be taken apart and the cod spread out, flesh side up, on the pebbles or *rances*. For the first night, the cod would be placed with the skin side up (*premier soleil* or "first sun"). At noon the next day, they would be turned over, flesh side up. In the evening, they would be piled up, three by three, one above the other, with the skin side up (*second soleil* or "second sun"). The following day, they would be spread out separately, skin side up in the morning, flesh side up in the afternoon, after which they would be very meticulously arranged in piles of eight, skin side up (*troisième soleil* or "third sun"), and so on. Each evening, the piles became larger, transforming themselves into piles called *moutons* until the *septième soleil* ("seventh sun"). The cod were then left in piles for 15 days before being given a *huitième soleil* ("eighth sun"), and again for a month in piles called *pattes* (or *fumier*, for the Saint-Malo fishers) before the *neuvième soleil* ("ninth sun"), and in the same way before the *dixième soleil* ("tenth sun"). It took close to three months to obtain the finished product.[20]

The actual fishing took place over a very short period, because the cod stayed inshore for a relatively brief time—only two months, from early June to late July. The dry-cod fishery was thus a veritable race against the clock, whence the obsessive fear of the passing of time, and the extraordinarily intensive nature of the work. The sedentary fishery, a pathetic struggle between low productivity and the implacable race against time, often forced crews to work up to the extreme limit of their strength. Night work was common, and there was no scheduled day of rest.

When the drying phase was finished in September or October, the cod would be loaded into the hold of the ship, piled up on dunnage of dry wood. At the time of loading, the cod would be counted and the weight of the cargo assessed. Thirteen cod were placed on each stretcher. Every eight loads going aboard the ship made a "hundred" (*cent*) (104 cod). The chaloupes were hauled up on shore beyond the reach of the highest tides and were marked for identification the next year. Any unused salt was buried in the ground and marked with the name of the ship that owned it.

Once a Newfoundland vessel was ready to sail, a phenomenon unique in maritime navigation would take place: two-thirds of the crew would board another vessel called a *saque*, which was specialized in taking the fishermen, utensils, and oils directly back to the home port. The rest, limited to the number of crews necessary for taking the ship to the port of unloading, would remain on board. The *saques* were either sedentary vessels that had not had good fishing and had turned over their meager production to the Newfoundland vessels and taken on their crews, or else ships especially assigned for the purpose by companies with several vessels. The overcrowding on these vessels was often frightful.[21]

The ships, loaded with dry cod, would leave the coasts of North America in September or October. Some of them left much earlier, in July or August; these vessels, which belonged to companies, had made up a cargo by collecting the first cod of their fellow crews. They would thus always arrive first at the markets. Thanks to the heavily concentrated structure of their fleet, the Saint-Malo fishermen retained a quasi-monopoly on these early and highly profitable deliveries.

The discrepancy between the geographical distribution of the home ports and ports of landing was much greater for dry cod than for green cod. If the green-cod trade was carried on within a "northern" system (i.e. north of the Loire), the dry-cod trade linked northern and southern France. The great landing ports for dry cod were located south of Nantes and on the Mediterranean. North of Nantes, there were two notable landing ports, Saint-Malo and Granville. Much more important were—south of Nantes—La Rochelle, Bordeaux, Bayonne-Saint-Jean-de-Luz, Sète, and above all Marseille, the great dry-cod market where over half the landings in France were normally concentrated.

The duration of the trip from Newfoundland to Marseille was highly variable. Unfavourable winds often hindered the passage through the "straits" and some vessels put in at Cadiz, Alicante, or Malaga after crossing the ocean. The 566 Newfoundland vessels from Granville that went to Marseille from 1714 to 1792 had an average voyage of 46 days.[22]

Hailings of Newfoundland vessels by Barbary pirates were very frequent throughout the 18th century.[23] They seem to have been regarded as routine incidents by captains, who were little inclined to take risks by resisting. Their fears resided chiefly in the consequences that physical contacts with Algerians would have on arrival in Marseille: any vessel "inspected" ran the risk of being quarantined in the same way as vessels coming from North Africa.[24] Thus the captain of the *Pressigny*, in December 1784, took care to note that "he showed his papers with caution, without any member of his crew going aboard the Barbary vessel, or any member of the latter's crew having come aboard his own."

The call at Marseille could last from a few weeks to several months, depending on the difficulty of finding return freight. The captain sometimes remained in Marseille to look after the sale of the cod, while the ship set sail again for the west; but this practice tended to fall into disuse over the course of the century, since most outfitters had consignees on the spot. The prospect of obtaining good return freight gave captains an incentive to arrive in Marseille as early as possible. Only the earliest arrivals were assured of getting any; the later ones often had to go home in ballast. The freight taken on at Marseille consisted of products from Provence (soap, olive oil, dry fruit, and wines) or the Levant (cotton fabrics, yarn wool). When Marseille had no more freight to offer, ships would head for Sète, Toulon, or the Italian ports of Nice, Genoa, Leghorn, or else to Alicante, Malaga, and Cadiz, in Spain. Only a small part of the freight picked up in the Mediterranean ports was destined for the home ports of the Newfoundland vessels. Most of it was delivered to Le Havre (for Paris), with smaller amounts going to Bordeaux, La Rochelle, Nantes, and Dunkirk. From these ports, the cod vessels returned to their home ports with new freight or else in ballast, about a year after their departure. If the ship completed its loop too late—after April—it could not be fitted out for the current year's sedentary fishing season. Here again, companies with several ships held the advantage: by using two, three, or four vessels at once, it was possible, to some extent, to escape the time factor. The different possible combinations in the use of several ships gave these fleets a flexibility that individual vessels could not match.

The dry-cod fishery, like the green-cod fishery, constituted an integrated industrial and commercial system. But a certain number of characteristics were peculiar to it. Its production units employed considerable labour: several hundred men, usually. In 1786, the Saint-Malo outfitter Fromy du Puy had 1,155 men working for him in Newfoundland, which was probably a record.[25] Such fleets and crews represented very large enterprises, technically much closer to the *fabrique* or factory than to the traditional workshop. With such numbers of employees, discipline, elaborate division of labour, maximization of mass production, mechanical repetition of the same actions, and frenzied work, the whole industrial universe seemed already to be taking shape. However, here again, the extraordinarily intensive character of the work was not dictated by the call of the markets but by the extremely tenuous match between the techniques and the available production time. The fishery was constantly threatened by two factors over which humans had little control: its very limited duration, and the low productivity of a technical system based entirely on human effort. Either of these threats could have been easily avoided had the other not existed. That they existed simultaneously created a fundamental problem for the sedentary fishery that could only be resolved through the extraordinary harshness of the work requirements.

The extremely cyclical character of production imposed its rhythm on the cod-trading circuit. The phases in the three-way cod traffic—transportation of workers, transportation of fish, transportation of freight—had to be carried out on a very precise schedule, and the whole cycle could not exceed one year. The circuit from France to the Newfoundland fishery, then to the West Indies and back to France, was too long. Most of the cod shipped to the French West Indies was produced by settlers on Isle Royale. The loss of Isle Royale in 1758 put an end to any hope of providing the French West Indies with an adequate supply of cod from a base in French North America. The Anglo-Americans took over. France either had to let them do so, or else provide public subsidies to support extremely costly circuits such as the one that involved re-exporting to the West Indies cod landed in Bordeaux or Marseille.[26] The dry cod circuit therefore fit into a rigid timeframe that severely limited its geographical extension.

The Mixed Fishery and Trading

Alongside the two major fishing techniques that I have described—and these were the ones most commonly practised—there were two other methods for obtaining cod in the 18th century: the mixed fishery and trading (*traite* or *troque*). The mixed fishery was practised by ships that went fishing for cod far from shore, on the banks, for several weeks, even several months. The fish was split and salted on board, just as on the bankers. But instead of returning to France with their cargo of green cod, these vessels would bring it to the coasts of Newfoundland or the North American continent and dry it there, as in the sedentary fishery. There was no need for stages: all one had to do was "wash" off the salt before spreading it on the beaches. The bank fishery with drying operations on shore was thus a non-sedentary or itinerant form of the dry-cod fishery. It made it possible to expand considerably the grounds used for the dry-cod fishery, which were normally limited to the inshore zone. Small, schooner-type decked craft of 30 to 60 *tonneaux* were used. The schooner had fore-and-aft sails that could be worked from the deck, and therefore required only a very small crew, in the order of 7 to 12 men; its manoeuvrability and speed accelerated the trips back and forth between the coast and the banks, which were repeated several times during the course of the season.

Fishing trips to the banks with drying operations on land were a novelty in the 18th century, and were a refinement of dry-cod fishing methods. But the advance resided in the original combination of the two great systems practised up until that time, and not in the techniques, which did not change. By broadening the geographical horizon of the dry-cod fishery, the mixed system also extended its time limits, freeing it to some degree from the rigid constraints of space and time which, as we have seen, characterized the sedentary fishery.

Some French ships that came back from North America laden with dry cod had purchased their cargo from resident fishermen in exchange for delivery of goods from Europe. This cod trade (*traite, troque*) was always a secondary means for French outfitters to obtain fish. The thin human settlement on the coasts of French Canada and the prohibition against trading with the English in Newfoundland blocked the development of this type of commerce. Trading was concentrated, then, on the two zones where the density of resident fishermen was high: Isle Royale and, after 1763, Saint-Pierre-et-Miquelon.

The economic activity of the few thousand colonists living on the shores of Isle Royale and Isle Saint-Jean, and later on Saint-Pierre-et-Miquelon, was based almost exclusively on the cod fishery: there was practically no agriculture, and the only industrial facilities were a few shipyards. The population depended almost totally on imported food and finished products for its subsistence. The only local resource that could be exported was dry cod, so the exchange of fish for imported products formed the basis of the trading system of these colonies. The population of Isle Royale and the islands of Saint-Pierre-et-Miquelon, a population "displaced" from Plaisance in 1714 and from Isle Royale in 1763, had received beach areas in compensation for losses suffered at the time of the expulsions. Since the colonists shared the best beaches, only the less favourable ones remained accessible to ships from France.

The inhabitants of Isle Royale and Saint-Pierre-et-Miquelon practised both types of dry-cod fishery: the sedentary fishery with chaloupes, as on the coast of Newfoundland (but without ships), and the itinerant fishery with schooners. It took at least two chaloupes per enterprise to reach the break-even point. Any fishing enterprise thus included at least ten men: six for the two chaloupes, and four for the beach. The schooner fishery required a minimum of 14 men: ten on board and four on shore. Since local labour was insufficient, the inhabitants would recruit fishermen from France, who hired on for one fishing season.

Saint-Malo, Bayonne, and Saint-Jean-de-Luz dominated this fishery throughout the 18th century. Trading fleets had less room to maneuver than fishing fleets. Since the production process was beyond their control, it was difficult for them to be the first to bring cod to the markets.

Crew Payment Systems

The way in which the crews of Newfoundland fishing vessels were paid was a very original aspect of the French maritime fishery under the Ancien Régime. The pay systems in commercial shipping and the navy were based on wages for time worked, each man being paid based on the number of days he had spent aboard. Two types of navigation were excepted from this arrangement: privateering and fishing.[27] In some cases, the crew's activity produced a certain amount of wealth in the form of vessels taken or in the form of fish. The profitability of voyages depended on the work done by the crews. The crew's pay was, therefore, but for exceptional circumstances, independent of the time spent aboard the ships. Crews received a lump-sum advance before setting sail, and when they returned, a share from the sale of their prizes or of their cargoes of fish.

The pay systems in the Newfoundland fishery were the most complex in the French maritime system. Except for the case of the Dunkirk cod trade, which was similar to conventional trading, the entire Newfoundland fishing fleet of France operated on a share basis.

The seven pay systems practised in the Newfoundland fleet can be divided into two distinct groups:

1) Systems without a premium (*pot de vin*)
   ◆ *Engagement au tiers franc*: western part of the English Channel, Isle Royale, and Saint-Pierre-et-Miquelon
   ◆ *Engagement au quart*: central part of the Atlantic coast (between the Loire and the Gironde Rivers)
   ◆ *Engagement aux deux cinquièmes*: southern part of the Atlantic coast (Basque country)
   ◆ *Engagement aux trois septièmes*: Isle Royale and Saint-Pierre-et-Miquelon

Under these systems, the crew would receive a share (one-third, *tiers*; one-quarter, *quart*, etc.) of the net value of the proceeds of the fishery (the sum remaining after deduction of costs borne in common by the outfitter and the crew). Any advance received before departure was deducted from their share.

2)   Systems with a premium
   • *Engagement au cinquième*: western part of the Channel
   • *Engagement à la mode du Nord*: western part of the Channel
   • *Engagement au last*: North Sea

Before departure, the crew would receive a sum called a *pot de vin* (premium) that was theirs definitively, whatever the outcome of the voyage. At the end of the voyage, they collected a small share, proportionate either to the value of the catch (*lot au cinquième*) or to its volume (*lot à la mode du Nord, last*). The *lot à la mode du Nord*, which were of very small value, were offset by large premiums.

The boarding of stowaways on vessels fitted out for the sedentary fishery at Saint-Malo (only) seems to have been directly linked to the pay system. Keeping stowaways on board posed no problem in the dry-cod fishery, since the presence of extra labour was more beneficial than detrimental. The stowaways received neither advances nor premiums before departure, but they were always assigned a share on their return, for to make their participation in the fishery effective, they had to be motivated by the certainty of receiving a share of the proceeds. *Engagement à la mode du Nord*, which enormously increased the premiums and reduced the shares to a ridiculously small sum, made it possible to keep the indispensable remuneration of the stowaways to a minimum without disorganizing the system as a whole.

The Newfoundland Traffic From Granville and Saint-Malo

Throughout the 18th century, Granville and Saint-Malo were the two main ports of the French Newfoundland fleet. In these ports an ever-greater share of the fleets became concentrated, especially after 1763.

From 1722 to 1792, the port of Granville fitted out 4,027 Newfoundland vessels (we do not know the number of vessels fitted out from 1713 to 1721 or from 1760 to 1762). During this period, Granville sent out 2,344 ships (58.2% of its Newfoundland fleet) to the itinerant fishery on the banks. Unlike that of Saint-Malo, the Granville fleet was essentially interested in green cod.

There were five distinct phases in Granville's presence in the Newfoundland fishery:
1)   from 1722 to 1750, a long period of expansion that accelerated suddenly after 1737, with a boom in the sedentary fishery;
2)   from 1751 to 1768, a period of stagnation and decline. The fishery appears to have reached its maximum size between 1751 and 1755. After 1763, the depression deepened and a withdrawal began;
3)   from 1769 to 1786, a new period of expansion, a more commercial one than in the 1740s. The very favorable years following the American Revolutionary War (1783-1785) set off a period of feverish activity that peaked in 1786.
4)   from 1787 to 1789, a short and brutal crisis;
5)   from 1790 to 1792, the beginnings of a recovery that was quickly interrupted by war.

The sudden expansion of the Granville sedentary fishery after 1738-39 was brought on by fiscal changes: not until 1739 were Granville sedentary vessels of 80 *tonneaux* or more authorized to "raise" their salt in Brittany, a region exempt from the *gabelle*.[28] The merchandise and supplies they carried were also dispensed from paying export duties.[29] At the same time, the Anglo-Spanish War (1739) gave Granville and Saint-Malo fishermen an opportunity to compensate for the suspension of English landings in Spain. The Granville sedentary fleet then expanded very quickly.

From 1713 to 1792, Saint-Malo fitted out 4,654 vessels for the Newfoundland fishery (66% of its total, the coasting trade excepted). Of this total, 1,131 ships (24.3%) were fitted out for the itinerant fishery—only half the number from Granville (2,344 vessels). Thus in the 18th century, Saint-Malo was a port centred mostly on the sedentary fishery. Eight successive periods can be identified in the Newfoundland fishery out of Saint-Malo:

1) from 1713 to 1724, sedentary fishermen abandoned the coast of Newfoundland and fell back on mainland Canada, Isle Royale, and Labrador, and starting in 1720, Gaspé. An abnormal drop in yields (noted for the English fishery) in Newfoundland between 1711 and 1720 probably triggered the exodus.[30] The effect of these phenomena was strengthened by the letters patent of April 1717 that exempted from export duties supplies for ships fitting out for the French colonies in America; Newfoundland, a British territory, was excluded.[31] This resulted in the withdrawal of Saint-Malo fishermen to the coasts of Canada.

2) From 1724-25 to 1733, there was an appreciable decline in the fishery on Isle Royale, at Gaspé, and on the banks, while it picked up again on the coasts of Newfoundland and Labrador.

3) From 1733-34 to 1749, a general decline in the continental fishery occurred, to the benefit of the Newfoundland coast, where extraordinary expansion occurred. The bank fishery continued to stagnate. The extension of the exemption from export duties to Newfoundland, obtained in 1733, accelerated the abandonment of the shores of Canada.[32]

4) From 1749 to 1755, the Saint-Malo itinerant fishery expanded rapidly (in 1739, the duties on Saint-Malo green cod entering Normandy had been reduced).[33] Along the Petit Nord of Newfoundland the expansion continued. Trading developed in Gaspé and on Isle Royale to the detriment of the migratory fishery. Labrador was abandoned. Well before the Seven Years War, Saint-Malo fishermen had almost completely left Canada.

5) From 1763 to 1767-68, the itinerant fishery continued its expansion.

6) From 1768-69 to 1774, a crisis abruptly cut off the astounding thrust of the itinerant fishery; trading expanded.

7) From 1774 to 1786-87, trading declined, the sedentary fishery stagnated and the itinerant fishery grew.

8) From 1786-87 to 1792, there was a general crisis in the Newfoundland fishing industry: the fishery collapsed, resulting in a return to trading.

The growth of the Granville Newfoundland fleet between the end of the 17th century and the end of the eighteenth far exceeded that of the Saint-Malo fleet: from the 1680s to the 1780s, Saint-Malo increased its Newfoundland traffic by approximately 50% (thanks to the itinerant fishery), Granville by nearly 400%.[34] The Granville expansion thus represents one of the major features of the history of the French Newfoundland fishery in the 18th century. Granville, unlike Saint-Malo, was not authorized to trade with the colonies. This "confinement" to the fishery was one of the bases of its astonishing expansion in Newfoundland. The Granville entrepreneurs had to concentrate their capital and energies on a single transoceanic destination.

Fishing and Politics

A thorough analysis of the sources relating to the French Newfoundland fishery reveals the existence of two simultaneous conflicts in the North American fisheries. The first conflict was between France and Great Britain, and had to do with maritime and colonial power in the world. In the eyes of contemporaries, the place of the Newfoundland fishery in maritime power gave this fishery considerable political importance. The other conflict, less visible but much more serious and dangerous for the French and British fisheries, was between fishermen who settled on the shores of North America (from Labrador to New England) and the Newfoundland fleets from the mother countries in Europe. This second conflict was technical and commercial: there was an incompatibility between the resident and migratory fisheries. The residents, since they occupied the ground permanently, had lower production costs. The British fisheries thus gradually fell into the hands of the residents, while ships from England were more and more limited to trading. The French practised such a system on Isle Royale and at Gaspé, but since the residents were less numerous there than on the British coasts, the incentive to abandon the fishery and turn to trading was not as strong. The transatlantic seesaw continued to dominate the French fisheries system after the Seven Years War, since France now had access (apart from Saint-Pierre-et-Miquelon) only to coasts that it was not entitled to populate. The massive installation of English resident fishermen on the French Shore after 1763 appeared as a grave menace for the fishery based in France, which was regarded as one of the pillars of that country's maritime power.

The Newfoundland fishery trained a considerable number of seafarers, not only because it mobilized a large fleet (300 to 400 ships on average), but also, and especially, because of the exceptionally large numbers of men who could be taken aboard ships and the low mortality rate among them. So far as relations between states were concerned, the role played by Newfoundland in the training of crews who could serve in the navy was a major issue, constantly present in the minds of the governments of the time. This explains the very special place reserved for the island of Newfoundland on the world chessboard.

Under Article 13 of the Treaty of Utrecht, France obtained for its subjects the right to catch and dry fish "from the place called Cape Bonavista to the northern extremity of the said island, and thence, following the western part, to the place called Pointe Riche."[35] Until 1755, no clashes between French and English occurred on this coast in peacetime. The only incidents between the two nations arose on the extreme southern part of Isle Royale, in the Canso area, in 1718-20. On the other hand, serious tensions developed at that time concerning the occupation of beaches between the residents of Isle Royale or the Gaspé and the crews of sedentary-fishery vessels that had come from France. The latter could easily fall back on the coast of Newfoundland, which was forbidden to the French colonists. That is what the Granville-Saint-Malo fishery did in the 1730s and 1740s. The mere threat of incidents turned the France-based fishing fleets away from inhabited regions.

Taking advantage of the absence of the French from 1756 to 1762, English residents on the north coast of Newfoundland continued their expansion to the west of the French Shore, for the first time entering zones traditionally frequented by the Saint-Malo and Granville fishermen. When the latter returned to the Petit Nord in 1763, they were surprised to see a large number of English ships there.[36] The English felt that they could fish on the French Shore because the Treaty of Utrecht did not mention that the fishing right of the French there was exclusive; the French countered that everything that restricted the free exercise of their fishing right undermined the strict application of the treaty; the presence of the English fishermen was therefore implicity condemned by the Treaty of Utrecht.

Five other problems arose:

1) The exact location of Point Riche, the southwestern boundary of the French Shore (until 1783), was set in different places by French maps and English maps.

2) The English did not go to the fish stage at Le Croc, but received their places from the admiral of each harbor, which gave them an advantage in the race for fishing rooms. The French therefore adopted the English system starting in 1764: each harbor would have two admirals, one French, the other English.

3) When were the French to leave the French Shore? After 20 September, according to them— and they won their case—since the right to fish also implied the right to dry the catch.

4) The English maintained that the Treaty of Utrecht permitted the French to leave only huts and fish stages on shore, and that it said nothing about chaloupes, salt or utensils. The French got their way so far as the chaloupes and utensils were concerned, but the question of salt remained unresolved.

5) The English government agreed to compensate French outfitters for the damage caused by English warships, but dragged its heels for ten years before paying.[37]

In the face of these difficulties, the Duc de Praslin, the Secretary of State for Foreign Affairs, threatened the English:

> I finally told the Duke of Bedford: I will make no difficulty about disclosing our policy to you. We set as the first condition for peace the conservation of the cod fishery; this was a *sine qua non* condition. If it had been refused by your Court, the war would perhaps be still going on. By renewing Article 13 of the Treaty of Utrecht, we thought we would enjoy the fishery as in the past, (...) but if our fishermen come home empty-handed and our fishery becomes illusory thereby, despite the good intentions of the King of England and the protestations of his ministers, then we cannot dispense ourselves from demanding and asserting our rights (...). (I have given) the English minister to understand that the maintenance of peace depended on the conservation of our fishery, and that if we should lose it, we would be obliged to resume the war.[38]

Clashes multiplied after 1765 on the eastern part of the French Shore, which the French (especially the Granville fishermen), who were cramped on the western part, were seeking to "reconquer." That is where, between 1767 and 1777, most of the clashes between French and English fishermen were to occur, and not on the coast traditionally occupied by the French, where the British were in a clear minority.

In the 1770s, the Court of France moved toward the solution advocated as early as 1763 by the Saint-Malo outfitter Chateaubriand: an exchange of coasts. The boundaries of the French Shore were to be moved in such a way that the two nations could enjoy an exclusive fishery. The failure of the negotiations on this question—among other factors— led Vergennes, in March 1776, to propose secret assistance from France to the American colonists in revolt, and then war against England. It regarded the settlement of the Newfoundland question as one of the chief objectives of France's entry into the conflict.[39] So at the end of the war, Article Five of the Peace of Versailles again took up the question of an exchange of coasts. The French Shore, with the coastline between Cape Bonavista and Cape Saint John removed, now extended from the latter point to Cape Ray. [40]

The vagueness of the terms of the Treaty of Utrecht gave France the room to manoeuvre that it required in order to have Article 13 legally interpreted in the same way as it had been applied until 1756. In reality, the French fishery on the French Shore in 1783 enjoyed the same conditions of exclusiveness and the same geographical boundaries (Cape Saint John-Cape Ray) as before the Seven Years War, but this time it was sanctioned by a precise text that blocked any installation of settlers. By transferring sovereignty over Newfoundland to England, the Treaty of Utrecht made possible an English colonization that ultimately led to the destruction of the migratory fishery based in England. The same treaty prevented the installation of French resident fishermen while giving France a legal claim for eliminating English residents. In this way, the French home-based fishery in Newfoundland was able to remain in existence throughout the 19th century.

Notes

1. J. Bosher, "A Fishing Company of Louisbourg, Les Sables d'Olonne, and Paris: La Société du Baron d'Huart, 1750-1775," *French Historical Studies*, IX (1975), 263-77.
2. Salt preserves cod by dehydrating its flesh and the bacteria it contains.
3. Ordonnance sur les Gabelles, 1680, titre XV, art. 1 & 2.
4. This, of course, has nothing to do with the duty-free status of ports.
5. AD Loire-Atlantique C744, fo 15, Lettres patentes, avril 1717.
6. APC, MG2 A1 32, fo 53, Ordonnance du 2 juin 1695.
7. Duhamel du Monceau, *Traité général des pesches* (Paris, 1769-1777), t. 2, 52.
8. Ordonnance sur la Marine, 1681, titre VI, art. 7.
9. Valin, *Nouveau commentaire sur l'Ordonnance de la Marine du mois d'août 1681* (La Rochelle, 1766), t. 1, 509, Ordonnance du 23 juillet 1745.
10. Very young boys were sometimes taken aboard the Newfoundland boats: Joseph Louvet, a boy on the *Bienvenu* of Granville, died in Newfoundland at the age of 10 in 1722; AMar, Cherbourg, C6 141, Rôles d'armements Granville, 1722. Most of these children's fathers were among the crew members.
11. Ordonnance du 22 décembre 1739, Valin, *Nouveau commentaire...* t. 1, 509; Ordonnance du 23 juillet 1745, AMun, Saint-Jean-de-Luz, EE5. The requirement to take apprentices aboard was abolished by the ordinance of 4 September 1786, Ch. de la Morandière, *Histoire de la pêche française de la morue en Amérique septentrionale des origines à 1789* (Paris, 1962) t. 1, 100.
12. AN, Marine, D2 54, Mémoire sur le commerce des Sables d'Olonne, 1760.
13. The cod were particularly fond of a shellfish called a *pitot*, a kind of large clam that 18th-century fishermen called a *vit de moine* (Bronkhorst, *La pêche à la morue* (Paris, 1927), 80, and La Morandière, *Histoire de la pêche...* t. l, 153.
14. AN Marine, C5 55, fo 8, and C5 58, fo 1, "Balance du Commerce: ports de France où se sont fait les retours de pêche, 1787 et 1788." The figure for Dunkirk relates to green cod from Newfoundland only.
15. Duhamel du Monceau, *Traité général...* vol. 2, 94.
16. AD Ille-et-Vilaine, 15 Rd 22, Registre armements Saint-Malo, 1784-1810.
17. AN Marine, C5 53, Etat des bâtiments expédiés pour la pêche de la morue en 1786.
18. AD Loire-Atlantique, C744, fo 108, Duc de Praslin à Chambre de Commerce de Nantes, 12 février 1769.
19. APC, MG1 F2C 5, fo 353 and A. Bellet, *La grande pêche de la morue à Terre-Neuve* (Paris, 1902), 61, Règlement du 11 mars 1770.
20. For the technical aspect of the fishery and the preparation of dried cod, see Duhamel du Monceau, *Traité général...*, 82-103.
21. APC, MG2 A1 21, fo 35, Arrêt du Conseil du 3 mars 1684.
22. R. Kaninda, *Les relations commerciales entre Granville et Marseille au XVIIIe siècle*, typewritten memorandum, Université d'Aix-Marseille, 1969, 39.
23. AD Bouches-du-Rhône, 200E, 474 to 548, déclarations des capitaines à l'arrivée, 1709-92.

24. Only ships coming from the Levant or North Africa were automatically subject to quarantine.

25. AD Ille-et-Vilaine, 15 Rd 22, registre armements Saint-Malo, 1784-1810.

26. On the question of supplying the French West Indies with cod, see J. Tarrade, *Le commerce colonial de la France à la fin de l'Ancien Régime. L'évolution du régime de l'Exclusif de 1763 à 1789* (Paris, 1972) 2 vol. and J. Mathieu, *Le Commerce entre la Nouvelle-France et les Antilles au XVIIIe siècle* (Montréal, 1976).

27. Both systems—time-based and share-based—were practised in the coasting trade.

28. AD Loire-Atlantique C744, 8, arrêt du Conseil du 13 janvier 1739.

29. AD Loire-Atlantique C744, 9, arrêt du Conseil du 27 janvier 1739.

30. Grant Head, *Eighteenth Century Newfoundland* (Toronto, 1976), chap. 4, 63-81. The author may have used sources for which there is no equivalent for the French fishery in the same period.

31. AD Loire-Atlantique C744, 9.

32. AD Loire-Atlantique, C744, 16, arrêt du Conseil du 31 octobre 1733 and AD Ille-et-Vilaine IF 1939, fonds Vignols 2127 bis, "Réponse des députés du Commerce au mémoire des Fermiers Généraux touchant l'exécution de l'article 4 des lettres patentes de 1717 pour le commerce des îles et colonies françoises de l'Amérique relativement au commerce de la morue sèche," 1733.

33. AN Fonds général, AD, VII (fonds Rondonneau), 7a, arrêt du Conseil du 14 juillet 1739.

34. Robert Richard, "Comptes et profits de navires terre-neuviers au Havre au XVIIe siècle," *Revue d'Histoire Economique et Sociale*, vol. 54, 1976 (4), table p. 81, and AN Marine, C5 53, fo 140, "Etat des bâtimens...."

35. Article 13 of the Treaty of Utrecht: "L'isle de Terre-Neuve avec les iles adjacentes appartiendront désormais absolument à la Grande-Bretagne, et à cette fin le Roy Très Chrétien fera remettre à ceux qui se trouveront à ce commis en ce pais-là, dans l'espace de sept mois à partir du jour de l'échange des ratifications de ce traité, ou plus tost si faire se peut, la ville et le fort de Plaisance et autres lieux que les Francais pourraient encore posséder dans la dite isle, sans que le dit Roy Très Chrétien, ses héritiers ou ses successeurs, ou quelques-uns de ses sujets, puissent désormais prétendre quoi que ce soit, sur ladite isle et les isles adjacentes en tout ou en partie. Il ne leur sera pas permis non plus d'y fortifier aucun lieu, ni d'y établir aucune habitation en facon quelconque, si ce n'est des échafauds et cabanes nécessaires et usités pour sécher le poisson, ni aborder dans la dite isle dans d'autres temps que celui qui est propice pour pecher and nécessaire pour sécher le poisson. Dans la dite isle, il ne sera pas permis aux dits sujets de la France de pecher et de sécher le poisson en aucune partie que depuis le lieu appelé cap de Bona-Vista jusqu'à l'extrémité septentrionale de ladite isle, et de là en suivant la partie occidentale jusqu'au lieu appelé pointe Riche." [The island of Newfoundland with the adjacent islands shall henceforth belong absolutely to Great Britain, and to this end the Most Christian King will cause those appointed thereto in that country, within the space of seven months from the day of the exchange of ratifications of this treaty, or earlier if it is possible to do so, to turn over the town and fort of Plaisance and other places which the French may still possess on the said island, without the said Most Christian King, his heirs or successors, or any of his subjects, being able henceforth to assert any claim whatsoever to the said island and the adjacent islands, in whole or in part. Nor shall they be permitted there to fortify any place, or establish any habitation in any way whatsoever, except fish stages and huts, necessary and customary for drying fish, nor land on the said island at other times than that which is favorable for fishing and necessary for drying the fish. On the said island, the said subjects of France will not be permitted to fish or to dry fish in any part except from the place called Cape Bonavista to the northern extremity of the said island, and thence, following the western part, to the place called Pointe Riche.] (Bellet, *La grande pêche de la morue,* 39).

36. AN Colonies, C11F 1, fo 87, armateurs de Granville à Choiseul, 9 november 1763.

37. *Loc cit.* and AAE, corresp. Angleterre, no 457, fo 9, Guerchy à Praslin, 3 mai 1764.

38. AAE, corresp. Angleterre, no. 456, fo 324, Praslin à Guerchy, 14 avril 1764.

39. D.D. Irvine, "The Newfoundland Fishery: a French Objective in the War of American Independence," *Canadian Historical Review*, vol. 13 (1932), 278-79.

40. AN Marine, C5 52, fo 15, note de Bretel, 1786 and fo 22, instructions au commandant de la Station de Terre-Neuve, 1785.

chapter
five

THE NEW ENGLAND FISHERY AND TRADE AT CANSO, 1720-1744

Judith Tulloch

The Canso Islands, at the northeastern tip of mainland Nova Scotia, have been an important fishing centre for centuries. Basque and French vessels first came there in the mid 1500s to fish on the rich offshore banks and to dry their catch on the islands' rocky beaches. In 1607, Marc Lescarbot encountered a Basque fishing captain at nearby Whitehaven who had already made 43 voyages to the region. French fishermen came regularly to Canso during the 17th century. By the 1690s, New Englanders were beginning to sail north to trade with them, but for the next 30 years, conflict between France and Britain made fishing at Canso a risky endeavour and slowed development of the industry.

The Treaty of Utrecht transferred mainland Nova Scotia from France to Britain in 1713. Although the capital of Nova Scotia was at Annapolis Royal, Canso was the economic centre of the colony. During the first half of the 18th century, Canso formed a corner of a busy transatlantic trading network. The tiny resident population was swelled each summer by hundreds of fishermen, and fish cured at Canso was transported to southern Europe and New England. Throughout the period, the fishery was dominated by New England entrepreneurs who sent fishing schooners and transport ships to Canso. The shoremen who managed their fishing establishments were also New Englanders. Similarly both the fishermen and the vast majority of on-shore servants were from the American colonies with only a few shorehands from Ireland. English participation in the Canso fishery was limited almost completely to the transatlantic trade.

The yearly routine at Canso was focussed on the busy summer fishery. The first boats to arrive were fishing vessels from New England, bringing both fishermen and shore workers. These schooners, usually of about 40 tons and with an average crew of five, arrived early in March. The shorehands set about readying the flakes for use while the fishing schooners departed for the banks. The schooners were usually at sea for periods of two or three weeks, occasionally up to six weeks, and continued to fish until late September. Transport, or "sack," ships arrived at Canso from Britain or New England late in May or early in June. These vessels carried the higher quality "merchantable" cod to markets in Spain, Portugal, and Italy. The remainder of the catch, the so-called "refuse" fish that had been damaged during curing, was carried by New England ships south to the West Indies.

During the summer months, Canso presented a scene of considerable activity. Fish flakes lined the shores of the islands, and the major centre of habitation was on what is now called Grassy Island. After the catch was unloaded from the fishing schooners, the shorehands washed the salt from the fish and spread it on the flakes to dry. Each night, they gathered it into stacks to keep it from getting wet and laid it out again in the morning. Finally the cured fish was made up into bales for shipping. The large transport vessels, capable of carrying several thousand quintals of fish, usually made only one voyage a year from Canso. Depending upon the weather and the size of the catches, the sack ships began to set sail to Europe late in the summer. The fishing schooners returned to New England somewhat later, usually in October. Most of the workers departed with them, leaving only a few people to maintain the flakes and storehouses over the winter. By early November, Canso was again reduced to its small resident population of 30 to 40, protected by a garrison of two or three companies of the 40th Regiment.[1]

The Canso fishery was only part of a far-flung trade for many of the entrepreneurs and vessels connected with it. Much of the merchantable cod was carried to Europe in ships owned by New England merchants such as Peter Faneuil, Joshua Peirce, and William Pepperell. Their vessels followed a regular transatlantic pattern based on seasonal trade. Cargoes of New England and West Indies goods were taken to Britain or to the Mediterranean late in the winter. There the ships were loaded with salt and miscellaneous supplies before sailing for Canso in the spring. After returning to southern Europe with Canso fish in the fall, the New England ships picked up cargo for either Britain or the American colonies but usually returned to home waters for the worst months of winter.[2]

The price of the fish recorded in the annual fishing returns fluctuated substantially over the quarter-century of the New England fishery. Late in the 1720s, it occasionally dropped as low as 7 shillings sterling per quintal, but by 1732 had risen to a high of 12/6 sterling. On average, the price remained between 8/ and 10/ sterling per quintal. The fishermen were usually paid in shares of their vessel's catch. Their proportion was one-half of the proceeds while the owner, who customarily provided provisions and equipment, retained the other half. Some of the fishermen's income was spent in purchases such as tobacco and rum from the New England merchants or their local agents. Bills of exchange were issued for the remainder, as was typical with most financial transactions of the period. The shorehands were paid entirely in truck supplied from the storehouses of their New England employers. Such goods as fabric, rum, tobacco, sugar, and molasses were common media of exchange. Paying in kind undoubtedly meant that many of the servants were perpetually in debt to their employers. As well, the heavy rum consumption that often accompanied this type of payment occasioned complaints about damages to the cured fish caused by drunken men.[3]

The Canso fishery had one major product: dried salted cod. Higher quality cod was exported to the Mediterranean market, which required vast quantities because of the dietary regulations of the Roman Catholic church. In 1755, the Iberian peninsula consumed 300,000 quintals a year and the populous Italian peninsula formed an equally reliable market.[4] Poor quality cod that had been damaged during curing, known as "refuse" fish, was exported to the Caribbean as food for the large population of plantation slaves. Refuse fish was usually transported by the New England fishing vessels on their voyage south in the autumn. A minor by-product was cod liver oil, used chiefly in the tanning industry. Only a relatively few barrels were made yearly, ranging from a low of 28 in 1729 to 486 in 1737. Most of the oil appears to have been shipped south to New England at the end of the fishing season.[5]

Provisions for seasonal and permanent inhabitants formed most of the import trade at Canso during the busiest years of the New England fishery. These provisions came both from overseas and from New England. Some of the sack ships from Britain carried supplies on their outward voyage in the spring. In 1732, for example, sack ships brought bread, flour, pork, and beef for sale. These supplies appear to have come principally from Ireland.[6] On the other hand, the dominance of the Canso fishery by New England suggests that the bulk of supplies were in fact imported from the American colonies. There were, for example, complaints in 1726 that Boston supplied all the goods required at Canso, to the detriment of the British traders.[7] As well, naval officers who reported on the state of the fishery regularly noted that New Englanders brought supplies to Canso.

New England's proximity to Canso permitted shipment of livestock and perishable produce. In 1726, Portsmouth merchant Joshua Peirce sent the sloop *Happy Returns* to Canso with at least 16 sheep and 15 lambs, as well as several cows and calves. The purchasers included permanent residents and captains of both fishing and sack ships. Similarly an account book kept by Peirce shows such foodstuffs as rum, sugar, tobacco, limes, molasses, and beef sold to residents of Canso.[8] A more detailed list of goods shipped to Newfoundland by Boston merchant Thomas Hancock, also active in Canso, probably represents the type of merchandise regularly exported northwards. Hancock's ship carried a cargo of 13 tierces of Indian corn, 3,500 hogsheads of staves, 99 barrels of flour, 66 loaves of white sugar, 13 hogsheads and tierces of tobacco, 4 hogsheads of Barbados rum, 5 boxes of soap, 39 barrels of port, and 6 barrels of cider.[9] Certainly the records of food consumed by Joshua Peirce's Canso household during the summer of 1723 support the availability of these sorts of goods. His living expenses included purchases of bread, pork, beef, flour, sheep, oil, and molasses.[10]

In addition to foodstuffs, the Canso population required fishing supplies. Of these, salt was the most obvious and requisite. The Canso fishery regularly used ten hogsheads of salt for every 100 quintals of fish.[11] Hence during the 1720s and 1730s, when production sometimes reached 50,000 quintals a year, as many as 5,000 hogsheads of salt could be consumed. Salt reached Canso from both Europe and New England. Sack ships sometimes carried salt on their western voyages in the spring. This salt usually had been imported to Britain from the Iberian peninsula and then shipped onwards to North America.[12] Salt also came direct to Canso from the Iberian peninsula, principally on New England ships coming out to Canso early in the spring. In 1729, Joshua Peirce ordered one of his captains to load a full cargo of good white salt at Cadiz and sail to Canso.[13] Salt also was brought from New England, which had long been allowed to import it direct from the Iberian peninsula as an exception to the customary restrictions of the English trade laws.[14] As well, ships on coastal trading voyages put in to Canso with salt. In only one case was the origin of the salt noted, i. e. the West Indies, although West Indian salt was generally considered to be of inferior quality.[15]

Other fishing supplies required at Canso included such items as nets, lines, and hooks. Since almost all fishermen were New Englanders, much of this type of material was probably brought from New England by the fishing captains in the spring. Inventories of Joshua Peirce's ships going to or returning from Canso mention such items as nails, nets, canvas, hooks, lines, anchors, and cables.[16] Similar fishing supplies were undoubtedly brought from England by the sack ships. Indeed Peirce himself noted that most ships' supplies like canvas and cables were cheap enough in England that it was worthwhile importing them for use in New England. Certainly in the case of the Newfoundland fishery, England was the principal source for these goods.[17]

Canso also served as an entrepot for both legal and clandestine trading ventures. The legitimate trade, between Europe and New England, does not appear to have been substantial but rather was a by-product of the fishery. In most cases, goods shipped through Canso appear to have been brought out by captains of sack ships, either as a speculative venture for sale to New England merchants or as delivery of goods previously ordered in Britain by the colonial traders. One well-recorded transaction by Joshua Peirce may be representative of this type of trade. Late in 1728, Peirce dispatched one of his ships to Cadiz with a cargo of fish and oak timber. There the cargo was to be sold and the ship loaded with salt that it would then carry to Canso where it was chartered to load fish. Peirce's letters to his European agents and to the captain of the ship indicate the kind of material that may have arrived direct from Europe as part of the transatlantic trade circuit. Some of the goods were clearly intended for re-sale. In addition to a full load of white salt, Captain Cate's shopping list included white wine, olives and olive oil, fresh lemons, and coarse Holland linen. Peirce also ordered goods for his own use: two china punch bowls and five gallons of liquor. Such mixed cargoes were not uncommonly brought to Canso by westward-bound vessels in the spring. Some of the goods thus imported may well have been re-exported to Louisbourg or New England. In this particular instance, however, the goods did not pass through Canso since Captain Cate remained in Cadiz till after mid-August 1729 and was therefore ordered to return direct to Piscataqua.[18]

The occasional presence of Mediterranean goods in Canso is confirmed by customs' records. In 1728, the brigantine *Gardiner*, which had arrived from Spain to load fish for transport to Portugal, was confiscated on the grounds that the captain had imported Spanish goods without paying duty. The cargo—olive oil, olives, wine, and soap—was sold at an auction.[19] Such transactions only rarely appear in the documents, and available evidence suggests that European goods were usually imported direct to New England rather than through Canso. A list of fishing vessels leaving Canso in September 1730 shows only four ships of 154 carrying foreign imports back to New England ports. The vast majority of returning ships carried only fishing stores, refuse fish, or ballast.[20]

The clandestine trade with Louisbourg seems to have been a significant part of the Canso economy. As early as 1726, an experienced New England trader reported that New England ships frequently carried such commodities as cattle, pork, boards, and shingles to Louisbourg. These cargoes had been cleared from New England ports as if bound for Canso but were instead intended for Louisbourg.[21] Louisbourg's chronic difficulties with obtaining sufficient food supplies made it a lucrative market for the New Englanders with their accessible supply of livestock and produce. Moreover, frequent shortages forced Louisbourg authorities to accept and, at times, to solicit New England trade. In 1743, for example, expected shortages prompted French officials to send to Canso for flour, biscuit, corn, vegetables, and cod.[22] One result of this ready market was a scarcity of provisions in Canso and a consequent rise in prices. The cost of a sheep at Canso in 1726 for example was noted to be double that at Louisbourg.[23]

The New England trade with Louisbourg had opponents on both sides. British authorities concluded that it was detrimental to Canso's development, while British merchants complained that because the French traded at Canso, legitimate traders could not purchase fish and were therefore shut out of the rich Mediterranean market.[24] Some French merchants were equally opposed to the New England trade. In 1725, a group claimed that the English smugglers interfered with their livelihood by importing such products as beef, pork, tobacco, salt, tar, and textiles.[25] Nevertheless, the illegal New England trade with Louisbourg flourished throughout the early-18th century. Despite risks of seizure by either English or French authorities, New England merchants found the rewards lucrative, especially access to the French West Indian market. Such West Indian goods as sugar, cotton, cocoa, molasses, and indigo found ready sale in Boston and provided additional profits for many merchants engaged in the Canso fishery. The extent to which Boston entrepreneur Peter Faneuil was involved in trade with Louisbourg is illustrated by his correspondence with his Canso agent, Thomas Kilby, over a period of two summers. During the summer of

1737, Faneuil sent at least three vessels to Canso en route to Louisbourg. Cargoes included building supplies such as board, plank, shingles, and brick as well as such food items as beef, biscuit, and flour. Kilby was instructed to purchase West Indian goods in exchange, specifically indigo, rum, molasses, and sugar. Kilby made several visits to Louisbourg during 1738 as well, taking goods sent by Faneuil to customers there and under orders to buy wine, sweetmeats, or sugar if they were available.[26] These commodities appear to have then been shipped through Canso to Boston.

During the 25 years of the New England fishery at Canso, catches and revenue fluctuated substantially. New Englanders were fishing at Canso soon after the cession of mainland Nova Scotia to Britain in 1713, and by 1718 John Henshaw and Giles Hall of Boston, the first English settlers, had approximately 50 workers fishing with them at Canso.[27] Seizure of French fishing vessels by a Royal Navy frigate from Boston in 1718 and the retaliatory attack on the New England establishment in 1720 by a combined force of French and Mi'kmaq served as only temporary setbacks to the growth of the fishery. Indeed the extent of losses indicates the value of the industry at this time. In 1718, for example, the French claimed a loss of goods valued at 200,000 *livres*; and two years later, New England losses, even after the return of some goods, were estimated at £9,000.[28]

The 1720s appear to have been the high point of the Canso fishery. A British naval officer reported that there were 96 English ships at work in the area in 1720.[29] The same year Governor Richard Philipps of Nova Scotia claimed that the New Englanders caught between 80,000 and 100,000 quintals per year, although statistics for later years indicate that the annual catch seldom exceeded 60,000 quintals.[30] Fear of renewed attacks by the Mi'kmaq early in the decade retarded the growth of the industry, and in 1722 a group of Canso entrepreneurs threatened to leave unless the British government provided adequate protection.[31] A measure of security was achieved in 1723 when a naval ship was dispatched to patrol the Nova Scotia coast, and a year later a small fort was built on Grassy Island to defend the shore establishment.[32] As well, signing of a peace treaty with the native people in 1726 lessened the threat.[33] Although statistics for the late 1720s are incomplete, the veteran Canso trader, Joshua Peirce, reported that 1728 was the best year in many at Canso, and the following summer he predicted that the catch would be as good as it had been in 1728.[34]

Direct overseas trade seems to have declined in the 1730s. Throughout the preceding decade, a substantial number of the sack ships loading for the Mediterranean had come from Britain with fewer from New England. From the early 1730s, New England transports began to predominate and by the end of the decade, as the possibility of war with Spain increased, there was little direct overseas export from Canso. Indeed, in 1740 no British sack ships came to Canso, and in the following year the fishing report noted that no sack ships had come since the outbreak of the war late in 1739.[35] Merchantable fish was taken back to New England for re-export to Europe or the West Indies.

The decline in overseas trade was influenced by several factors. British merchants and Royal Navy officers sent out to protect the fishery complained that much of the catch was sold to the French trading from Louisbourg, rather than shipped for the Iberian market.[36] Canso fish also gained a reputation for poor quality as a result of bad curing practices. By the late 1720s, consumption of Canso fish in the lucrative Iberian market had dropped and British consuls and merchants there complained of the inferiority of the fish. Board of Trade officials warned that unless greater care was taken, the Mediterranean market might be lost.[37]

As an offshore bank fishery, the Canso fishery experienced inherent problems. Freshly caught fish was salted and stored in the holds of the schooners on the banks throughout the length of the fishing voyage, between two and five weeks. By contrast, in the inshore fishery, fishermen could return to land every night so that the day's catch could be washed and laid out to dry almost immediately. Hence, particularly in summer, Canso fish could suffer salt burns and deteriorate to the extent that it could be sold only as refuse fish rather than the higher-priced merchantable fish. Another complaint against the Canso fish was that in an attempt to win the lucrative early market, New England merchants shipped the cod before it had dried properly, again lessening its quality on arrival in Europe.[38]

The Canso fishery, closely tied to the Boston merchant community, was also affected by commercial rivalry among the New England seaboard towns. During the 1720s and 1730s, entrepreneurs in the coastal towns of Essex County, including Marblehead and Salem, strove to break the overwhelming domination of Boston over Massachusetts Bay shipping. In 1737, Peter Faneuil warned his agent at Canso that the entrepreneurs must be careful to improve the quality of their product, pointing out that the fishing industry at Marblehead would take advantage of any problems with Canso fish to destroy the industry there and strengthen their own trading position.[39] Moreover, fish markets in both Europe and the West Indies experienced a serious glut in the 1730s.[40] Hence, insecurity of transport because of the war with Spain, lower demand for fish, and powerful competition all combined to erode Canso's position in New England's cod fishery. The settlement's failure to revive after the French attack of May 1744 merely confirmed its loss of status.

A short-lived aspect of the Canso fishery was a whaling industry that lasted barely five years. First reported in 1732, whaling was prosecuted on the Grand Banks, primarily by Nantucket sailors. The most successful year was 1732 when 22 whales were landed at Canso, representing 2,000 tons of oil and 6 tons of bone for a value of about £4,500. Fewer whales were taken in subsequent years: in 1738, none of the vessels were successful and the official returns for 1740 noted that there had been no whaling that year. Whales were at first reported to be numerous, but within a few years their numbers declined as they had earlier along the New England coast. In 1737, for example, only nine were taken, two of which were described as "small." Most of the vessels that had formerly put into Canso had moved further north to hunt in Davis Strait, where that year there were between 50 and 60 whaling ships from Massachusetts alone.[41]

During the years of the New England fishery at Canso, the settlement was guarded by a detachment of the 40th Regiment. One company of about 36 soldiers was stationed there late in 1720 after the French and Mi'kmaq attack of that summer, and in later years the garrison increased to as many as 120 soldiers. Their presence necessitated regular provisioning by military authorities. Goods were supplied by British contractors, usually through New England agents. The contractors were required to provide bread, beef or pork, butter or cheese, and flour or rice at a weekly rate per person of 7 lb. of bread, 7 lb. of beef or 4 of pork, 3 pints of pease, 6 oz. of butter or 1 lb. of cheese, and 1 lb. of flour or 8 oz. of rice.[42]

The dominance of the Canso market by New England traders probably meant that at least some of these goods were supplied from Boston, as was the case with provisions for the detachments of the 40th stationed at Annapolis Royal.[43] On the other hand, closer ties with Britain through the annual voyages of sack ships may have brought more of these contract supplies from overseas. This was almost certainly the case in 1736 when a group of officers at Canso reported on the quality of the provisions recently brought by a British sack ship, the *Sir Thomas Pinck* of Bristol. The cargo appears to have represented a substantial portion of the garrison's provisions: bags of bread (probably hardtack biscuit), casks of beef, pork, flour, pease, and firkins of butter.[44] Similarly in 1742, the military engineer, J. M. Bastide, reported that the garrison at Canso depended upon annual provisioning from England.[45]

Fresh provisions and livestock were regularly brought from New England. The records of New England merchants show officers of the garrison among their Canso customers. Joshua Peirce for instance provided a wide variety of goods to clients there, among them the lieutenant-colonel of the 40th Regiment, Lawrence Armstrong. Late in the 1730s, Christopher Kilby's daybook contained numerous references to the officers, including the sale of sugar, tea, and wine glasses to Captain Patrick Heron at Canso.[46] These transactions were clearly part of a regular trade between New England and Nova Scotia. The officers at Canso, like the permanent residents, also supplemented their diet by cultivating small vegetable gardens. Bastide's 1742 map of Canso shows garden plots along the south shore of what is now Grassy Island, while his accompanying report notes that the officers still required fresh goods brought by the New Englanders in the summer, despite having "some little Gardens."[47]

Canso played a significant role in the New England economy of the early-18th century. Fish cured there represented a major source of revenue for the colonial entrepreneurs who dominated the industry. On the other hand, Canso's chief importance in the Atlantic trading network was the opportunities it offered for trade with Louisbourg. The settlement functioned as a conduit through which French goods, especially products of the French West Indies, were imported for transshipment to New England. During the summer, Canso represented a useful market for New England traders supplying food and fishing equipment, but the settlement was too small to sustain year-round trade. Indeed, in 1732 Governor Philipps reported that the only manufactured goods consumed in Nova Scotia were cloths of various sorts while the only commodity exported was Canso fish.[48] Nonetheless, as a source of a valuable staple and as a vehicle for camouflaging illicit trade with Louisbourg, Canso served as a useful adjunct to the New England commercial network.

Notes

1. Great Britain, Public Record Office, Colonial Office 218, v. 2, ff. 53-54. Council of Trade and Plantations to Lords Commissioner of the Admiralty, 16 February 1726. (Colonial Office hereafter CO). For details of the fishery, see the annual "scheme of the fishery" submitted to the Board of Trade, CO217, passim.
2. New England Historic Genealogical Society (hereafter NEHGS), Pepperell Family Papers, folder 13, Elias Pearse to Pepperell, 21 May 1724; Massachusetts Historical Society (hereafter MHS), Miscellaneous Bound Documents, 1728-33, Samuel Appleton to Joseph Brandon, 24 June 1728. For analysis of the trans-Atlantic pattern of New England trade, see W.T. Baxter, *The House of Hancock: Business in Boston 1724-1775*, (New York, 1965) and Byron Fairchild, *Messrs. William Pepperell, Merchants at Piscataqua*, (Ithaca, 1954).
3. For an excellent review of the financial system common in New England and hence at Canso, see Baxter. For site specific information, see CO217/6, f. 17, scheme of fishery 1729 and ff. 26-27, scheme of fishery 1730.
4. James G. Lydon, "Fish and Flour for Gold: Southern Europe and the Colonial American Balance of Payments," *Business History Review*, v. XXXIX, 1965, p. 173.
5. See, for example, CO217/5, f. 231 ff., list of ships clearing Canso, September 1730: no oil transported eastwards.
6. CO217/6, f. 143, scheme of fishery 1732; ff. 26-27, scheme of fishery 1730; f. 17, scheme of fishery 1729. Irish foodstuffs were common in the Newfoundland fishery. See C. Grant Head, *Eighteenth Century Newfoundland*, (Toronto, 1976), pp. 109-10.
7. CO194/8, f. 140, Commodore John St. Lo to Council of Trade and Plantations, 30 September 1727.

8. Harvard University, Baker Library, Wendell Family Papers, II A-1, Joshua Peirce Daybook, sundries sold at Canso, sloop *Happy Returns* 1726 and miscellaneous entries.

9. NEHGS, Hancock Family Papers, Box 13, bills of lading for government supplies, 1734-62.

10. Baker Library, Wendell Family Papers, II A-2, Joshua Peirce Ledger, House expenses, Canso, 1723.

11. CO217/6, ff. 26-27, scheme of fishery 1730.

12. Head, p. 107.

13. Baker Library, Wendell Family Papers, II A-1, Peirce Daybook, Peirce to Captain Cate, 9 February 1729.

14. NEHGS, Pepperell Family Papers, Folder 17, Gibbs, Lenven and Potter to Pepperell, 16 April 1728. For laws on salt importation, see Charles M. Andrews, *The Colonial Period of American History, England's Commercial and Colonial Policy*, (New Haven, 1938), p. 109.

15. See, for example, Baker Library, Wendell Family Papers, II A-1, Peirce Daybook and II A-2, Peirce Ledger, miscellaneous entries and NEHGS, Hancock Family Papers, F-4, Peter Faneuil Letterbook 1737-39, Faneuil to Thomas Kilby, 20 June 1737 and same to same, 1 September 1738; Andrews, p. 109.

16. Baker Library, Wendell Family Papers, II A-1, Peirce Daybook, inventory of ships sent fishing to Canso, 26 October 16725, and II A-2, Peirce Ledger, memo of stores on *Dolphin* at Canso, 31 August 1724.

17. Baker Library, Wendell Family Papers, II A-1, Peirce Daybook, Peirce to unnamed correspondent in Britain, 10 September 1728; Head, p. 102.

18. Baker Library, Wendell Family Papers, II A-1, Peirce Daybook, Peirce to Francis Wilks, 1 February 1728 [sic-1729], Peirce to William Jacks, 3 February 1729, Peirce to Captain Cate, 9 February 1729 and 14 August 1729, Peirce to Jacks, 15 August 1729.

19. CO217/5, ff. 159-60, evaluation of ship and contents, 20 August 1728.

20. CO217/5, f. 231ff., list of ships clearing Canso, September 1730.

21. CO217/4, f. 306, John Bradstreet to Lords of Trade received 24 March 1726. Similar problems, including the method of circumventing customs' officials, were described 17 years later by the customs' collector at Canso. CO217/31, f. 199, Newton to Lords of Trade, 1 September 1743.

22. Donald F. Chard, "The Impact of Ile Royale on New England, 1713-1763," Ph. D. thesis, University of Ottawa, 1976, p. 64.

23. CO217/4, f. 306, Bradstreet to Lords of Trade, 24 March 1726.

24. CO217/6, f. 92, petition of Matthew Mauger *et al.* to Lords of Trade, n.d. [ca. 1731-32].

25. Harold A. Innis, "Cape Breton and the French Regime," *Transactions of the Royal Society of Canada*, 1935, sect. 2, p. 67.

26. NEHGS, Hancock Family Papers, F-3, Faneuil invoice book 1725-29, miscellaneous lists of goods received from Louisbourg, and F-4, Faneuil letterbooks, Faneuil to Thomas Kilby, 13 and 20 June 1737, 15 July 1737, 12 and 18 August 1737, 8 September 1737, 21 August 1738 and 18 September 1738.

27. CO217/2, f. 188, copy of letter from Cyprian Southack to lieutenant-governor at Annapolis Royal, 7 January 1718.

28. *Calendar of State Papers*, v. 31, (1719-20), p. 99; CO217/3, ff. 128-30, memorial of Henshaw, Taylor *et al.*, 29 August 1720 for loss of goods valued at £18,000 and ff. 212-15 for revised total.

29. CO217/3, ff. 13-14, memorandum of Captain Benjamin Young, 21 October 1720.

30. CO217/3, f. 107, Governor Philipps to Lords of Trade, 26 August 1720.

31. CO217/4, ff. 142-43, Taylor, Richards *et al.* to Armstrong, 10 August 1722.

32. CO217/4, f. 180, Armstrong to Lords of Trade, 23 March 1732; CO217/38, p. 37, Armstrong to Newcastle, 5 September 1725.

33. George A. Rawlyk, *Nova Scotia's Massachusetts: A Study of Massachusetts-Nova Scotia Relations 1630 to 1784*, (Montreal, 1973), p. 130.

34. See Baker Library, Wendell Family Papers, II A-1, Peirce Daybook, Peirce to Tidmarsh and Appleton, 11 July 1728; Peirce to unnamed correspondent, 25 April 1729 and Peirce to Tidmarsh & Co., 28 July 1729, for good fishery.

35. Scheme of fishery for 1730s for decline in overseas shipments, specifically CO217/8, ff. 101-03, scheme of fishery 1740, "no sack ship loading . . . this year," and CO217/8, ff. 128-33 for 1741 report.

36. CO194/8, f. 140, St. Lo to Lords of Trade, 30 September 1720; CO217/6, f. 92, petition of Mauger *et al.* to Lords of Trade, n.d. [ca. 1731-32].

37. CO218/2, ff. 237-46, Popple to Burchett, 30 March 1731.

38. For complaints of bad curing practices, see scheme of fishery for 1730s, e.g. 1730, CO217/6, ff. 26-27, see also CO194/8, f. 139v, St. Lo to Lords of Trade, 30 September 1727.

39. NEHGS, Hancock Family Papers, F-4, Faneuil Letterbook, Faneuil to Kilby, 15 July 1737.

40. James G. Lydon, "North Shore Trade in the Early Eighteenth Century," *The American Neptune*, v. XXVIII, 1968, pp. 272-73.

41. For details of whale fishery, see scheme of fishery, CO217/6, f. 143 (1732), CO217/7, f. 28 ff. (1733), CO217/7, ff. 155-56 (1735), CO217/7, f. 199 (1736), CO217/7, f. 27 (1737), CO217/8, f. 38 (1738), CO217/8, ff. 101-03 (1740). George Brown Goode, *The Fisheries and Fishery Industries of the United States*, (Washington, 1884-87), Section V, Vol. II, p. 94.

42. Nova Scotia Archives II, *A Calendar of Two Letter Books and One Commission-Book in the Possession of the Government of Nova Scotia, 1713-1741*, A.M. MacMechan ed., (Halifax, 1900), Philipps to Armstrong, 4 April 1721, p. 73.

43. Brenda Dunn, "Food for Philipps' Regiment, Annapolis Royal, 1717-1757," unpublished manuscript, Atlantic Regional Office, Parks Canada, p. 2.

44. MHS, Belknap Papers, 61 A 130, John Jephsen, John Bradstreet, Christopher Aldridge, Thomas Armstrong and James Gibson to Paul Mascarene, 3 August 1736.

45. Great Britain, Privy Council 1, v. 58, B3, Report for Canso in Nova Scotia by J.M. Bastide, Engineer, 1742 (microfilm from Fortress of Louisbourg National Historic Site).

46. Baker Library, Wendell Family Papers, II A-1, Peirce Daybook, miscellaneous accounts, including goods sold to Armstrong from sloop *Happy Returns*, July 1726; NEHGS, Hancock Family Papers, TH-1, Christopher Kilby's Daybook, account of Patrick Heron, 21 September 1738.

47. Great Britain Privy Council 1, v. 58, B3, Report for Canso in Nova Scotia by J.M. Bastide, Engineer, 1742.

48. Account of trade and revenue in Nova Scotia prepared by Governor Philipps, enclosed in Philipps to Lords of Trade, 24 January 1732, in Public Archives of Canada, *Documents Relating to the Currency, Exchange and Finance in Nova Scotia with Prefatory Documents, 1675-1758*, rev. and ed. Gustave Lanctot, (Ottawa, 1933), pp. 180-81.

Nineteenth-Century Fishery

chapter
six

FAMILY ECONOMY AND THE FISHING COMPANY, MISCOU 1841-1847

Sheila Andrew

In the 1840s, fishing companies based in Jersey dominated the economy of New Bruns-wick's northeast coast. Fishers bought goods on credit at the company store and sold their fish to the company. The amount spent almost invariably exceeded the amount earned, so they had to bring in next year's catch to pay for last year's purchases. The company also gave positive credit for produce and wage labour but never paid in cash. When fisheries' inspector Moses Perley visited the island of Miscou in 1849, he wrote that the fishers were in "a state of poverty and bondage...a worse position than southern slaves."[1] Historians also have blamed the fishing companies for the continuing poverty in Acadian coastal communities.[2]

There is limited information on the effects of this system on individual families. Roch Samson and Rosemary Ommer have studied relations between the fishers and the compa-nies as capitalist ventures and analysed the experiences of individual men.[3] Nanciellen Davis, Hilda Murray, and Ellen Antler have emphasised the women's share of labour in fishing communities, but information on 19th-century fishers' families has been hard to obtain.[4] Gaspé companies did not usually give women their own accounts and their pur-chases were only listed in day books.[5] However, the Fruing company's Miscou station kept separate accounts for some women and even for girls as young as ten. They show who worked for the company on Miscou between May and October when the store was open, what they bought and how they paid for it. A complete set of ledgers for male and female workers has survived for 1841-47.[6] Then the women's accounts were moved to a separate book that has not survived.

This paper examines the fortunes of nine families who worked for the company be-tween 1841 and 1847. The picture is not necessarily complete as transactions outside com-pany books could modify conclusions on the fortunes of these families.[7] However, the results show variations in income, available credit and spending patterns at the company store that were influenced by the stage of a family's life cycle, family structure, and oppor-tunities for individual initiative.

Most of the Miscou community depended on the fishing companies for employment and provisions. Moses Perley said the island was made up of willow, caribou bogs, swamps, small lakes, salt marsh, and sand plains, with only small tracts of land fit for agriculture.[8] He listed 18 families living there and a total population of 125. About 80 men worked for the Fruing and Lebouthilier fishing companies at their stations on the point of the island; a total of 97 people lived in this area. Perley thought they were squatters on crown land and described their houses as "built of logs and poles...small and very ill fitted to resist the severity of the climate."[9]

The families studied were chosen from a possible 23 identifiable family units in the company's books between 1841 and 1847.[10] They represent the middle section of Miscou's community and a comparatively stable group. At least one member of these families worked for the company each year and all but the Chiassons can be identified in the first surviving census for Gloucester County in 1861. Their origins and family structures showed the diversity of Miscou's community.

Table 1

Families Studied: Origins, Age in 1841, Children born before 1841

Parents	Origins		Age		Children
	Husband	Wife	Husband	Wife	
Beseaux, P & G	N.B.	N.B.	29	20	2
Blanchard, O & M	Caraquet	Caraquet	28	26	2
Brown, G & M	N.B.	Ireland	40	35	1
Chiasson, G & M	Miscou	Tracadie	49	44*	1
Gauthier, L & M	France	Miscou	47	50	6
Harper, R & S	Scotland	N.B.	57	30	6
Sivret, G & A	N.S.	Miscou	35	39	5
Ward, J & A	N.S.	Canada	24	40	3
Ward, W & M	Canada	Canada	28	31	2

* Estimated from the record of their marriage in 1817.[11]

These diverse origins were typical of the community. Perched at the northeastern tip of New Brunswick, the island was open to the sea lanes and men also came from Prince Edward Island, the Gaspé, Cape Breton Island, and Jersey.[12]

At the top and bottom of the island's economy, there was a marked difference between families of different origins. The most prosperous families were the Wilsons and the Marks who were both from Scotland.[13] They are not included in this survey because few family members worked for the company. However, Moses Perley explained the secrets of their success.[14] They had been granted the only productive agricultural land on the island where they raised cattle, sheep, oats, wheat, and potatoes. Wilson had eight sons and two daughters, all living at home and all contributing to the family economy. They produced enough food for their families, and because they owned schooners they could trade for cash where the prices were best. At the other end of the scale the Mi'kmaq couple, Nicholas Marthe and his wife, rarely appeared in the books. He occasionally shared in a fishing trip and she was paid by other men and women for unspecified help.

Table 2

Value of Family Purchases From the Miscou Ledger*

Parents	1841-2	1843-4	1845-6	1847
Beseaux, G & P	378.25	315.83	1001.92	522.66
Blanchard, M & O	0.00	302.83	1464.58	1093.16
Brown, G & M	158.75	347.16	838.58	380.08
Chiasson, M & G	889.78	949.33	2866.00	5.25
Gauthier, M & L	465.75	844.66	1866.33	1036.62
Harper, R & S	409.83	52.50	432.42	98.82
Sivret, A & G	323.91	174.83	1033.92	153.83
Ward, J & A	716.27	1087.25	1896.25	1415.66
Ward, W & M	480.57	321.33	884.08	597.66

* Amounts are in shillings. To simplify calculations, the decimal system has been used rather than pounds, shillings and pence. The figures are correct to the nearest two decimal points.

Among the families studied, spending power (measured by the company's books) was not strongly influenced by origins or language of origin.[15] The Browns and the Wards had similar expenditures and income levels to the Beseauxs. Robert Harper the Scot had less spending power than the Acadian Chiassons.

A good education did not necessarily bring more prosperity. Records are not adequate for a full investigation of the relationship between literacy and income, but the company did not use the skills of educated residents. All the clerks were brought out from Jersey. Georges Sivret had a better education than most men on the island and was a man with some standing in the local community. He had been brought up in a home with books where accounts were kept, as his father had been a clerk from Jersey who lost his job when he married an Acadian.[16] Georges taught the local children when time allowed.[17] He was employed by the parish as a constable but this did not increase his credit substantially. The ledgers show one payment of 15 shillings for an arrest. Whatever small sums he earned from the local government office of "tidewaiter" did not pass through the company books. They did not benefit the family much, as he left his widow and children in poverty.[18]

Family spending power was influenced by life cycles. The company needed fishers and it needed male and female workers for salting, drying, and farm work.[19] As the clerks were not allowed to bring wives or to marry local women, they also needed women for domestic work. Couples in their late forties or early fifties were thus in a strong situation. This was the most prosperous period in the lives of many rural parents, if they stayed healthy.[20] Most of the children were either independent or earning part of the family income. The Gauthiers and Chiassons were both in this age group and consistently among the big spenders. In the Gauthiers' case, this reflected the number of family members paying into the books and other families showed the same pattern. More accounts meant more spending power. The Chiassons had three children old enough to contribute to the family income and one baby born in 1845.

Table 3

Nuclear Family Members with Company Accounts.

Parents	1841-2	1843-4	1845-6	1847
Beseaux, G & P	1	1	2	1
Blanchard, M & O	0	1	2	2
Brown, G & M	1	2	2	2
Chiasson, M & G	2	2	3	1
Gauthier, M & L	5	4	5	6
Harper, R & S	1	1	2	2
Sivret, A & G	1	2	3	3
Ward, J & A	2	2	2	1
Ward, W & M	1	1	2	2

The family with the oldest husband in this survey was also the poorest. The retired Scottish soldier, Robert Harper, was 57 in 1841 and his wife was 30. He had only one son old enough to help in a boat or work for the company, and his wife had to look after six small children. Neither of them could earn much money. In 1842, they did not even have a home of their own, and company books show that they were paying 80 shillings for room and board.

Daughters often had their own accounts, so their contribution to the family's spending power is clear. The Gauthiers, Chiassons, and Sivrets all had daughters working for the company with their own accounts. The books did not explain what work was done by the younger girls who were paid 1.25 to 1.33 shillings a day. Caroline Sivret was taken on for 15.75 days and again for three days in September 1845 when she was 10 years old. Elisabeth, the Gauthier's third daughter, was already working in 1841 when she was 13. Between 1841 and 1842, she earned 49.75 shillings. The Chiasson's daughter had her own account when she was 16. Older girls were paid for the same work their mothers did and at the same rate. The Gauthier daughters were still part of the family economy until they were 19.[21]

The contribution of sons is harder to measure. Only two of the identified sons had accounts before setting up their own households. Peter Sivret earned 24 shillings when he was 11. Seven shillings were for occasional days' work for the company and the rest was his share of the fish catch.[22] Lisée Gauthier got his own account when he was 14 and a recognised member of his father's cod-fishing enterprise.[23] He earned 29 shillings as his share of the catch. Even those over 14 did not necessarily have their own accounts. The Chiassons' two sons occasionally earned small sums paid into their mother's account. The fish catch of James Ward's two boys was credited to their father's account. Robert Harper's son did not have his own account until he was 21 years old. The purchases were those of a dependent son rather than a household head, although Roch Samson found that young men on the Gaspé started their own families around the age of 21.[24]

The younger families lost spending power because the women had to look after children. As the population was mobile, extended families were unusual, and the women studied here normally joined the paid work force when the eldest child was around ten years old and able to look after younger siblings.[25]

Table 4

Women's Earned Income in Shillings

	1841-2	1843-4	1845-6	1847
Beseaux, G	0.00	0.00*	27.00	0.00
Blanchard, M	0.00	0.00*	237.25*	143.33
Brown, M	0.00*	6.00*	110.50	40.50
Chiasson, M & S	95.00	94.50	253.66*	5.25
Gauthier, M, M, E, S & J	238.25*	159.25*	384.33	232.00
Harper, S	0.00*	0.00	18.00	0.00
Sivret, A & C	0.00	46.33*	27.33	30.25
Ward, A	81.50**	15.00*	10.50	0.00*
Ward, M	0.00	0.00*	0.00*	36.50*

* Indicates a child born that year.

In the year when children were born, some women were able to continue earning money. Although Marie Chiasson had a child in 1845, she worked for the company every year until 1847, when she and Guithome disappeared from the books. Marie Gauthier also worked in the years when her three sons were born, but her income dropped below that of her 19-year-old daughter in the year of the last birth. Younger women with small children rarely earned much. Geneviève Beseaux, for example, was 20 in 1841 and had two children, and a third in 1843. Her only positive credits were five shillings salary and 22 shillings for sounds and cod oil in 1845-46.

Older or unmarried women who worked for the company made a useful contribution to the family's spending power. They were hired for 20-35 shillings a month or for occasional work during peak periods, such as potato planting and harvesting, at two shillings a day. They also did mending and washing for the Jersey bachelors and received some payments from local residents for unspecified services.

Women also gained credit from other transactions with the company. The most important of these was cod oil. The manufacturing process was less labour intensive than drying fish and the company paid one and 1.75 shillings a gallon. Cod livers were left in barrels in the sun and the oil was tapped off as it separated.[26] Sometimes women processed fish for the family instead of for the company, earning more than the company paid for fresh fish. Marie Chiasson, Mary Ward, and Appoline Sivret earned up to six shillings income from dried fish in some years, but none produced it regularly. Processing took time and required the right weather. Every stage could increase the profit but also increased the danger of losing money through spoilage.[27] Women employed by the company made very little of their credit from manufactured goods, but those not on salary occasionally made scoop nets and sounds.

Table 5

Sources of Income for Women

Salary	67.36%
Cod Oil	13.87%
Paid by Men	6.19%
Fresh Fish	5.91%
Paid by Husband	2.93%
Dried Fish	2.00%
Farm Produce	0.90%
Returned Goods	0.36%
Son's Salary	0.32%
Manufactures	0.15%

The earning capacity of women was intermittent and limited. A survey of the 49 women who had accounts with the company shows only four worked for the company every year. No woman on Miscou was employed on salary for the full season from May to October.[28] There was competition for jobs, as summer residents moved into the area from other communities.[29]

Husbands earned more than their wives and fishing provided a more consistent source of income. Most fishers relied on round and fresh fish as their major source of income.[30] This provided faster credits but less profit than processed fish. Green fish were salted enough to get them back from a longer fishing trip, requiring more provisions but possibly providing a bigger catch. Sivret, Chiasson, and James Ward provided most of the green fish. Those families fortunate enough to have labour resources to split or dry fish made the best profits. This favoured the Gauthiers, who could do the work themselves, and James Ward, who paid others to do it. However, the company kept a close eye on the quality of dried fish and poorer product gained less credit and was sent to feed slaves in Jamaica.[31]

Ward was not the only man in the community to use his credit to finance ventures that might contribute to the family economy. Guithome Chiasson bought supplies for a schooner voyage in 1841. This meant paying more to the store for provisions, but also gave potential access to other fishing grounds and increased his income from fish that year. Louis Gauthier kept to hand-lining, but he did decide to hire a boat of his own when his son Lisée was 14. Ironically, the extra expense cost him more than the extra money earned from fish sales.

Almost a quarter of the men's income came from sources other than fishing. They worked for salary at salting, repair work, construction, or transportation jobs that paid between 2 and 7.25 shillings a day or 60 shillings a month. Unlike the women, they were sometimes paid to work in the winter, maintaining the buildings or preparing fishing gear for spring. Other men hired help occasionally, but most payments between men were for a share of the fish catch. Women who did not have accounts occasionally had credits put into their husbands' accounts. Some women also transferred credits from their accounts to their husbands'. By selling oars, yokes, and firewood to the company, men earned more than the women did from manufactured goods.

Table 6

Sources of Income for Men

Round Fresh Fish	46.97%
Split Fish	14.40%
Salary	11.47%
Paid by Other Men	7.33%
Green Fish	6.59%
Paid by Women	3.72%
Produce	2.06%
Manufactured Goods	2.05%
Transferred	1.70%*
Dry Fish	1.75%
Jamaica Fish	1.10%
Cod Oil	.71%
Items Returned	.15%

* Unexplained credits were sometimes transferred from the Lebouthilier station or the Shippagan station.

Farming would have been a great help to the family economy, but the poor soil on Miscou made this difficult.[32] Perley described the Gauthier's farm as "part of an acre of ground in cultivation, but weeds and thistles were more abundant than crop."[33] The Gauthiers bought potatoes, meat, beans, and peas from the store every year. Other families managed to avoid this. The Harpers and Sivrets did not buy potatoes from the store and Guithome Chiasson sold three barrels of potatoes to the store in 1844. However, potato crops were unreliable in the 1840s and Chiasson was buying potatoes again in 1845. A bushel of potatoes cost about as much as one day's pay for working in the company potato fields, so the economic benefits of growing your own were not immediately obvious, particularly when the crop might fail because of disease.[34] The Sivrets and Browns also kept cows. Georges Sivret sold one to a neighbour in 1847 and had 103 shillings credited to his account. George Brown slaughtered one of his and sold it to the company for 294.16 shillings. Mary Ward once paid with a pig and got eight shillings credit. Pork sold for 10 pence a pound so the company made a good profit on this as all the families bought pork from the store occasionally.

Men also contributed to the family economy by hunting. Wild geese and black ducks were frozen to eat in the winter and feathers were used for down. However, this represented significant capital expenditure as a bag of shot cost 14 shillings; a pound of powder was 1.75 shillings and a flint was 1.5 pence. The more prosperous families went hunting. The Wilson family claimed to have shot 400 geese one winter.[35] Olivier Blanchard and the two Ward families also bought shooting supplies regularly. However, George Brown and the Gauthiers rarely bought them and Guithome Chiasson, Georges Sivret, and Peter Harper never bought them. Peter Beseaux, one of the younger husbands with little credit, spent more than anyone else on hunting supplies.

This suggests at least one source of income was not reflected in the books. We have no way of measuring the barter trade. The very limited unidentified transactions between individuals on the company books may have covered some barter items, but an unofficial exchange network seems likely. We can surmise that Beseaux's above-average purchase of shot and powder would have allowed him to supply others with game.

Business ability and other resources may have modified the pattern suggested by life cycles, as there were exceptional families. James and Amelia Ward were 40 and 24 respectively in 1841. She did not earn much money from the company as their eldest son was ten in 1841, and they had seven younger ones by 1847. However, they consistently maintained one of the highest expenditure levels and were allowed credit for the largest debt in 1847, even when the company seemed to be requiring some reductions in debt load. James was a good businessman; he claimed refunds for excessive charges more often than any other member of the families studied. He controlled family expenses by paying for his sons' needs through his own account and he bought broken items at the store at reduced prices. He was the only man to buy a compass, suggesting that, like Guithome Chiasson, he had access to vessels able to go some distance from the island. His clothing expenses suggest he maintained a high profile in the community. Monique and Olivier Blanchard were also exceptions. They were a young married couple who had two children during this time period, but they still enjoyed more spending power than others in their age group. Olivier also spent more on ready-made clothes than any man except James Ward. Like James Ward, Monique claimed her money back for over-charging or inferior goods. Marie Chiasson showed similar business ability as she was the only other woman asking for her money back, and she earned more than any other individual woman.

Family spending patterns show they needed to use strategies to improve their situation because necessities took most of their credit.

Table 7

Spending Patterns

Male		Female	
Essential Food	33.78%	Cloth	49.44%
Clothes	12.99%	Shoes	14.44%
Fishing	12.73%	Clothes	8.35%
Small Luxuries	8.15%	Unreadable	6.66%
Cloth	7.86%	Small Luxuries	6.50%
Shoes	6.24%	Essential Food	6.40%
Other Stores	6.20%	Household	4.36%
Paid to Men	5.78%	Sundries	4.24%
Household	4.70%	Other Stores	3.92%
Sundries	1.10%	Cash	.79%
Unreadable	.13%	Paid to Men	.78%
Cash	.09%	Paid to Women	.42%
Paid to Women	.04%		

The foods labelled as essential in this table were those any 19th-century rural house-hold might have bought including flour, oatmeal, cornmeal, pease, pork, biscuits, butter, lard, and molasses. Perhaps because the women were rarely sure of an income, the men bought most of these. Sometimes, there was money left over for a few luxuries. All the men bought tobacco and pipes. James Ward bought liniment for his rheumatism. Family members of all age groups bought raisins, figs, and quantities of tea. Sometimes the men bought pepper, cocoa, and small amounts of coffee, and all the adults bought a variety of alcoholic drinks. Rum was the favourite with the men and some of the women, but both sometimes bought cider. Men also bought brandy, port, and, by 1847, gin. James Ward, for example, bought 1.25 gallons of rum, almost 2 gallons of gin, 1.25 gallons of cider, and half a pint of brandy in 1847. A quarter gallon of gin cost 2.5 shillings. A gallon of rum was ten shillings. A considerable number of the "sundries" marked down to the ladies cost 7.5 pence, exactly the same amount as half a pint of rum. This may have been coincidence.

Cash was rarely bought. Even the annual tax of 1.25 shillings paid by Guithome Chiasson and James and William Ward between 1845 and 1847 was paid through company accounts. Some of the more business-minded families were involved in cash transactions. Monique Blanchard borrowed five shillings in 1846 and repaid a quarter of the loan in cash. Guithome and Marie Chiasson bought 20 and 16 shillings worth of cash for unspecified purposes and did not apparently pay back in cash or pay a premium, so the company policy in this case is not clear.

None of the families spent large amounts on household goods, but expenditures varied according to taste and income. Perley described the Gauthiers' home as little more than a shack and said that it was typical of the settlement.[36] However, the books show considerable diversity. Even the Gauthiers bought a few plates and spoons, two tins of paint, a pane of glass, and some planks and nails. Every family bought soap from the store. In most cases, purchases were related to income; from 1842 to 1847, Robert and Sarah Harper, for example, did not pay for room and board, and they bought two pocket knives, one chest lock, two porringers, and four plates. George and Mary Brown had far more available credit than the Harpers and still bought only one tin dish, one tea kettle, and three porringers. Guithome Chiasson, who usually had more credit than James Ward, spent far less than Ward on household goods. Chiasson bought basic cooking and eating utensils, a lamp, a counterpane, and a chamber pot. Predictably, James Ward and Olivier Blanchard bought more household goods than other families. Ward's expenditure rose every year and included 18 "plates," six white plates, six blue plates, and six "fancy" plates with a matching teapot and basin.

Cloth was the biggest single expenditure for most of these families. Women bought most of it, but men also bought cloth for their wives. The Gauthiers, for example, bought 110.66 yards between them in 1845. In 1846 they bought 117.25 yards, and in 1847 they bought 86 yards. None of this was canvas or duck that would have been suitable for sails. Even with five daughters, three small sons, and an adolescent boy to provide with clothes and linen, this suggests the family enjoyed dressing well.[37] Other salaried women were buying comparable amounts of cloth. Marie Chiasson, with only one daughter and a new baby at home, bought 99.75 yards of cloth and 19 yards of lace in 1845. Most of the women on salary and even the younger girls also bought dress patterns; at 12.5 shillings each this represented about two weeks' salary. Women also bought necklaces at one and a half shillings or hair bands at 1.25 shillings, which cost almost a day's wages. Lace and ribbons at four pence to six pence a yard sold very well.

Women spent little on ready-made clothes, restricting themselves to shawls, handkerchiefs, and caps. Men had to spend more for work clothes such as trousers, drawers, braces, heavy sweaters, jackets, and sou'westers. After these essentials, purchases varied. Boys with their first pay credits usually bought a new cap and sometimes ribbon or lace for a woman. William Ward, Georges Sivret, and Louis Gauthier rarely bought more than the essentials. Robert Harper bought only the essentials except for one "spring bonnet," for his wife in 1845. Guithome Chiasson spent more and treated himself to a "fine cloth jacket" worth 45 shillings as well as new work clothes in 1841. George Brown bought three shirts and two pairs of trousers that year. Olivier Blanchard and James Ward spent most on clothes. As well as the usual one pair of trousers, drawers, braces, cap, handkerchiefs, and a monkey jacket, Blanchard bought three cravats, two pairs of socks, and a plush cap in 1845. Ward's 1847 purchases for himself and his son included six shirts, five pairs of trousers (including three for work), one pair of socks, two caps, and a "fine" waistcoat.

While some of these families had a comfortable standard of living between 1841 and 1847, their future prospects were not necessarily good. Increasing family income was rarely enough to keep up with the expenses of another family member with the right to credit. The amount of debt carried by most families increased faster than the family income. Although the company gave individuals separate accounts, they recognised them as members of family units. Men occasionally returned goods or transferred money from their wives' accounts to lower the debt at the end of the season. Women and children did not run up large debts, which reflects their lower earning power. They had to return goods to balance the account and got transfers from their husbands. However, family debts continued to rise as if the permitted amount was part of their earned income. As Roch Samson has pointed out, most fishing families did not intend to pay off their debts.[38] Bigger permitted debt meant a better standard of living. There was some reduction of debts in 1847, but the company books do not include any threats to prosecute or even to cut off credit.[39]

Table 8

Family Earned Income in Shillings

Parents	1841-2	1843-4	1845-6	1847
Beseaux, G & P	157.42	138.56	35.70	409.00
Blanchard, M & O	0.00	161.83	1340.08	1145.13
Brown, G & M	185.16	306.00	741.75	160.50
Chiasson, M & G	631.33	778.66	2791.76	5.25
Gauthier, M & L	515.50	694.57	1401.91	749.50
Harper, R & S	185.90	253.00	70.92	64.00
Sivret, A & G	237.70	422.25	766.63	170.75
Ward, J & A	655.50	726.16	1506.41	335.42
Ward, W & M	700.92	146.75	717.80	488.16

Table 9

Family Debt Carried Forward in Shillings

Parents	1841-2	1843-4	1845-6	1847
Beseaux, G & P	-194.58	-371.92	-951.50	-842.80
Blanchard, M & O	0.00	-149.00	-236.92	-216.16
Brown, G & M	- 52.00	- 93.16	-190.00	-409.58
Chiasson, M & G	-345.30	-516.50	-590.75	-590.75
Gauthier, M & L	- 97.50	-247.58	-711.96	-999.08
Harper, R & S	- 1.00	- 28.50	- 90.00	- 72.82
Sivret, A & G	-131.92	-115.50	-382.80	-365.92
Ward, J & A	-111.75	-472.83	-863.25	-943.50
Ward, W & M	-216.40	-391.00	-557.20	-666.75

Some of the people of Miscou knew that the companies were taking advantage of them. When Perley visited them, the fishers were organising a petition against company charges for the use of salt marsh hay meadows, and they complained of high prices at the company store.[40] Company policies also caused longer term damage. As they did not employ local men in administrative positions, there was little incentive to get an education. There was no record of families buying books, pens, or paper that might have helped them educate themselves. The growing family debts also looked alarming. By 1848, the Robins in Caraquet were annotating their accounts with chilling comments like "To have nothing this winter" and "interest on debt, twenty shillings."[41] It was impossible for fishers to save because cash was a commodity bought like any other, not a medium of exchange within the Miscou community. This meant families had no support in times of trouble. Widows were particularly vulnerable. However hard they worked, women did not command the same income or credit amounts as men. The only widow on the Miscou books was Appoline Sivret, whose husband apparently died in 1845 or 1846.[42] The family income dropped sharply from 766.63 shillings to 170.5 shillings. She made courageous efforts to keep up: the record for 1847 shows a payment to her for driving a horse and sleigh over the ice to Shippagan. However, buying all the necessities on wages available to a woman and her teenage children was difficult. By 1848, she was selling fish to the Robin's Caraquet store and by 1849 that company refused to give her credit. She only owed 70 shillings, but her earning ability had diminished to 43 shillings and the company judged her to be a bad risk.[43] Old age was also a problem. Spending power could only mask the declining earning power of Robert Harper for a few years longer. By 1847, he was 63 years old and earned only 64 shillings.

A longer study is needed to establish the significance of the varying fortunes of Miscou families when they reached later phases of their life cycles. The series of Miscou ledgers is broken after 1847, so we are left with a brief snapshot of an economy where fishing families had to use individual initiative to make the best of a difficult situation by taking advantage of family structure, access to distant fishing grounds, and even a small patch of fertile agricultural ground.

Notes

1.   Moses Perley, "Report on the Sea and River Fisheries of New Brunswick," *Journal of the House of Assembly* (Fredericton, New Brunswick: Queen's Printer, 1850), cxxvii. See also Rev. Jean-Baptiste Ferland, "Journal d'un voyage sur les côtes de la Gaspésie," *Soirées Canadiennes*, Oct.-Nov., 1861 and F. W. Remiggi, "La lutte du clergé contre le marchand de poisson," L. R. Fischer and E. W. Sager eds. *Enterprising Canadians* (St.John's, Newfoundland: Memorial University Press, 1979), p. 183.

2.   Bernard Thériault, *Les Robins: presence jersaiase en Acadie* (Caraquet: Historical Resources, 1975); Gary Hughes, *Two Islands: Miscou, Lamèque and their State of Bondage 1849-1861* (Saint John, N.B.: N.B. Museum, 1979); R. Mailhot, "Prise de conscience collective et comportement de la majorité anglophone au Nouveau-Brunswick, 1860-1891," Ph.D. thesis, Université de Montreal, 1973, pp. 39-40.

3.   Roch Samson, *Fishermen and Merchants in Nineteenth Century Gaspé* (Ottawa: Parks Canada, 1984). Rosemary Ommer, *From Outpost to Outport: A Structural Analysis of the Jersey-Gaspé Cod Fishery, 1767-1886* (Montreal and Kingston: McGill-Queen's University Press, 1991) and "The Truck System in Gaspé, 1822-1877," Ommer ed. *Merchant Credit and Labour Strategies in Historical Perspective* (Fredericton: Acadiensis Press, 1990), pp. 42-73.

4.   Nanciellen Davis, "Women's Work and Worth in an Acadian Maritime Village," Naomi Black, ed. *Women and World Change* (Beverley Hills: Sage, 1981); Hilda Chaulk Murray, *More Than Fifty Percent* (St. John's, Newfoundland: Breakwater, 1982); Ellen Antler, "Women's Work in Newfoundland Fishing Families," *Atlantis*, Spring 1977, pp. 106-13.

5.   Samson, *Fishermen and Merchants*, p. 51; Ommer, *Outpost to Outport,* p. 129.

6.   Fruing Company. MC 82A, MS 3 "Ledgers Miscou Office," 3/1 3/2 3/3. Public Archives of New Brunswick (hereafter PANB).

7.   The Chiassons and Sivrets had access to schooners before this period and may have been able to use them in 1841-47. Guithome Chiasson shared one with his father and brothers in 1827 and Sivret was captain of a 36-ton schooner in 1831. Fidèle Thériault, *Les Familles de Caraquet* (Caraquet N.B.: author, 1985), pp. 87, 432.

8.   Perley, "Report," cxxvii.

9.   Ibid.

10.   Genealogical information from Thériault, *Les Familles* and Donat Robichaud, *Le Grand Chipagan: histoire de Shippagan* (Montreal: 1977), the 1861 census and the parish records of St.-Pierre aux Liens, Caraquet, St.-Urbain de Lamèque, and St.-Jerome de Shippagan.

11.   Thériault, *Les Familles*, p. 82.

12.   W. F. Ganong, "A History of Miscou," *Acadiensis*, VI, no. 2 (April 1906), pp. 79-94 gives the origins of anglophone families.

13.   Ganong, "History of Miscou," p. 82.

14.   Perley, "Report," cxxix.

15.   The author wishes to thank Dawn Betts for her help with data entry and research on the accounts of the husbands of these families.

16.   He willed both the library and the account books to his wife when he died. Thériault, *Les Familles*, p. 427.

17.   Thériault, *Les Familles*, p. 432.

18.   Robichaud, *Le Grand Chipagan*, p. 360.

19.   For the role of women in preparing the fish for export, see Marilyn Porter, "'She Was Skipper of the Shore Crew': Notes on the History of the Sexual Division of Labour in Newfoundland," ed. Arlene Tigar Maclaren, *Gender and Society: Creating a Canadian Women's Sociology* (Toronto: Copp Clark Pittman, 1988), pp. 169-83; Murray, *More than Fifty Percent*; Antler "Women's Work in Newfoundland Fishing Families."

20.   J. T. Main, *Society and Economy in Colonial Connecticut* (Princeton: Yale University Press, 1985), p. 373.

21. For details on this family, see my "The Gauthier Girls: Growing Up on Miscou Island, 1841-1847," *The Child in Atlantic Canada*, Hilary Thompson ed. (Dundurn: Children's Literature Press, forthcoming).

22. Samson, *Fishermen and Merchants*, p. 52, notes that young boys went along in the boats to help out and learn the trade.

23. Robichaud, *Le Grand Chipagan*, pp. 172-73, gives an account of the methods.

24. Samson, *Fishermen*, p. 87.

25. There may have been extended families that do not show up on the books. However, the 1861 census shows only eight extended households from a total of 118 households on the neighbouring islands of Miscou and Lamèque.

26. Jean-Claude Dupont, *Histoire populaire de L'Acadie* (Montreal: Leméac, 1978), pp. 352-53.

27. Robichaud, *Le Grand Chipagan*, pp. 172-74.

28. The company's station on the neighbouring island of Lamèque station employed a female cook for the whole season in 1848. Robin Papers, MC 82A MS 3, 3/4 "Prudent Maillet," PANB.

29. *The Gleaner and Northumberland Schediasma*, 22 July 1850.

30. Some fishers also went to the ice for seals in April. Perley, "Report," cxxix.

31. Most ledger entries list this product as Jamaica, suggesting a rum trade, but others specifically call it Jamaica fish. I thank Fidèle Thériault, Department of Historical Resources, for identifying this as inferior quality dried fish. Personal communication, 18 December 1995.

32. Even the best farm land on the island had "severe limitations," and the area where the Sivrets were farming in 1861 was mapped as having "no capability for arable culture or permanent pasture" in 1971. *Soil Capability for Agriculture: Bathurst* (Ottawa: Department of the Environment, 1972).

33. Perley, "Report" cxxii.

34. Mr Wilson explained this problem to Perley, "Report," cxxvii.

35. Perley, "Report," cxxix.

36. Perley, "Report," cxxiii.

37. Samson, *Fishermen and Merchants*, p. 52, suggests that Gaspé women made clothes to sell and barter. This idea is discussed in Andrew, "The Gauthier Girls..." where I suggest the market on Miscou was limited and the Gauthiers could have been using the eight to ten yards per family member per year themselves. Abbé Nérée Gingras, "Impressions de Gaspésie en 1857," cited in Samson, *Fishermen and Merchants*, p. 52 complained that the fishers' wives on the Gaspé "are lazy and do no housework and are spendthrifts. They spend their days strolling up and down roads showing off and depriving themselves of nothing during the fishing season, even though they suffer later. Their greatest pleasure is drinking tea and the only thing that bothers them is to be without it."

38. Samson, *Fishermen and Merchants*, p. 79. He cites the report of a fisheries officer in 1872.

39. Samson suggests that companies had little to gain from prosecution of fishermen with no capital. He found only one example of prosecution for debt and that was a man who had moved his business to another company without paying off his debt. *Fishermen*, p. 37.

40. Perley, "Report," cxxxiii.

41. See, for example, Widow Damien Lebouthilier, Robin Company Papers, MC 82 A, MS 3, 3/4.

42. According to Thériault, *Les Familles*, p. 432. It was Appoline's father-in-law, Georges senior who died in 1845, but the account book in Miscou called her a widow in 1846.

43. Robin Company Papers. Caraquet, 1848-49. PANB.

chapter
seven

SOCIAL HISTORY IN A NEWFOUNDLAND OUTPORT: HARBOUR BRETON, 1850-1900

D.A. Macdonald

Migration to Newfoundland and the origins of permanent settlement in the 18th and 19th centuries have been well studied, most notably by the writers in Mannion's volume and by Gordon Handcock.[1] Less attention, however, has been paid to the following stage of development, from frontier work-camps to settled outports. Early settlements were little more than seasonal work-camps with fluctuating, largely transient populations, mostly of single men. The settled outport with church, school, and web of kinship was a creation of the 19th century and can seldom have emerged before mid-century.

In this paper I have used the Anglican church records of Harbour Breton, a small town on the south coast, to examine this transition in the second half of the 19th century. It is largely a descriptive work of social history based upon the Anglican parish records and the reconstitution of families from them.[2] They enable us to examine the course of and reasons for the growth of the permanent population, the life-cycles of the inhabitants, and the pattern of social stratification. Though descriptive, the account does allow of some conclusions regarding the heterogeneity of Newfoundland communities (and their inhabitants) and suggests that there was a distinctive regional sub-culture in Harbour Breton during the period.

By 1800 St. John's was the largest town and on its way to becoming the administrative and commercial capital of Newfoundland. Smaller towns in each district became local or regional centres; these had more diversity of occupations than the majority of settlements, which remained fishing villages. An indigenous way of life had arisen in the fishing villages by 1800. Outporters made a living by using the products of both land and sea for subsistence and sale. Saltfish production, however, remained the main export industry until well into the present century.

Harbour Breton is in Fortune Bay, a sparsely-populated south-coast district far from St. John's; it was one of the frontier districts of internal settlement in 1850. Harbour Breton was used seasonally by French fishermen in the late-17th century and English settlement had been founded there by 1763.[3] For more than a century thereafter the town was a mercantile rather than a fishing centre. It is doubtful that most of the workforce was ever engaged in fishing, though colonial censuses may seem to show otherwise.[4]

Early in the 19th century, Newman and Company (of port-wine fame) made Harbour Breton their principal depot for the Newfoundland fish business and the town grew thereafter largely in response to their need for a supply of labour and fish. Newmans did little fishing on their own account but were still large employers of labour on their premises and vessels. Until late in the century the resident population could not provide them with enough workers, so they imported most from England for short terms; they were still importing workers in the 1890s. There was thus a transient population in addition to the permanently-resident. As public services were extended during the 19th century, Harbour Breton became the administrative centre of the district and the parish seat of Anglican and Catholic churches.

Newmans' main trade was in the cod fishery, which was prosecuted for the most part by small, inshore boats. There was a winter as well as a summer cod fishery on the south coast, unlike more northerly parts. After 1850 a second marine industry grew in the supply of herring and other bait to visiting American and French bank-fishing ships. The value of the bait trade (which was banned in the 1880s) may at times have rivalled that of cod to Fortune Bay fishermen. The bait trade sponsored the construction of hundreds of schooners in the district and led to the growth of a class of prosperous fishermen who were independent of supplying merchants. Newmans did not engage in the bait trade, which they resented because it drew planters' attention from the cod-fishery.

The Settled Population

The 1836 colonial census showed Harbour Breton to have fewer than 150 inhabitants more than a century after its founding. Most of these were servants of the merchants and there were not more than 15 households. The population showed the usual frontier characteristics of many young men and few women and children.[5] By 1901 the population had more than trebled to 493, though the settled population grew faster; the transient population, Newmans' servants, had not grown. Table 1 shows two measures of the number of settlers—the number of families and twice the number of females at each census.[6] It also shows that the settled population grew by almost twice as much as the total population, though

Table 1

**Population, number of families and settled population,
Harbour Breton, 1836-1901**

Census year	a. Population	b. Number of families	c. Settled population*	c as % of a
1836	149	15**	86	58
1845	241	20**	144	60
1857	271	29	200	74
1869	361	47	314	87
1874	386	54	350	91
1884	456	82	440	96
1891	484	93	466	96
1901	493	94	482	98

Source: Censuses of Newfoundland and Labrador, 1836-1901.
Notes: *Twice number of females enumerated in census.
　　　**Number of dwelling houses.

---

there was little increase in either after 1884. The total population, in fact, grew little more than the colonial average, which is surprising in a frontier district. Part of the explanation is that Newmans needed to import fewer workers as the settled population grew; loss of transients partly offset increase of settlers. The other part of the explanation is that the young emigrated steadily throughout the period. Their places were taken by immigrants but not, for the most part, by newcomers to the district. Most (at least Anglican) immigrants to Harbour Breton came from its own and nearby parishes.

Harbour Breton lost 56 Anglican families through death and departure from 1850 to 1899 but gained 63 by immigration. Families moved in at a fairly steady rate during the period and this, not natural increase, caused the settled population to rise. The south coast was the last area of Newfoundland outside St. John's to receive substantial immigration from Europe,[7] and about one-third of incoming Anglican heads of households were from England. The main sources of immigrants to Harbour Breton were, however, its own and adjoining parishes.[8] Immigration contributed more families than natural increase after 1850, at least as far as Anglicans are concerned. Of Anglican male heads of households whose place of birth is known and who are listed in the 1898 directory, fewer than half were born in Harbour Breton.[9]

Once resident, families seldom moved on, most departures being caused by death or by the transfer of officials. With the unmarried it is a different matter—once married, most stayed in Harbour Breton, but only a minority stayed long enough to marry. Table 2 shows that, during the period in question,[10] most of the town's young emigrated before marriage.

Table 2

	Emigration of children baptized to Harbour Breton residents,* 1850-89**				
	Decade baptised	**Died unmarried**	**Married and settled***	**Emigrated or unknown**	
Boys	1850-9	5%	58%	37%	(N=19)
	1860-9	29%	12%	59%	(N=17)
	1870-9	24%	26%	50%	(N=46)
	1880-9	31%	31%	38%	(N=58)
	1850-89	25%	31%	44%	N=140
Girls	1850-9	18%	23%	59%	(N=22)
	1860-9	7%	26%	67%	(N=27)
	1870-9	23%	20%	57%	(N=30)
	1880-9	21%	26%	58%	(N=53)
	1850-89	18%	24%	58%	N=132

Source: Anglican parish registers, Harbour Breton.
Notes: *Whether born in Harbour Breton or not if parents immigrated during child's minority. **1889 rather than 1899, or too many children would have been unmarried at 1921 census. ***Settlement at Harbour Breton need not have been permanent—baptism of first child there counted as settlement. As we have seen, few moved after marriage.

Typically there is a record of confirmation in the middle or late teens and then they slip from view. Given the mean number of births per marriage, this rate of retention would not have replaced, let alone increased, the settled population. Table 2 makes it seem that before 1860 Harbour Breton held its men though not its women, but this may not have been so.[11] Girls settled less often than boys because they were more prone to move upon marriage (though most girls who married settled in Harbour Breton). A few of those unaccounted for can be found elsewhere in the parish or in nearby parishes,[12] but most left for parts unknown. The great majority of emigration was of single people and very little was of families; family persistence co-existed with high and constant emigration of single people.

Two groups of immigrants deserve special mention—English servants and fishermen. It was work that brought servants to Harbour Breton but marriage that led them to settle. I estimate that at least 800 of these English servants passed through the town during the 50 years of study, but not more than two dozen of them married and settled in Harbour Breton, and perhaps two or three dozen elsewhere in the parish. Of English servants who settled in Harbour Breton, 74% married townswomen and, conversely, 88% of those who married townswomen settled in the town.

Most transients served a single tour of duty and then left.[13] Our only knowledge of their social and geographical origins comes from the few who married locally. This evidence conforms with Handcock's[14] conclusions drawn from south-coast marriages of immigrants during the period. Most immigrants came from Dorset and Somerset—in fact from an area adjoining the borders of the two counties—not from Devon, though Newmans were based there. Most were from small towns rather than villages, and were of the poor— though not necessarily the poorest—class. Where father's occupation was stated, in 46% of cases was he a labourer. The rest had a trade of some kind—carpenter, shoemaker, butcher, gamekeeper, gardener, and farmer being examples.

In the case of immigrant fishermen, it is less clear that it was either work or wedlock that caused them to settle in the town. Probably it was some combination of the two, for many lived elsewhere between marriage and immigration. A clear majority (79%) of fishermen-settlers had married townswomen, but only a slight majority of fishermen from other communities who married townswomen settled in Harbour Breton on marriage. Almost half of fishermen-settlers moved to Harbour Breton after a period of post-marital settlement elsewhere, usually in the groom's community. It seems to have been less the case that marriage meant immigration than that those who wished to move did so to a place where they had affinal kin. Most fishermen-settlers came from nearby Sagona Island, which was losing population in the last quarter of the century; all but two others were from other communities in the parish.

In summary, the population of Harbour Breton grew briskly, if unevenly, from 1850 to 1899, and the rate of growth was higher if transients are excluded. Persistence of male heads of households co-existed with steady immigration of families and with high and steady emigration of the unmarried, in the face of which natural increase would not have maintained the population. Harbour Breton's population grew largely from immigration by English servants who married local women and from fishermen from the same and adjacent parishes who had married townswomen, often some years before immigration. As far as can be told, Catholic families contributed much less to the growth of the resident population than Anglican ones during the study period and very little after 1869.

The Household Cycle

Unusually detailed parish registers allowed the reconstitution of virtually all Anglican families. Catholic families could not be reconstituted, but rough comparisons between Anglicans and Catholics were sometimes possible using other sources. From the parish records I have identified 112 Anglican marriages from 1850 to 1899[15] and most of the following data are drawn from the reconstituted families of these marriages.

The literature suggests that, in 19th-century Newfoundland, men married later than they do today and to wives who were considerably younger than themselves. Handcock reports mean ages at first marriage of 26.7 and 19.4 years for men and women respectively in Trinity in the early-19th century.[16] Thornton recorded even later marriage of men to young brides in the Strait of Belle Isle.[17] Nemec found mean ages of 31.2 and 24.2 for men and women respectively in an Irish-Catholic fishing village.[18]

In Harbour Breton, by contrast, men and women married little later than they do today and on average there were only a few years difference in age between the spouses. Harbour Breton marriage ages, in fact, resemble those which Laslett found to be typical of Tudor England—a man in his middle or late twenties marrying a woman a few years younger.[19] By Nemec's ages three-quarters of townspeople had already married. The median ages at first marriage in Harbour Breton from 1850 to 1899 were 25 and 21 years respectively. These do not vary appreciably with decade of marriage, occupation, birth order within families nor, seemingly, with religion,[20] though foreign birth correlates in some cases with later marriage.

Marriage usually took place within denominational boundaries, though there were some mixed marriages. Six per cent of Anglican marriages involved a Catholic spouse, usually the bride. If most mixed marriages were celebrated in the Catholic church they would have been more frequent than the Anglican records show. The denominational boundary persisted, however, despite a modest flow of persons across it, and the Catholic minority remained distinct. In Trepassey,[21] by contrast, an English protestant majority converted to Catholicism in the early-19th century. Nemec attributes this to the need for co-operation in the inshore fishery, the lack of protestant clergy and the equal status of the two religions. In Harbour Breton there was less need for co-operation, as most were land workers rather than fishermen; the merchants (and most officials) were Anglicans; and there was no Catholic priest before the late 1850s. In light of the Trepassey case, the Harbour Breton Catholics' failure to assimilate is difficult to explain.

Whether townspeople practised residential as well as religious endogamy is difficult to establish; at times the marriage registers record places of residence of the spouses and at other times those of the parents. Table 3 shows that it was more common for brides and their parents than for grooms and their parents to live in Harbour Breton before (the children's) marriages. Altogether 95% of brides and 77% of their parents lived in Harbour Breton (or had done so when alive). Conversely, it was rather likely that a groom would be resident in the town at marriage but rather unlikely that his parents would; if the foreign-born are excluded, then it was little more than an even chance that either grooms or their parents had lived in the town before the marriage.

Place of residence after marriage is a different matter. We do not find here the pattern of viri-patrilocal residence[22] reported in some other areas of Newfoundland in the 19th and 20th centuries. In fact, it is hard to show that there was any pattern of post-marital residence in the present case. Brides and grooms were about equally likely and unlikely to live in the same communities as their parents. This mobility after marriage, coupled with the high rate of emigration of single people, must have worked against the formation of extended families of any genealogical depth. Marriage partners were usually, if not always, chosen from within the same social stratum.

Table 3

Residence of Harbour Breton marriage partners
(1879-99) and of their parents (1850-99)

1879-99:	spouses resident in Harbour Breton at marriage—	
	both	68%
	bride only	27%
	groom only	5%
		100%
1850-99:	parents of spouses resident at Harbour Breton at (children's) marriage—	
	both spouses' parents	23%
	bride's parents only	54%
	groom's parents only	20%
	neither spouses' parents	3%
		100%

Source: Anglican parish records, Harbour Breton

Marriage was closely followed by children, the median interval between marriage and baptism of the first child being 12 months. Illegitimate births were rare (only 3% of births) but premarital conception was frequent. In 23% of cases, the first child was baptized less than nine months after marriage and, as delaying baptism would have removed the evidence from the registers, we might look askance at more cases. It is most unlikely, however, that in Harbour Breton as in Cat Harbour,[23] marriage was the result rather than the cause of children in 71% of cases. In fact, the rarity of illegitimate birth taken with the surplus of unmarried males pays tribute to the chastity of local women. It seems that rural English courting customs, in which conception fixed the date of a union that was already intended, had crossed the Atlantic.[24]

Typically, then, a year or so after marriage a couple embarked upon a career of child-bearing and child-rearing that would occupy them for most of the rest of their lives. Most accounts depict Newfoundland families in the past as having been large, sometimes improbably so. Among Anglicans in Harbour Breton, the mean number of children per completed marriage was seven and the mean number of children for all marriages, including those broken by early death, was 5.5. The largest number of children born to one union or to successive unions was 12. These figures are similar to those considered typical of pre-industrial Europe, where seven or eight children have been reckoned per completed marriage. My mean of all marriages is higher, which perhaps means that premature death was less frequent in Newfoundland than in England or France. If the parents married at the median ages, their last child was born on average when the wife was 39 years old and the husband was 43. Both would be well into old age before all their children were married and they might well not live to see this happen.

Household Size and Composition

Evidence is drawn from two censuses of Anglican church members (dated from internal evidence to 1853 and 1866) and the raw data of the 1921 government census, the earliest for which such data is available. All this must be treated with caution, as no definition of household is given and it is not stated that the church censuses list families by household, though this seems to have been the case. Census summaries of the 19th century are unreliable guides to household size because of the presence of a large body of transients employed (and housed) by the merchants, though this evidence can be compared with that from more reliable sources.

Even taking all these difficulties into account, we find the evidence yielding a consistent pattern of household size and composition. The pattern is consistent within and after the study period and is very similar, with a few exceptions, to that described by Laslett for Tudor England. Households were small and contained one nuclear family each; under ordinary circumstances no two married couples occupied the same household and couples were prepared to delay marriage so as to live by themselves. Laslett claims that it was this rule, rather than the need to wait upon inheritance, that accounted for delayed marriage in pre-industrial England.[25]

The evidence suggests that the mean size of Harbour Breton households approached five persons from the early-19th to the early-20th centuries,[26] close to the 4.75 persons per household that Laslett says was typical of pre-industrial England.[27] There were few cases of extended-family living in Harbour Breton; only 13% of families in 1853 and 1866 and 18% in 1921 including kin outside the nuclear family. Measures of household size and composition in Harbour Breton suggest that the rule was of one conjugal family per household and there is evidence that, where the arrangement differed from the norm, it was usually to care for the widowed.[28]

A puzzling feature of the 19th-century church census is the complete absence of servants. In rural England servitude was an accepted stage in the life-cycle of the poor and more than a quarter of households had resident servants. It is known that some Harbour Breton families of the period had domestics or fishing servants, so their absence from the census is doubly puzzling. Moreover, in the 1921 census 12% of households kept resident servants—more if some unmarried kin were really servants in disguise. It may be that servants were kept but were not resident, or else that they were Catholics (as were most servants in 1921) and therefore not recorded in the Anglican censuses.

Marriage, once joined, was often broken by early death. One wife in six and one husband in four died before the age of 45; one-tenth more died later but with minor children still at home. Of all couples joined within the study period, only 40% lived to see the majority of all their children; only 25% of husbands and 40% of wives are known to have lived to see the marriage of all their children. In view of late marriage, early death, and lengthy child-rearing, a three-generational household would have been exceptional.[29] Old age as a distinct stage of the life-cycle seems, in fact, hardly to have existed.

Although early death occurred, it seems not to have been particularly common in Harbour Breton—in fact, by the standards of the day, life seems to have been quite a healthy one.[30] The numbers are perhaps too small for accurate comparison, but death rates and infant-and-child mortality rates were below those found in other pre-industrial countries, though they rose late in the century.[31] To judge by the age at death of those buried during the period, a man who had survived to the age of 20 had on average 38 years of life remaining, a woman 45 years—which is longer than they would have had in Tudor England.[32]

Social Stratification

The completeness of the Harbour Breton parish registers allows some analysis of social stratification among Anglicans during the period. Parish registers yield information on choices of marriage partners, social mobility within and between generations, and literacy (in ability to sign the marriage register). Any discussion of class differences must necessarily simplify a complex subject and I avoid any theoretical discussion; as used here the term relates to life-chances. Moreover, little information is available on several important aspects of the matter—family earnings, for instance, or differences in power and influence—and we will leave open the question of how far status was judged by personal qualities and other non-class attributes.

I have borrowed a model of social inequality in 19th-century Newfoundland from Nemec's work on the Irish Catholic town of Trepassey, on the Southern Shore. Nemec found at least three social classes; fishermen were in all of them, not confined to the lowest. A rural middle class grew during the century because of the creation of public offices tenable by fishermen and because of the introduction of more productive vessels and fishing gear, which only some could afford. The upper class consisted of the priest, some officials, professionals, and the richest fishermen. Their wealth only partly explained the high status of the elite, for they also functioned as political brokers—in fact, official appointment often resulted from strong ties to political patrons.[33]

Fishermen who held minor public offices belonged to the middle class though, being manual workers, they ranked well below the elite. Nemec concluded, from collating census data with the memories of the oldest inhabitants, that there was a crucial distinction between fishermen on the basis of size of vessel and amount and type of fishing gear owned and, to a lesser degree, on the personal attributes of the fisherman and his family. Nemec distinguished three classes of fishing technology (and therefore of boat-owner) during the late-19th and early-20th centuries. Ranked below all types of boat-owner were sharemen, usually single, transient outsiders who owned no boats or gear.[34]

From Nemec's model I have derived some hypotheses, set out below, to be tested on the Harbour Breton data. There were in Harbour Breton some occupations that Nemec did not mention—labourers, seamen, ships' captains, skilled manual workers, and small traders and dealers. Membership of the elite is easily distinguished by the degree of literacy that it required. To distinguish between the middle and lower classes I have adopted Gagan's[35] concept of social improvement, which he used to distinguish between successful and unsuccessful farmers in 19th-century Ontario, but I have used measures of material improvement more appropriate to my case. Ownership of a schooner, tenure of a minor government office, petty trading,[36] a skilled manual trade or command of a vessel places a subject in the middle class.[37] Those without any measure of improvement are in the lower class, which therefore consists of labourers, sailors, and small-boat fishermen.

The hypotheses to be tested on the Harbour Breton data are:

(1) There were three social classes, whose members differed in their life chances.

(2) The upper class consisted of merchants and their agents, clergy, professionals and senior public officials, whom collectively I call the official class. It is defined by requiring a high degree of literacy, then a rare accomplishment.

(3) Fishermen were found in all three classes according to the size of their operations. The crucial distinction is between large-vessel owners—who can be identified from the shipping registers—and small-boat fishermen.

(4) Tenure of a minor government office ranked fishermen above the lowest class because it gave them a security of income denied to the poor.

We have an index, albeit a flawed one,[38] of literacy in the ability to sign a name on the marriage register, either as spouse or witness. Inability to sign at least shows illiteracy, though some in signing may have reached the limits of their letters. Table 4 below summarizes the evidence from the marriage registers by decade of birth and by occupation of parents.[39] It shows that literacy did follow class lines, with some interesting sub-group variations in the middle ranks. The ability to sign was always universal among the children of officials and almost so among those of skilled manual workers. It spread to skippers' families after 1860, more slowly to improved planters' families after 1870, and in small measure to the children of the lower class near the end of the century.

Table 4

Ability to sign by decade of birth and occupation of parents

Decade born	Officials	Skilled manual trades (SMT)	Skippers	Improved planters	Others*	All
before 1850	100% (N=29)	78% (N=9)	12% (N=17)	------------	29% (N=65)	49% (N=120)
1850-9	100% (N=6)	73% (N=11)	50% (N=4)	33% (N=3)	30% (N=27)	49% (N=51)
1860-9	100% (N=7)	86% (N=7)	100% (N=4)	43% (N=7)	30% (N=40)	54% (N=65)
1870-9	100% (N=8)	100% (N=3)	100% (N=4)	67% (N=9)	28% (N=40)	50% (N=64)
1880-9	100% (N=4)	100% (N=6)	100% (N=5)	77% (N=13)	49% (N=39)	66% (N=67)
1890-9	100% (N=1)	100% (N=4)	--------------	100% (N=6)	43% (N=21)	67% (N=32)
1850-99	100% (N=55)	85% (N=40)	50% (N=34)	68% (N=38)	34% (N=232)	53% (N=399)

Ability to sign by sex and occupational group, 1850-99

	sons	daughters
officials	100%	100%
SMT	76%	100%
skippers	33%	69%
improved planters	72%	65%
*unimproved planters & c	32%	37%
All	52%	54%

Source: Anglican parish registers, Harbour Breton.
Note: *Unimproved planters, servants, sailors, others and no occupation.

Illiteracy was certainly widespread. No more than half of the population can have been literate during the study period, and no more than two-thirds in the last decade of the century, although schooling was available throughout. It also seems that education was considered more important in the last quarter of the century, the effects of which would have appeared in the marriage registers after 1900. During the study period, literacy among even young adults was not especially common, and 50 years of schooling had failed to make it so. Literacy did not make for a more rewarding career for men, unless it helped them to emigrate; for those who stayed, there was little social mobility for literate and unlettered alike. The same was not true of women of the lower class, whose marriage prospects were improved by the ability to sign. A large majority (73%) of those able to sign married into the middle and upper classes but only 10% of those unable to sign married into the middle (and none into the upper) class.

Social mobility within generations can be measured by comparing men's occupations at marriage with those at the peaks of their careers—in this case I have used the highest status attained after the age of 40. This necessarily yields a small population concentrated in the later years of the century. (Table 5).

Table 5

Final status of Harbour Breton grooms who married 1855-99*
and reached age 40 by 1899

Status at marriage	Highest status after age 40					
	Officials	SMT**	Skippers	Improved planters	Lower class	N
Officials	3	-	-	-	-	3
SMT	-	6	-	-	-	6
Skippers	-	-	3	1	-	4
Improved planters	-	-	-	-	-	-
Working class***	-	1	2	5	16	24
	3	7	5	6	16	37

Sources: Anglican parish records, Harbour Breton; 1898 directory of inhabitants; shipping registers.
Notes:
*Occupation at marriage not given before 1855.
**SMT = skilled manual trades.
***Lower class = unimproved planters, seamen, labourers, servants, others and not stated.

The numbers in each category are small, but the conclusions are clear. Membership of the official class was acquired young or not at all and the same was close to being true of skilled manual and skippers' positions. A degree of social mobility was possible into the middle groups from the lower; in fact a third of such men at marriage enjoyed (albeit fairly modest) social mobility and 44% (8/18) of eventual middle class position-holders had so risen. In all, however, 73% (26/36) of the total other than officials experienced no mobility after marriage and we shall see below that the apparent mobility out of the lower class is largely illusory.

With respect to mobility between generations we have again a small population and the requirement of comparing highest status of father and son after the age of 40 limits us to births before 1860.

Table 6

Occupational mobility between generations: highest statuses of men born before 1860 compared with those of their fathers

Status of fathers	Officials	SMT	Skippers	Improved planter**	Lower class	N
				Status of sons		
SMT	-	2	-	2	1*	5
Skipper	-	-	4	2	1	7
Improved planters	-	-	-	-	-	-
Lower class	-	-	-	2	21	23
	-	2	4	6	23	35

Sources: Anglican parish registers, Harbour Breton; 1898 directory of inhabitants; shipping registers.
Notes: No sons of an officials married and remained to the age of 40.
*Improved planter after 1900.
**Traders and general dealers included with improved planters.

The official class was filled from outside and no middle or lower class resident managed to rise into it. The middle classes were in large measure self-recruiting—very often within specific trades, son following father. Comparing Tables 5 and 6 shows that the apparent career mobility within generations from the lower to the middle class was illusory. Mostly it related to different stages in the same career and the need to acquire capital before a schooner could be purchased or a small shop stocked, while minor government positions were not achieved until middle age. Classes as we have defined them were largely filled from within and the chance of a man changing his station was slight.

Marriage patterns reveal that classes were to an extent endogamous, though the picture is clearer at the top and bottom of the social scale than in the middle. Most children of officials and of the lower class married within their stations, though about a third of lower-class women married up. Fewer than half of children of the middle class, however, married within their class, though this is still double the number that random choice of marriage partners would have produced. What influence demographic scarcity may have had upon choice of spouse is not known. Actually, most cases of anomalous marriage in class terms involved members of just five families. Three families were open to accepting partners from a lower class, while two families of planters[40] often married above their station. The possibility cannot be discounted that factors other than class entered evaluations of status within the community, though they were much less important.

Table 7

Occupations of fathers of marriage partners, Harbour Breton, 1850-99

Groom's father's occupation	Bride's father's occupations					
	Official	SMT	Skipper	Improved planter	Lower Class	N
Official	8	1	-	-	3	12
SMT	2	2	2	2	7	15
Skippers	-	2	2	2	3	9
Improved planter	-	-	-	1	7	8
Lower class	1	1	6	5	42	55
	11	6	10	10	62	99

Source: Anglican parish records, Harbour Breton.

The parish registers in general bear out the three-class hypothesis well, though they do not reveal some aspects of social inequality—levels of earnings and differences in power and influence[41] in particular. As to earnings, wages of Newmans' servants and public officials are on record and they bear out the three-class hypothesis—there was barely any overlap between the salary scales of the three classes. What is missing is information about the earnings of fishermen. To guess from the very scanty evidence that is available, the earnings of small-boat fishermen did not exceed those of labourers; the earnings of large-vessel owners, by contrast, would alone have placed them in the upper class. We have, however, seen above that they do not fare so well on other indicators. It is in the position of these large planters that the Harbour Breton data most depart from those of Nemec, for the literacy barrier was too high to be scaled in the 19th century. It is probably in respect of power and influence that Harbour Breton least resembled pre-industrial England. The local elite, whatever their powers, were salaried officials rather than hereditary landowners. Though local officials' powers exceeded those of their present-day counterparts, they certainly did not speak with one voice and probably held more sway over their poorer than their richer neighbours.

Conclusions

In this paper I have examined the growth of a settled community as it left the frontier stage behind. In 1850 Harbour Breton had still, owing to Newmans' operations, one of the characteristics of a frontier town—a preponderance of men. In other ways, however, it had ceased to be a frontier community. Most of the population was now native-born,[42] settlement from overseas had slowed to a trickle, and there was no scarcity of women. The population thereafter was largely self-recruiting, though less by natural increase than by immigrants being aided by marriage (and perhaps other kinship) ties to move from nearby fishing villages.

Some aspects of migration to and from Harbour Breton are difficult to explain. Staveley has analysed the process of internal settlement in Newfoundland in the 19th century and found it to conform to a pattern, repeated successively in older and newer-settled districts. About mid-century the population of the oldest-settled districts of the east coast grew rapidly, which led to migration to frontier districts, including Fortune Bay. The frontier districts, by contrast, were still in the population build-up phase in the 1870s.[43]

We do not, however, see evidence of this process in Harbour Breton. The town received almost no immigrants from older-settled districts, while high and persistent outmigration (to unknown destinations) appears already to have been longstanding. Moreover, the rate of emigration of young people, both men and women, from Harbour Breton seems to have been constant from the 1830s to the end of the century. Emigration, in other words, was as high when there were 15 families in the harbour as it was when there were close to 100, so it seems that overcrowding was not the cause. It also seems, from the case of Harbour Breton, that the states of stability and a "great shaking loose" of the rural population need not have been successive but could have been simultaneous.

The Harbour Breton data have been derived from only the Anglican records, not those of the Catholic minority, but where other sources allow us to compare their characteristics it does seem that both denominations conformed to the same pattern. From this it appears that, somewhere in mid-century, a distinctive regional sub-culture had arisen. It resembled that of pre-industrial England in many respects, the distribution of power and influence and the keeping of servants being the most obvious areas of difference. The local sub-culture differs in some respects from that which Nemec has described for the Catholic Southern Shore. Marriage was earlier (though still late), a religious minority and (local) immigration persisted, and clergy played a lesser role in Harbour Breton than in Trepassey. Extended families in the male line formed the basis of the fishery in Trepassey and lived near to as well as worked with each other. Patrilocality was not found in Harbour Breton, and it seems unlikely that extended families could have had the importance there that they did in Trepassey. The constant emigration of youth must have attenuated kinship networks in Harbour Breton and made the use of family labour in the fishery short-lived in those cases where it did occur.

To generalize from such limited material is perhaps unwarranted, but it seems that we must allow of a certain heterogeneity both within and between 19th-century Newfoundland communities, even if most of them were based upon the fishery. Fishing communities differed to a degree in their economic bases—in whether local or distant resources were exploited, the variety of species taken, the size of boats and type of gear used, and the industries with which fishing was combined. Some communities—like Trepassey and Harbour Breton—were long established by mid-century, while others had only recently been founded; some were occupied only seasonally, while others lost part of their population in winter. Communities also differed in commercial function and in size (though nearly all were small). Some outports were simply fishing stations whereas others, like Harbour Breton, were local or regional centres with a greater diversity of occupations.

Notes

1.   J.J. Mannion (ed.), *The Peopling of Newfoundland: Essays in Historical Geography.* (St. John's: Institute of Social and Economic Research, Memorial University of Newfoundland, 1977); G. Handcock, *Soe Longe As There Comes Noe Women: Origins of English Settlement in Newfoundland* (St. John's: Breakwater Books, 1989).

2.   These records are unusually complete and allowed the reconstitution of nearly all Anglican families, who were about two-thirds of the population. They are continuous through the study period and some records exist from the 1830s and 1840s. They also contain two censuses of church members, which can be dated by internal evidence to 1853 and 1866. Other sources of data were summaries of colonial censuses from 1836 to 1901 and Newman and Company's papers (held in the Provincial Archives), which were especially useful for data upon the transient population.

Catholic records for the period have not survived, but the manuscript copy of the 1921 census gave some information about the Catholic population as well as the Anglican. Censuses make it plain that Catholics can have contributed little to the growth of the town, except perhaps early in the period. The Catholic population increased eightfold from 1836 to 1869, which shows immigration. For the rest of the century it grew little, though there are signs of new immigration late in the century - the 1921 census records 53% of Catholic heads of households to have been born elsewhere, along with 40% of their wives.

3.   D.A. Macdonald, "Really No Merchant: An Ethnohistorical Account of Newman and Company and the Supplying System in the Newfoundland Fishery at Harbour Breton, 1850-1900" (Ph.D. Thesis, Department of Sociology and Anthropology, Simon Fraser University, Burnaby, 1988), chapter 3.

4.   Macdonald, "Really No Merchant," pp. 101-07.

5.   Men were 70% of the population in 1836 and 1845, but the proportion of men then fell steadily and after 1869 it was never more than 55%.

6.   Assuming all transients to have been men. The 1911 census, after Newmans had left, shows this measure to have underestimated the population by 3%.

7.   G. Handcock, "English Migration to Newfoundland" in Mannion (ed.), q.v.

8.   Anglican male heads of households came from England (32%), nearby Sagona Island (21%), elsewhere in Fortune Bay (22%), elsewhere on the south coast (17%), elsewhere in Newfoundland (3%) and abroad other than England (3%). The longer-settled areas of the island, therefore, contributed very little to population growth in the second half of the 19th century.

9.   42% of Anglicans in 1898. *McAlpine's Newfoundland Directory* (Halifax: McAlpine Publishing Co., 1898). The corresponding figure for Catholics in the 1921 nominal census was 47%.

10. It appears to have been a chronic condition. A sample of births in the 1830s and 1840s was taken from the few recorded baptisms and from those children in the 1853 church census that can be presumed to have been born in Harbour Breton or to have moved there during their minority. Of this sample 33% of boys and 39% of girls married and settled in Harbour Breton (n=36&28). This method underestimates the frequency of early deaths and departures.

11. The 1850s may have been exceptional; see previous note.

12. In most parishes the places of residence of the marriage parties, rather than those of their parents, are shown. Former townspeople who emigrated long before marriage might therefore not be identified as such in the parish registers. Mixed marriages that were celebrated in the Catholic

church would also not appear in my data. If there had been many such cases, however, they would have been found in directories and other sources.

13. For more details of the transients see Macdonald, "Really No Merchant," pp. 90-101.

14. Handcock, "English Migration," pp. 36-37.

15. The places of residence of their parents, not those of the spouses, are usually given in the marriage records, so that Harbour Breton marriages cannot be distingushed this way. I have counted as Harbour Breton marriages those in which one of the spouses had previously lived in the town or in which the couple settled there after marriage. The usual evidence for each is a baptismal record, either of one of the spouses or of their children.

16. Handcock, *So Longe*, p. 140.

17. P.A. Thornton, "The Demographic and Mercantile Bases of Initial Permanent Settlement in the Strait of Belle Isle" in Mannion (ed.), q.v., p. 167—means of 31.3 years for men and 19.9 years for women. After the early days of settlement, however, men married younger, though the median age of marriage of women did not rise: Thornton, "Newfoundland's Frontier Demographic Experience: The World We Have Not Lost," *Newfoundland Studies*, Vol. 1, No. 2, (1985), p. 153.

18. T.F. Nemec, "An Ethnohistorical and Ethnographic Study of the Cod fishery at St. Shotts, Newfoundland" (Ph.D. Thesis, Department of Anthropology, University of Michigan, 1980), p. 129.

19. P. Laslett, *The World We Have Lost* (London: Methuen, 1965), p. 85.

20. A rough check on Anglican and Catholic marriage-ages was possible using the 1921 census. I assumed marriage to have occurred one year before the birth of the oldest resident child and excluded cases in which the wife was over 40 (and might therefore have older, married children). This yielded median ages at marriage of 25 and 24 years for Anglican and Catholic men respectively and 21 years for women of both faiths.

21. T.F. Nemec, "Trepassey, 1505-1840: the Emergence of an Anglo-Irish Newfoundland Outport" *Newfoundland Quarterly*, Vol. 70, No. 1, (1973), pp. 17-28.

22. Meaning that after marriage men settle near their fathers, women marrying in; see Nemec, "St. Shotts" and J. Faris, *Cat Harbour: a Newfoundland Fishing Settlement* (St. John's: Institute of Social and Economic Research, Memorial University of Newfoundland, 1972).

23. Faris, *Cat Harbour*, p. 79.

24. Laslett, *The World We Have Lost*, pp. 147-50.

25. Laslett, *The World We Have Lost*, p. 94.

26. The 1836 census, the only one to distinguish families and servants, yielded a mean household size of 4.75 persons, assuming one family per dwelling. The two church censuses yielded means of 4.94 and 5.38 in 1853 and 1866 (n=16 and 24). The 1921 census showed a mean of 4.93 (4.84 Anglican, 5.10 Catholic). The 1911 census, the first after Newmans' withdrawal had removed the complication of transients, showed a mean of 4.86.

27. Laslett, *The World We Have Lost*, p. 93.

28. Wrigley states that if 15% of families were multi-generational then it was usual for widowed parents to live with their married children. E. Wrigley, *Population and History* (New York: McGraw-Hill, 1969), pp. 131-35.

In Harbour Breton in 1921 slightly more families (16% of Anglican and 22% of Catholic) had extra kin. Of these about half were aged or widowed parents, a quarter were nephews or nieces and the remainder were grandchildren, unmarried brothers and sisters and widowed sisters, and their children. From the available evidence, it seems unlikely that aged parents went to live with their married children unless infirmity overtook them.

29. In 1921 only 6% of households had three or more generations.

30. There was a resident doctor throughout the study period.

31. Crude death rates (excluding transients) were below 10 per 1,000 from 1850 to 1879 (n=43) and 17 per 1,000 for the remainder of the century (n=104). Infant and child mortality rates (per 1,000 live births) were below 100 in the 1850s and 1860s but rose above 200 after 1880. Even at their peak, these figures are rather below those typical of pre-industrial Europe.

32. Laslett, *The World We Have Lost*, p. 98.

33. Nemec, "Trepassey, 1840-1900: an Ethnohistorical Reconstruction of an Anglo-Irish Outport," *Newfoundland Quarterly*, Vol. 69, No. 4, (1973), p. 19; and Nemec, "St. Shotts," chapter 5.

34. Nemec, "Trepassey, 1840-1900," pp. 19-20.

35. D. Gagan, *Hopeful Travellers: Families, Land and Social Change in Mid-Victorian Peel County, Canada West* (Toronto: University of Toronto Press, 1981), pp. 99-100.

36. I have classed traders, general dealers, and shopkeepers as improved planters, as they were usually larger planters and engaged in trade for only a short time - of such persons listed in the 1898 directory in Fortune Bay barely half were listed in the 1904 directory.

37. Large and small planters were not distinguished in the parish registers but other categories are drawn from contemporary usage, at least that of Reverend W.K. White, who spent 40 years on the coast. Captains and skilled manual workers are marked as such in the parish registers, as are traders and minor government officials even when they were also fishermen. These categories are distinguished in the parish registers from servants, which might include foremen, ships' mates, etc.

38. Some confirmation comes from censuses. Taking the numbers able to read and write (or in some cases just to write) in censuses as a proportion of the population over 10 years old yields figures very consistent with (though a little higher than) mine. It also confirms the fall in literacy rates during the late-19th century and their revival after 1900.

39. Except own occupation for those born outside the south coast. Population is all spouses and witnesses 10 years old or more.

40. In one of these cases there is reason to suspect an unregistered schooner, which would have made them improved planters.

41. I have reviewed the evidence in Macdonald, "Really No Merchant," pp. 114-17 and 125-32.

42. 69 percent at the 1857 census.

43. M. Staveley, "Population Dynamics in Newfoundland: the regional patterns" in Mannion (ed.). q.v.

chapter
eight

GOOD DEBTS AND BAD DEBTS:
GASPÉ FISHERS IN THE 19th Century[1]

Roch Samson

The Jersey merchants run all the businesses and have an almost total monopoly. It is they who set the price of fish. "This year" says the merchant,
"we will give you so much per quintal of cod" and there is no use complain-
ing. The Robin company fixes the prices for the whole coast and the
"habitants" have no choice but to accept it — every last one of them is in the
merchant's debt....
— Abbé Nérée Gingras, missionary from Percé, 1857.

even if you were in debt, they gave you credit all the same, just to keep you
there....
— Louis Fortin, 94, retired fisher from Cap-aux-Os, Forillon, 1976. [2]

These words, uttered a century apart, take us right back to the world of the Gaspé fisher
in the 19th century. In 1857, Abbé Gingras was quite fatalistic in his observations of the
social conditions of the fishers who confided in him as a missionary in Percé.[3] In 1976, Louis
Fortin, a Cap-aux-Os fisher in his 90s, gave a very lucid summary of the conditions under
which he had worked on the Forillon peninsula at the beginning of the century. The durabil-
ity of these conditions, still lodged in the collective Gaspé consciousness, underlines the
fact that the commercial practices inherited from 19th-century export merchants were still in
use in the Gaspé peninsula in the 1930s, when fishers were just starting to experiment with
the cooperative model.

The persistence of this era, so decisive for the people of the Gaspé, explains why the
conditions that were prevalent then are still relevant in this coastal region whose history is
now one of its main tourist attractions. The above testimonials prompt us to examine the
historical conditions that prevailed during the formation of this 19th-century society of
Gaspé fishers. Like Abbé Gingras, many visitors to the region denounced the fishers' dependence on the Anglo-Norman merchants from the Channel Islands who reigned over
Gaspé society. Nowadays, the people of the Gaspé look back on this period as one of hard
times and deprivation. Although this is not all that far from the truth, it has given rise to a
somewhat simplistic take on commercial fishing conditions and an overly homogeneous
view of Gaspé society. The fishers of the Gaspé coast produced dried cod that was considered among the best in the world. "Gaspé cure" was renowned for its whiteness, nutritious-
ness, and long "shelf life." But these people who were so skillful at their work did not have
any control over their working conditions: they could not embark on a fishing season unless their merchant gave them credit in the form of "advances."

Based on a study of the production relations binding a Forillon peninsula merchant to his fisher customers, this article will show how the credit system instigated by the merchants and the resultant debts accumulated by the fishers were decisive factors in fishery viability. It will also show how they led to the fishers' chronic dependence on these merchants who exported dried cod.

Gaspé Coast Fishery

In the 19th century, the Gaspé coast fishery covered 248 coastal miles from Miguasha Point in Bonaventure County to Cap Chat on the western border of Gaspé County. There were three major fishing areas made up of fishing grounds: Chaleur Bay, with 72 miles along its north coast; Gaspé Bay and vicinity, with 52 miles of coast; and the north coast of the Gaspé peninsula between Cap Gaspé and Cap Chat—124 miles, but only sparsely dotted with fishing establishments. The inshore or coastal fishery extended two miles from shore, but many fishers went beyond these limits when cod was scarce. They would fish within sight of shore, taking bearings on "marks" such as church steeples, lighthouses, or homes.

Technical Aspects

Boats

In the 19th century, the Gaspé fishing fleet was made up primarily of small undecked sailboats that were approximately 25 ft. long and shaped like whalers, sharp at both ends. Referred to as "barges," these fishing boats had schooner rigs: two masts, a mainsail, foresail and jib, and often a mizzen added to the stern. They were designed to accommodate two fishers, one fore and one aft, and could hold from three to five drafts, or 1,000-1,600 lb. of cod. When they reached the fishing grounds, the fishers would drop sail, throw their killicks (hand-made anchors) overboard, then bait and set their hand-lines.

Bait Fishing

The actual cod fishing was not at all complicated, but fishing for bait called for more equipment. Herring was the most common bait, but caplin, sand eel, smelt, and squid were also used. Herring was caught with special gill nets. These herring nets, which floated on the water's surface, were held in place by a beam called a "mooring" set a short distance from the fisher's beach. This method was fairly effective early in the season when the herring came close to shore to spawn, but was not as successful later on. The fishers would go out at night and drift, without sails, a series of nets attached to the bow of their boat, to their fishing grounds, then haul in and separate their nets and prepare their bait.

Later on in the season, when herring was harder to catch, they used large seine nets, which were hauled in from shore to catch sand eel. For the purchase and use of these nets, they would join together in a "seine club" headed by a "seine master."

Hand-Line Fishing

Hand-lines were used to catch cod. The 50-fathom hemp line had a two-to-five-pound lead sinker at the end with a swivel to which two snoods and their hooks were attached. Each line was wound around a reel. The two fishers on board would constantly raise and lower their lines alternately on either side of the fishing boat. The boats cast off around 3 a.m. and returned late in the afternoon, when the cod would be prepared immediately by the fishers themselves working on a family basis or by the specialized shore crews at the companies' fishing premises.

Cod Processing

Dried cod was prepared, or processed, in four phases: splitting, salting, washing, and drying. First the cod was dressed on a table, with the fish being passed from a "throater" to a "header" and then on to a "splitter." The split cod was then salted, i.e. kench-cured or soaked in tubs of brine. After three or four days, when it had "taken its salt," the cod was washed in fresh water and drained. The drying process came next, with the fish being laid out in the sun on "flakes." The drying cycle lasted ten days ("ten suns") plus an extra day on the beach itself before the cured cod was ready to be shipped overseas. At the end of each day in the wind and sun, the cod was piled up to remove as much moisture as possible. By the end of the drying operation, the cod was amber-coloured and very hard and dry, down to a mere 35% water content, and had lost 80% of its original weight.

A Typical Fishing Establishment or Plant

The specialized production of dried cod gave rise to fishing establishments with certain typical spatial and architectural characteristics.[4] The first company fishing plants, sometimes referred to as "fishing rooms," appeared at the end of the 18th century, particularly at Grande-Grave on the Forillon peninsula where the Janvrin Company from Jersey built a typical Gaspé coast fishing establishment (Fig. 1). Grande-Grave was situated in a generally wind-swept cove with a pebble beach or shingle (the origin of its French name *grave*). There were a number of wooden buildings used as tenements, as workplaces, and for storage. The owner or company agent's home, called the *grand' maison*, was set in the middle of the plant, usually on a slope or plateau overlooking the beach. Surrounding the house, there were buildings used as offices, stores, and warehouses. There were also "cookrooms" —huts that housed fishers and seasonal workers in the summer. The production-related buildings were generally built along the shore: the salt house or *"saline,"* washhouse, sheds for fishing gear, and dried cod warehouses. Also situated on the shore were workshops such as a cooperage, a sail loft, and sometimes a blacksmith's shop. Here too were the net-drying racks and the flakes on which the cod was dried. On the beach itself, there was a very typical building, the "stage." Some fishing rooms had two or more stages. The stage was a large building with a "stagehead" on stilts extending into the harbour. The fishing boats would draw up to the stagehead every day and unload their catch. The cod was then immediately dressed and split by a separate shore crew on the tables set up for this purpose on the stagehead.

Main Fishing Plants

The most important fishing establishments in Chaleur Bay were located at Paspébiac, Grande-Rivière, and, especially, Percé. Although there were many merchants in other locations, the powerful Charles Robin and Company dominated, and has lived on in the minds of the people of the Gaspé to the point where the others have been all but forgotten: Le Boutillier Brothers, Robin's biggest competitor, and also De La Parelle, Savage, Hamilton, Hammond, Forrest, etc., whose activities and influence were more local than regional. In Gaspé Bay, although cod production was on a smaller scale than at Percé for instance, the most important plants were at Pointe-St-Pierre, Douglastown, Grande-Grave, and Gaspé. In the middle of the 19th century, the Collas, Fruing, John Le Boutillier, and Hyman companies were the most important ones, alongside Fauvel, Veit, Dumaresq, Lowndes, Lyndsay, etc. On the north shore of the Gaspé peninsula, the most active establishments were at Anse-au-Griffon, Rivière-au-Renard, and Grande-Vallée; they were operated by Fruing,

109

Figure 1: The Janvrin establishment in Grande-Grave in 1809. (NAC, RG1, L3L, Vol. III, fol. 54457)

Hyman, and John Le Boutillier, and a number of others, including a few relatively important French-Canadian merchants such as the Blouin brothers at Rivière-au-Renard, and Michel Lespérance at Grand Étang near Cloridorme.

Fishing Effort

In 1830, the population of the area referred to in legal parlance as the "inferior district of Gaspé," which included the counties of Bonaventure and Gaspé, was 7,500. By 1860, this figure had more than tripled, to almost 25,000.[5] This population increase was related to the growth of the commercial fishery. From a little over 1,800 fishers in 1830, the numbers had grown to over 4,000 by the end of the 1860s, without including the 2,000 or so shore workers—mostly seasonal workers who dressed, split and dried cod in company establishments. The fishing fleet also tripled, reaching almost 2,000 fishing boats and about 30 schooners. Over this same time span, dried cod production increased from 55,000 quintals to over 100,000.[6] In the 1860s, the Gaspé coast fishery accounted for 60-70% of all the dried cod produced in the Gulf of St. Lawrence excluding western Newfoundland.[7] The increased fishing effort stemmed from an expanded work force, and not from improved productivity brought about by the introduction of new fishing methods. This conservative management was to influence the Gaspé fisheries up to the middle of the 20th century. The fishing companies, most of which were managed from Jersey, were run by merchants, and not by industrialists. Their profits were only partly reinvested in the region to open new establishments and increase the number of fishers; they were not used to revamp boats, improve fishing methods or diversify the fishery, which was almost exclusively centred on cod.

This merchant-based way of managing the fisheries had a direct impact on the number and attributes of the fishers. To boost production, the merchants needed to increase numbers of boats and fishers. The Anglo-Norman merchants therefore prompted many of their

Channel Island compatriots from Jersey and Guernsey to settle among the Acadians who had found refuge in Chaleur Bay after 1755. As of 1800, more French Canadians were drawn in from the cramped and overcrowded seigniories of Montmagny and Kamouraska. It was because of this seasonal influx of people that the French Canadians came to settle in the Gaspé. The growth in the Gaspé's population in the first half of the 19th century was therefore the result of a high birthrate[8] and immigration.

The almost exclusive marketing of dried cod had forced Gaspé fishers to specialize in this one species. This contrasts with American fishers who, owing to the 1854 reciprocity treaty, caught thousands of tons of mackerel, which the Channel Island merchants were not interested in harvesting or marketing.[9] The specialization of the Gaspé fishers increased their dependence on the merchants who controlled the sale and marketing of dried cod. Morover, this specialization, handed down through five generations of fishers, had become a tradition by the 20th century. Later it would be considered a major obstacle to fishery development.

Dried Cod Exports

In the 1860s, dried cod represented 65-80% of the value of the entire Gaspé fishing industry output. It was followed, in order of importance, by herring and cod liver oil. Salmon and mackerel together did not even account for 5%. Whale oil production represented a mere 2-4%, with whaling practised by only a few Gaspé Bay families, most of whom lived at Penouille on the Forillon peninsula. These percentages[10] make it clear just how specialized the Gaspé fishery was. It is only recently that mollusks and shellfish, such as scallops, crabs, and lobsters, have been systematically harvested.

Dried cod from the Gaspé was exported in huge quantities, mostly to the Mediterranean but also to the Caribbean and Brazil. Merchant ships would regularly leave the port of Gaspé with 2,000-3,000 quintals of cod and sail for the ports of Cadiz in Spain; Naples, Ancona, and Civitavecchia in Italy; Rio de Janeiro, Bahia, and Pernambuco in Brazil; St. Thomas and Bridgetown in the Caribbean; and Jersey in the English Channel where fish was trans-shipped to other markets.[11] Merchants based in Naples and Civitavecchia in Italy controlled buying and selling operations for all the dried cod shipped to the Mediterranean.

The Trade Triangle

Source of Capital

It is easier to understand how the 19th-century Gaspé fishery worked if the commercial organization behind it is clear. The fishery was incorporated within a trade network that made it viable and profitable. The operating capital came from the Channel Islands of Jersey and Guernsey. Merchants there set up a Joint Stock Bank and other financial institutions to run their operations. These merchants also belonged to the Chamber of Commerce of Jersey, the main purpose of which was to supervise and protect the business interests of Channel Islanders.[12] This European capital was beyond Canadian control, and the economic health of the fishery was dependent upon the health of the financial institutions set up by the Channel Island merchants, as well as conditions in foreign markets. This was made clear to those on the western side of the Atlantic in 1873 when Europe was hit by an economic crisis and the merchants' financial institutions, including the Jersey Joint Stock Bank, went bankrupt. In under 20 years, all Channel Island capital disappeared from the Gaspé fishery.[13]

Circulation of Capital: Commodities, Cod, and Ships

The Gaspé merchant capital circulated basically in the form of goods or commodities. Apart from locally built boats, all the equipment required to catch and process cod was imported from Europe. Fishing gear such as lines and nets was imported from England and Scotland, and the salt that was such an essential ingredient came from Spain. Many of the other items needed by the Gaspé families were controlled by the merchants: they imported these supplies from Europe, the Caribbean, and the cities of Halifax, Québec, and Montréal. All of the dried cod produced by the fishers was bought by these same merchants and sold on the Mediterranean and other markets. They picked up other goods, such as wine, fabric, and molasses, from these markets, then sold them in England, where they bought fishing gear to complete the trade triangle.

This triangular trade model included a fleet of ocean vessels owned by Channel Island shipowners that constantly sailed between three parts of the globe. The whole process, from the circulation of these ships to the sale and purchase of cargo, was coordinated from Jersey by brokers who represented the interests of the Channel Island merchants. The Channel Island merchants invested in dried cod production and in annual supplies for the Gaspé fishers. In this way, they managed a mercantile system that incorporated the means of production, the product, and the goods required by the producers. They therefore controlled the circulation of their capital at both ends of the production spectrum. The operations of the merchant William Hyman of Grande-Grave on the Forillon peninsula illustrate this phenomenon.

William Hyman and Sons Company

Founder

William Hyman was a Jewish merchant of Russian origin who settled in the Gaspé in the early 1840s. He was 38 years old in 1845 when he bought his first fishing establishment from the Channel Island captain Francis Ahier for £220.[14] This plant was in Grande-Grave on one of the best beaches in Gaspé Bay. For over 30 years, this beach accommodated the headquarters of the Jersey firm Mourant and Denis, and Hyman also shared it with the powerful Jersey-based Janvrin company whose establishment was sold in 1855 to William Fruing and Company, also of Jersey[15] (Fig. 2).

Grande-Grave Plant

The name Grande-Grave is made up of the French words meaning "big pebble beach or shingle" for drying cod. The word *grave* was used figuratively to denote an entire fishing establishment. Grande-Grave is located in Gaspé Bay on the south shore of the Forillon Peninsula. Along with Percé, Gaspé, and Penouille, it is one of the oldest fishing sites in the region and was used by fishers from France as early as the 17th century. During his campaign to conquer New France, General Wolfe set up camp here in 1758, before ordering everything along the bay's shores burned down. In 1777, two Anglo-Norman establishments were located here or close by. They were owned by a Lemesurier and a Bonami, and they employed 112 workers and operated 25 fishing boats.[16] In 1792, it was the site of a plant owned by brothers Philip and Francis Janvrin from Jersey, for which we have a plan dated 1809. By 1820, almost all of the "fishing lots" on the Forillon peninsula were occupied by fishers and small independent establishments. Most of these "beach masters" or "planters" dealt with the Janvrin company.[17] William Hyman was to build up his customer base primarily from among these people 25 years later.

Company's Importance

Until his death in 1882, William Hyman built a locally important company. By the beginning of the 20th century, after his son Isaac Elias had taken over the operation from him, it had attained the size and influence of a small regional empire. Isaac's son Percival Gerald then ran the family business until it closed in 1967.

In the second half of the 19th century, the Hyman firm exported an annual average of close to 6,000 quintals of dried cod, for an average annual value of $24,000, or £6,000. Hyman's sales figures were only half those of his main competitor and neighbour Fruing, however, and only one-sixth of those of the powerful Robin Company in Paspébiac. Hyman's production volume nevertheless made it the fourth largest exporter from the port of Gaspé,[18] excluding the Robin Company since it shipped directly from Paspébiac. Throughout his lifetime, William Hyman bought and resold many fishing beaches and other properties. When he died, he passed on six fishing establishments, without counting the many others on which he held mortgages. He also left a wharf, warehouses, a store, and a hotel in Gaspé. He had stocks and bonds in Québec City and Montréal banks and owned a home on St. Catherine Street in Montréal, where he spent winters with his family. In 1926, William Hyman and Sons was estimated to be worth $165,000.

Company Archives

The archives of the mercantile companies, consisting of accounts ledgers and correspondence (letterbooks), are especially vital to the study of the 18th- and 19th-century fisheries. On the one hand, they constitute a daily record of the countless business transactions that linked a given merchant to local people, and on the other they provide information on how the merchant worked his way into the business and trade network. Few such archival collections survive, and those that do are often not complete enough for historians to derive substantial help from them. The Gaspé is particularly well endowed in this regard, however, with the William Hyman and Sons collection which documents the history of the Gaspé Bay, and especially the Forillon peninsula fisheries, as well as the huge and almost complete archival collection of the famous Robin Company which traces the fisheries history of the greater Chaleur Bay area.[19]

Behind the multiplicity and complexity of the transactions recorded in the ledgers one sees the web of relations linking fishers to merchants, and also the fishers to each other. The individual accounts allow us to observe the situation of many fishers over extended time periods, and even bear witness in some cases to the transfer of a retired fisher's debts to the account of a son signed on by the same company. These ledgers are indisputably the best testimonials of the social and economic conditions of Gaspé fishers because of the detailed information that they provide for thousands of individuals over three or four generations. Moreover, these archives shed light on the differences in status among the members of the fishing community who all worked within the same commercial framework built on a vast credit system. The clarification of this social differentiation gives us a more complex and finely shaded picture of 19th-century Gaspé society. The customers of the Hyman company serve as a good example.

William Hyman's Fisher-Customers

> Father giving just now a good deal of credit to Cape folks & others, they give none at Fruing's....
> —Isaac Elias Hyman, Diary, 21 April 1869.

The Canadians, 6 of them, came down on board "Gaspé," they came down to Grand Grève in a boat today & began working....
—Isaac Elias Hyman, Diary, 5 May 1870.

Migrant Seasonal Workers

These two brief notes taken from the diary of William Hyman's eldest son establish the two categories of fishers that this merchant dealt with during the fishing season. The first were the permanent residents in Hyman's area of operations, the Forillon peninsula; the second, the *Canadiens*, who lived in the Gaspé for only four or five months of the fishing season. Most of the latter were peasants and farm workers from what was then referred to as the "parishes" of Montmagny and Kamouraska counties. Every year, they would migrate to the Gaspé's fishing beaches to find work or to make themselves some extra money in the lull between planting and harvesting. They were hired in the wintertime in their home towns by a local agent on William Hyman's payroll. Hyman reimbursed the agent for these workers' return ship fare and provisions. For many years, a Montmagny shoemaker named Fabien Guilmet hired workers for Hyman and other Gaspé merchants.

The seasonal workers were hired on as fishers paid on a wage or piecework basis, or as salaried shoreworkers, i.e. splitters, salters, general labourers, and even cooks. They would work on one of the Hyman company beaches or for one of Hyman's local customers or dealers who had a beach and a few boats. In the 1850s and 1860s, Hyman hired an average of 30 seasonal workers a year, two-thirds of whom worked in his own establishments or on his schooners, and the other third with his dealers. In 1860, his seasonal fishers made $12-$20 a month, splitters $20-$24, salters $16-$20, and shoreworkers $8-$12. Thirty years later, near the end of the 19th century, the wages he paid his employees had not increased.[20]

Local Fishers

Most of William Hyman's customers were heads of families from the townships of Cap-des-Rosiers, Baie-de-Gaspé, and Douglas. These resident customers formed the merchant's true base of operations. For the year 1860, Hyman's cod producers can be broken down as follows:

William Hyman's Dealers in 1860

CATEGORY	NUMBER OF PRODUCERS	DRY COD PRODUCED (QUINTALS)	PRODUCTION %
independent	148	2,850 q	49
merchant	3	1,301 q	22
beach master/planter	9	1,023 q	18
seasonal	21	346 q	6
schooners	2	276 q	5
TOTAL	183	5,796 q	100

Independent fishers therefore accounted for 80% of Hyman's customer base and 50% of his production or output. Most of them lived on small beaches that they often shared with another family for processing cod. The small merchants, who produced 22% of the dried cod, were actually intermediaries between Hyman and fishers who lived in areas where Hyman did not have a fishing establishment or station. The beach masters, or planters, who accounted for 18% of the dried cod, owned larger beaches equipped with many fishing boats, used by members of their families and seasonal workers. Generally speaking, these beach masters did not fish themselves: they supervised their shore workers. These beach masters always had fields of crops that their employees worked when fishing was poor.[21] Finally, with regard to the two schooners, irregular fishing was done by certain sailor/fisher crews belonging to a "club." They dressed and salted their cod on board ship, to be dried on the Hyman company beach.

How the Hyman Company Operated
Salaried Fishers
Seasonal workers were almost always hired on as wage earners and their terms of employment were determined before they arrived. These workers did not deliver a finished product at the end of the season. They either worked at sea as fishers or on land as shoreworkers. The number of seasonal workers varied from year to year in keeping with the fishing effort the merchants were willing to finance. Seasonal workers enabled merchants to intensify the fishing effort at low cost, and to determine this cost in advance.

Fishers Living on Credit
The situation was very different for local fishers. Basically, these fishers lived on credit and their working conditions were predetermined for life. They could not embark on the fishing season unless the merchant gave them advances, or credit, beforehand. These advances were usually received in the spring and consisted of salt, fishing gear, and provisions for the fisher's family's subsistence. At the end of the season, when the fisher delivered dried cod to the merchant, he would pay the merchant back for some—usually not all —of the things the merchant had sold on credit. For a long time, this practice was considered to be a form of barter. But since this implies an exchange between equals, it is really more accurate to refer to it as a credit system. From the outset the advance or credit created an obligation on the part of the fisher to reimburse the merchant. In other words, the fisher did not own the cod. Its price was unilaterally set by the merchant. Because of the shortness of the fishing season, the value of the dried cod produced was rarely high enough to cover the credit received at the beginning of the season. The fisher was therefore left in a permanent state of debt. In the autumn, the fisher had to appeal for more credit to get through the winter. The merchant would comply if satisfied with the fisher's output during the fishing season. If not, the fisher's beach or property could be mortgaged or seized.[22] This was a frequent occurrence after poor fishing seasons as this excerpt from a letter written in the winter of 1867 by William Hyman illustrates:

> ... the closing of last year's accounts does not improve anyone's mind or health. I had my share of it. I have done but very little else since the beginning of December than executing Mortgages, Bills of Sale and acknowledgements, and in cases had to resort to Summonses....

And at the end of the following season, he wrote: "This is the third season the fishers are doing very poor around here, in consequence large amounts of debts are accumulating and remain on the books...."[23]

Debts varied greatly from one fisher to the next. The accounts of individual fishers show that some producers, especially beach masters, had very few debts or none at all—owed to the Hyman company at least. They also show that some customers, most of them important ones, dealt with different competing companies from season to season. It should also be pointed out that, on the merchant's receipt, he himself received advances from European creditors, and that he also felt the consequences of a poor fishing season when he tried to get new advances on his cargo. Nevertheless, he could accumulate and invest profits elsewhere, while the slight surplus that a fisher might obtain after a good season was carried over in the merchant's books as a guarantee to cover future advances. The clerks at the Robin company called these amounts that were carried over the *fonds mort* (literally "dead fund").

The Fishers' Debts

The widespread use of credit causing indebtedness of a large part of communities in a region entirely controlled by European merchants for almost a century and a half is no small matter. It has left deep and lasting scars. By carefully studying the dependency of the people of the Gaspé one begins to understand why it has been said that these people were still "bound by tradition" in the 20th century. In the 19th century, travellers to the Gaspé who were appalled by the commercial control of the fishing companies and the outright indebtedness of the communities did not dare attribute the entire blame to the merchants. They suggested reckless spending as one cause, i.e. that the fishers were encouraged by the merchants to buy luxury items. Indeed, Abbé Ferland claimed in 1825 that the young ladies of Paspébiac were better dressed than those of the Saint-Roch district in Québec City.[24] Even the famous Captain Fortin, who was the superintendent of fisheries in the Gulf of St. Lawrence for 15 years and to whom we owe a number of excellent detailed reports on the fisheries, admired the Channel Islanders' entrepreneurial spirit and complained about the lack of interest in this area on the part of Canada's business people. His successor, Théophile Têtu, was somewhat more critical, however, and this undoubtedly lost him his prestigious job, since he only wrote two reports. This is what he had to say in 1868:

> ... it was when fishing was the best that the debts piled up. This may seem strange if one is unaware of the reasons. Let us take the example of a fisher who made $100 during the season. From this amount, he had to take let's say $40 to pay his old debt, because one way or another there was almost always an old debt. So this leaves our man with only $60 to get through the winter, which was not enough. Since he had done well the previous season, was a good fisher and could be expected to do well the next year, the merchant would let him buy on credit. And once his account was opened, he did not really look at it very much and every day the amount grew. This is why the best fishers have still not become prosperous; this is why, after working their whole lives, they have nothing left for their old age; this is why they have debts and why, virtually without exception, this is the case of almost all our fishers who have no crops and have made fishing their sole means of survival.... (author's translation)[25]

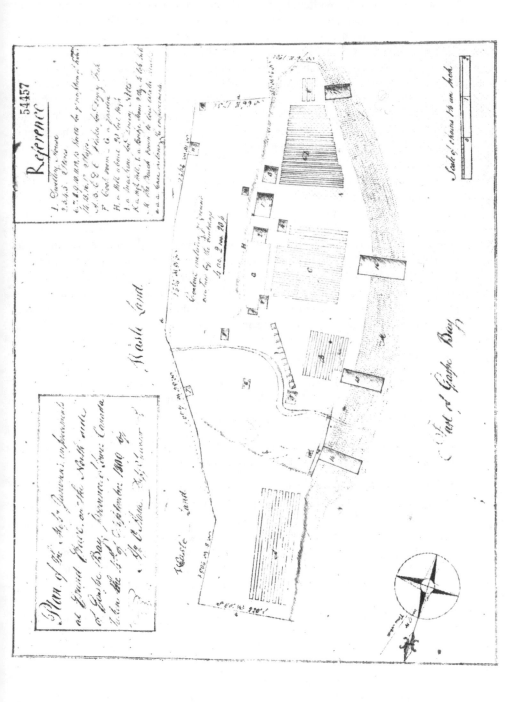

Figure 2: The Fruing and Hyman establishments in Grande-Grave in 1864. (Parks Canada, Quebec)

The personal accounts of William Hyman's fisher customers over 10 consecutive years (1854-63) support these perceptive and daring remarks by Superintendent Têtu, who knew that they would be read by the very merchants whose practices he denounced. It is important to understand that it was the larger amounts of salt, fishing gear, and other necessities obtained on credit with a view to intensifying fishing the following season that increased the fishers' debts the most, and not luxury or special items, as Têtu in fact pointed out. A fisher who did well was therefore up against greater risks, such as having his establishment mortgaged after a poor fishing season. As Abbé Gingras pointed out (quoted above), the planters "who have fishing stations generally [with] 2, 3, 4 or 5 boats" and who hired men to crew their boats, could better provide themselves against these risks.

When the fishers' debts are placed in the global perspective of the commercial organization of the fisheries, it becomes clear that this accumulation of debts played a fundamental role in controlling producers. First and foremost, it tied fishers to their merchants, ensuring a clientele of loyal producers. It was also the best way of boosting production within the framework of a credit system, by making the producer bear a large share of the risks. The producer could always be replaced if his debts got out of hand. Some merchants kept a special annual book of debts in which customers were rated as productive or non-productive, i.e. labelled "good debts," "bad debts," or "doubtful debts."[26] All these debts were therefore not necessarily seen as a sign of poverty but as an investment with variable risks for both the merchant and the fisher. Nevertheless, it is the difficult conditions resulting from a commercial system subject to these internal necessities that has lingered in the minds of the people of the Gaspé. The main actors have preserved, each in their own way, the memory of their social conditions, as the following two viewpoints illustrate:

> He [William Hyman] advanced provisions to the poor on the Gaspé coast in the winter, especially when they could get none anywhere, and as the people in their poverty were unable to repay him, he incurred great losses through his generosity.
> —Biography of William Hyman, published in Arthur Daniel Hart's *The Jew in Canada,* (Montreal, Jewish Publications, 1926).

> You know, in spring nobody had a cent. Me, I had no money. So, I went to Grande-Grave, got my fishing gear, got stuff for the family—I had an account there, right. Well, comes around August 15, if I had enough to pay my account, well I gave what I had on it. Then I finished in the fall. Then if I had ten dollars or around there, well I had a good summer. You know, we didn't count money in the hundreds or even in the fifties, no, no, no.... When a man finished his summer, if he could clear twenty-five bucks, dammit, he was rich!
> —Arthur Fortin, 82, retired fisher from Cap-aux-Os, Forillon, summer 1976.

Notes
1.   This article is based on the findings of a detailed study of the archives of the William Hyman & Sons Company of Grande-Grave, made up for the most part of ledgers and letterbooks. This study was published under the title *Fishermen and Merchants in 19th-century Gaspé* (Studies in Archaeology, Architecture and History, National Historic Parks and Sites Branch, Parks Canada, Environment Canada, Ottawa, 1984). The study was carried out as part of the research for the heritage preservation of the national historic site of Grande-Grave in Forillon National Park on the Gaspé

peninsula in Québec. It was submitted to Laval University in 1981 as a thesis for a master's degree in anthropology.

2.   These remarks were recorded by the author in 1976 in conjunction with an ethnographic study published under the title *Fishing at Grande-Grave in the Early 1900s,* History and Archaeology, No. 41, Parks Canada, Ottawa, 1980.

3.   Nérée Gingras, "Impressions de Gaspésie," en 1857, *Le Canada-Français*, Vol. 26 (janv. 1989), pp. 483-97.

4.   It was the superintendent of fisheries for the Gulf of St. Lawrence who noted and gave a detailed description of this in his 1859 report. *Annual Reports of the Superintendents of Fisheries for Upper and Lower Canada*; together with the Overseer for the District of Lakes Huron and Superior; also that of Pierre Fortin, Esq., Magistrate, and W.F. Whitcher, Esq., for the Year 1859. (Quebec, Thompson & Co., 1860) pp. 66-67.

5.   Bélanger, Jules, Marc Desjardins, and Yves Frenette, *Histoire de la Gaspésie* (Montréal, Boréal Express/Institut québécois de recherche sur la culture, 1981), pp. 149, 297.

6.   Joseph Bouchette, *The British Dominions in North America: or a Topographical and Statistical Description of the Province of Lower and Upper Canada....* (London, Longman et al., 1832), T.1, p. 329; Appendix to the XLth Volume of the Journals of the House of Assembly of the Province of Lower Canada, First Session of the Fourteenth Provincial Parliament, 1831, Appendix GG, A. 1831. Exports in 1830, from the Ports of Gaspé and New Carlisle; Canada, Annual Reports of the Gulf of St. Lawrence Fisheries Officer, 1860-69, in the Appendix to the Journals of the House of Assembly of the Province of Canada and in the Sessional Papers of the Department of Marine and Fisheries.

7.   This percentage is based on the overall figures for the years 1862-64, taken from the fisheries officer reports cited above. The Gulf of St. Lawrence fishing territory referred to in these reports covered Saguenay County, including the coast of Labrador, the north shore of the St. Lawrence and the Gulf of St. Lawrence, and Anticosti Island; Rimouski County; Gaspé County, including the Magdalen Islands; and Bonaventure County.

8.   Bélanger, Jules, Marc Desjardins and Yves Frenette, pp. 149-50.

9.   The fisheries officers complained repeatedly about this in their reports.

10.   See references in note 6.

11.   National Archives of Canada, RG 16, A2, Vol. 476, Ship Registers Outwards 1851-94.

12.   For an early history of the Chamber of Commerce of Jersey, see A.C. Saunders, *Jersey in the 18th and 19th Centuries....*(Jersey, Channel Islands: J.T. Bigwood, 1930).

13.   The powerful Robin and Le Boutillier Brothers companies of Paspébiac survived this crisis until 1886, when the Jersey Banking Co. went bankrupt, taking several other Jersey companies in the Gaspé with it. William Hyman's broker in Jersey, Edward de la Perelle, was stunned by the collapse of the Robin Company: "To learn of such things happening to the firm of C.R.C. (Charles Robin and Co.) now 120 years old could hardly be realized in one's mind." Letter from Edward de la Perelle to William Hyman, 19 January 1886, taken from the William Hyman & Sons Letterbook and cited in André Lepage, *Le banc de Paspébiac, site commercial et industriel* (Centre de documentation et d'interprétation sur la pêche à Paspébiac, Ministère des Affaires Culturelles, Direction générale du patrimoine, 1980), pp. 31-32.

14.   Percé, Bureau d'enregistrement, Registre B, Vol. 1, p. 252, no. 292 (1845).

15.   Percé, Bureau d'enregistrement, Registre B, Vol. 2, p. 567, no. 1159 (1855).

16.   National Archives of Canada, MG 21, 862, 1774-86, Mic. A. 773. Haldimand Papers (N. Cox census, 1777: "Gaspé within the Capes").

17.   I discussed the early fishing establishments in the Gaspé, and on the Forillon peninsula in particular, in the article: "Gaspé 1760-1830: l'action du capital marchand chez les pêcheurs," *Anthropologie et Sociétés*, (1981), Vol. 5, no. 1, pp. 57-85 .

18.   According to my compilations of exports from the port of Gaspé. National Archives of Canada, RG 16, A2, Vol. 476, Ship Registers Outwards 1851-94.

19. A systematic study of these collections was conducted in the 1970s, adding to and clarifying our knowledge of the commercial organization of the Gaspé fisheries and the consequent social and

economic conditions. Other researchers have followed suit extracting information from similar archives for other areas of the Gulf of St. Lawrence and northwestern Atlantic.The doctoral thesis of André Lepage, who made systematic use of the Robin archives, is an important study that is essential to the understanding of how the mercantile companies operated generally, and the Robin Company in particular. David Lee's work on Charles Robin is also important. See also the recent work of Rosemary Ommer. The Robin Company archive collection is filed and classified with the National Archives of Canada, along with that of the Hyman Company, bequeathed by Percival Gerald Hyman, whom I met at Gaspé a few weeks before his death in 1976. There are also other archive collections at the Gaspésie Museum, such as that of the Collas Company of Pointe-Saint-Pierre, although they are incomplete. André Lepage, *Le capitalisme marchand et la pêche à la morue en Gaspésie: la Charles Robin and Company dans la baie des Chaleurs (1820-1870)* (doctoral thesis, PhD in Anthropology, Laval University, 1983); David Lee, *The Robins in Gaspé, 1766 to 1825* (Markham, Ont: Fitzhenry & Whiteside, 1984). Rosemary E. Ommer, *From outpost to outport : a structural analysis of the Jersey-Gaspé cod fishery, 1767-1886* (Montreal : McGill-Queen's University Press, 1991).

20.   Roch Samson, *Fishermen and Merchants,* Table 6, p. 55.

21.   Nérée Gingras, pp. 490-92.

22.   I have shown elsewhere how the property transactions of the Hyman and Fruing companies on the Forillon peninsula in the second half of the 19th-century were an integral part of the exchange dynamic between fishers and merchants. The introduction of bonds and mortgages to guarantee the reimbursement of advances, or credit, created countless legal records on the properties. See Roch Samson, "La Gaspésie au XIXe siècle: espace maritime, espace marchand" in *Cahiers de Géographie du Québec,* Vol. 28, nos. 73-74 (avril-septembre 1984), pp. 205-21.

23.   Canada. Parks Canada, Québec Region, Archives of William Hyman & Sons, Letterbook 1866-68, Letter from William Hyman to John Hardeley of Jersey, 18 February 1867; Letter from William Hyman to Robert V. Tancrête Winter of Douglastown, 2 October 1867.

24.   Ferland, J.-B.A., "Journal d'un voyage sur les côtes de la Gaspésie," *Les soirées canadiennes* (1861), pp. 410-11.

25.   Canada. Department of Marine and Fisheries, Sessional Paper No. 12, 1868; Report of Théophile Têtu, pp. 62-63.

26.   Canada. Parks Canada, Québec Region, Archives of William Fruing and Company Ltd, Clorydorme, 1904: Good Debts Paid, Bad Debts, Doubtful Debts, Mic. M-68.

chapter
nine

ACCOUNTING FOR CHANGE:
UNDERSTANDING MERCHANT CREDIT STRATEGIES IN
OUTPORT NEWFOUNDLAND[1]

Robert C.H. Sweeny

The cod fishery at Newfoundland was among the first American resources to be exploited by European capital. This 500-year-old industry has also been one of the most favoured grounds for academic research. Until quite recently, a consensus has characterised much of the work: a merchant-dominated staple trade so thwarted Newfoundland's internal dynamics that the inshore fishery remained traditional. Fortunately, in response to the present ecological and social crises, this traditionalist image has come under increasing attack.[2] This article is a part of that larger critical re-evaluation. My subject is quite specific, an analysis of merchants' accounts for a particular place and time. The place is Bonavista, a centuries-old fishing community on the northeast coast of the island. The time is three years at the end of the 19th century: 1889 to 1891. The sources are the general ledgers of two merchants: James Ryan and Philip Templeman.[3]

In the human sciences, the choices of where, when, and what one looks at are all significant. If one chooses poorly, then the research results can be misleading or even meaningless. The danger is all the more important here, because my primary purpose is not the study of these particular Bonavista firms from 1889 to 1891. My purpose is much more general: to understand historically the complex social relations within the inshore fishery at a significant turning point in its long history. So, first, we must address the question: were these firms representative of something larger? The answer is ambiguous. These particular firms were not typical, but I know of no reason to consider them to have been exceptional. Throughout the inshore fishery's long history, families, fishers, and firms made choices. It is logical to assume, therefore, that the patterns and processes discussed here would vary, perhaps considerably, in the differing communities and regions of Newfoundland.

This idea, that what was typical was variety, is at odds with the inter-disciplinary, academic consensus on the inshore fishery. The consensus holds that in a context of scarce resources, merchant capital became so dominant through credit and truck that for the fishing families neither substantial capital accumulation nor significant social differentiation was possible. What little change there was came from outside. Historical research has, therefore, focused on strategies of merchant capital and the influence of market forces, while within the social sciences ethnographic research, which examines particular communities in the present, has generally been considered sufficient for understanding outport history.

Sources, Subject and Structure

Accounting records are complex documents. The merchant or his clerk kept a written, running account of each sale or purchase in what are known as day books. Usually at the end of the business day, this information was transferred into books known as journals. In some firms, as appears to have been the case with both Ryan and Templeman, the day book served as the journal. Periodically, summaries of the journal entries were written down or posted in general ledgers. Each journal entry of a purchase or a sale generated not one but two entries in the ledger: one on the debit side of an account and one on the credit side of another account. This technique is known as double entry bookkeeping. A sale of saltfish to a firm, for example, would be posted as a credit to the account of the person selling the fish and as a debit to the firm's cod account. Thus, the number of accounts in a general ledger is larger than the number of people doing business with the firm, because a firm maintains a large number of internal accounts to keep track of assets, inventories, supplies, and advances. By linking each sale with a purchase and *vice versa*, this double entry system allows a firm to know not only how each individual's account is doing but to find out relatively quickly how well the firm as a whole is doing by striking a balance. Both firms we examined struck their balances only once a year—on 31 December. In this article, I will be focusing on the external accounts of the firms' suppliers and clients.

In late-19th century Newfoundland, accounting was not yet standardised. Accounting techniques varied from firm to firm because they were almost invariably learned on the job. In developing our approach to these complex series of records, we thought of them as sets of concentric circles, like an onion. The analysis was done layer by layer, from the outside in. We started with the last record generated—the indexes to the ledger—and worked backwards to the first layer represented by the journal. Historical research often works likes this, backwards in time. This article reports on two layers still relatively close to the surface of our conceptual onion; both are based on an analysis of the yearly balances. The first explores the relationship between active and inactive accounts over time. The second goes beyond this to examine the varying patterns in the active accounts of suppliers and clients.

This article builds on and revises the findings from an earlier stage of our work which analysed the indexes to the general ledgers.[4] Those findings can be summarised as follows. Families could maintain multiple accounts and most families maintained accounts with more than one firm. The clientele was not stable. People appearing on the books in one year were often absent the next, but it was those clients with an account at only one firm who were the least persistent. There was clear evidence of social differentiation among clients. A small number of clients employed numerous non-family members and it was this rather select group who appeared to have been most tied to their merchant suppliers. These results highlighted the potential significance of varying and, in all likelihood, quite selective access to credit. This complex question of credit is at the heart of the present article. Credit is not simply an economic issue; indeed, it is perhaps best conceived in terms of differing strategies.

Ryan and Templeman kept detailed accounts because credit was central to the commercial inshore fishery. In an economy where relatively little cash circulated, credit was essential to the functioning of both exchange and production. Credit, as Tom Paine once remarked, is "suspicion asleep." So, it is interesting to note that neither firm chose to maintain an alert watch over credit. Although individual client accounts could be monitored on a continuous basis, it was only once a year, in the early weeks of January, that these merchants could possibly evaluate where the firm as a whole stood. Even then the situation remained at best ambiguous; although striking the year-end balance revealed the extent of total outstanding loans, it could not reveal how secure those loans were. Understanding this apparent paradox will take us to the heart of the contradictory credit strategies in outport Newfoundland.

First, I examine if there was a credit problem. After all, it is possible merchants did not bother to monitor credit closely, because, as the historiography has long assumed, the lines of credit tied fishers to their merchant. Second, I discuss the differing relations these firms had with their suppliers. Finally, I explore in some detail the extent and distribution of outstanding credit to active clients at year-end. The analysis here is at a certain level of abstraction; nevertheless, the discernible patterns strongly suggest that there was a complex social structure within these fishing communities. In conclusion, I propose a new way of thinking of the strategic aim of merchant credit in outport communities.

This brief introduction has introduced two problematic categories, that is terms whose meaning I have taken for granted, when in fact they need to be critically evaluated. In the context of the Newfoundland fishery, what did the distinction between supplier and client mean? After all, as fish merchants both Templeman and Ryan bought substantial amounts of fish products from their "clients" which they sold to their "suppliers." Here, clearly, the intermediary function of the merchant worked both ways. Nor is this just a semantic problem, for how one defines these terms is important in understanding where in the potentially complex web of production exchange and credit a particular firm stood.

Historians have not generally considered this lack of a clear definition to be a problem, because for them the situation was sufficiently clear without one. Outport merchants purchased goods from Water Street merchants on credit, which they in turn sold to fishing families on credit, who then paid off their debts by remitting fish. In short, the lines of credit not only served to distinguish supplier from client, but preceded and dictated the contours of the relationship. This conceptualisation of the economics of the fishery was based less on historical research than on an ideological choice and an a-historical assumption. The choice was to consider exchange relations more important than productive relations, because the wealth of nations was thought to stem from the operation of market forces, not from the creation of value by productive labour. Belief in the legitimacy of this choice, to accord analytical primacy to circulation over production, was buttressed by the assumption of a uniformity in the organisation of production. If a kin-based, small-scale, "traditional" inshore fishery predominated everywhere, then significant qualitative change could only come from the outside. Thus, against an unchanging backdrop of outport life, the history of the fishery became largely the history of merchant capital.

To challenge this perception of the fishery, it is essential to recognise the analytical and historical centrality of people's productive labour as the primary source of wealth in this society and the potential variety of social relations to appropriate the wealth created by that labour. In an historical analysis based on merchant records this is particularly difficult, because ledgers are organised in terms of accounts, not people, and chronicle relations of exchange, not production. So, what I think is central only appears here in an alienated form, once—or twice—removed if at all. Fishermen all appear first as consuming clients and then some reappear as sellers of their produce and others as sellers of their labour power, while extremely few women who were active in the fishery appear at all, despite or perhaps because of the importance that their labour in curing had for the creation of value.

In this context, no single definition of supplier or client seemed appropriate. Instead, for each firm I selected accounts as potential suppliers based on their name, volume, or location. I then examined in detail the patterns of debits and credits charged these accounts.

Confirmation of my initial selection was provided by the manner in which the firms kept these accounts, because in a double entry system a supplier's account is the mirror image of a client's; the normal postings are inverted: i.e. fish or cash are debits and merchandise is a credit. With the exception of a firm affiliated to Ryan in King's Cove, discussed below, all active external accounts that were not classified as suppliers are treated here as clients. So, in this study the terms supplier and client are historical categories based on the manner the account was kept in the ledgers.

Debt, Truck, and Agency

Over the three years of our study period, neither Ryan nor Templeman systematically monitored the credit-worthiness of clients. In 1891, Ryan established belatedly a ledger account for doubtful debts, while neither firm maintained a bad debt account. So any assessment of the size of their non-performing loan portfolios is necessarily the product of an imposition of categories developed in the present to make sense of the past. In an economic relationship where extensions of credit were essential and ongoing, merely owing money to a merchant for a long time did not mean the client was in arrears. As long as merchants allowed clients to purchase goods on credit, presumably, the merchants considered them to be good risks. If a client made any payment on the account at all, it may have been considered by the merchant sufficient proof of his or her good intentions. Aware of these ambiguities and the potential for misreading what was clearly a social as much as an economic relationship, I chose to err on the side of caution. I defined a non-performing loan as a situation where a client owed the firm money at the end of one year and did absolutely no business with the firm over the course of the ensuing year.

Credit advanced which became non-performing was not necessarily lost. Many clients, who chose not to carry on any business with a merchant in a given year, even though they owed them funds, could and did return to trade another year. At Ryan's 48% of the non-performing accounts in 1889, representing 59% of the total accumulated outstanding debt, recorded at least some business by 1891. Templeman was somewhat less fortunate; there, 35% of the 1889 non-performing accounts, representing by value a third of the portfolio, were once again active by 1891. Furthermore, it is important to remember that this category of non-performing loans is my own construction. Some clients may well have reached informal agreements for repayment with their merchant and would not therefore have considered themselves in arrears. Family ties were also important. So, even when no such agreement existed, the merchant might well have felt secure in the knowledge that a particular client's family would, if need be, cover the debt. Nevertheless, it is reasonable to assume that at both firms the risks involved in credit advances would have been an accepted, albeit difficult to calculate, cost of doing business.

Table 1

The amount of new non-performing loans each year.

	1889	1890	1891
James Ryan	7,775	13,768	14,758
Philip Templeman	1,607	2,950	2,484

Table 1 shows the amount of credit that had been advanced in the previous year and was non-performing by year-end. My definition of non-performing loans underestimates, probably significantly, the extent of doubtful or bad loans on these firms' books. In 1891, when Ryan established a doubtful debt account, following a major write-off of unproductive accounts, he calculated the value of all doubtful debts as being a third higher than that revealed by my method. Yet, my deliberately conservative method reveals a problem of major proportions. Given the scale of this problem, the cumulative weight of unproductive credit advances undoubtedly constrained and influenced merchant strategies.

Graph 1 highlights the long-term, cumulative impact of years of non-performing loans. Differing histories led to different situations for each firm. In part, as the graph's construction suggests, this was a question of scale. By the end of 1889, Ryan had advanced $73,289 in credit to its 673 active clients, compared to advances of $7,614 to Templeman's 434 active clients. So, on the graph, amounts are in thousands for Ryan, but in hundreds for Templeman, while the scale used for the number of accounts was doubled. Understandably, the length of time a firm had been in business influenced the situation. The long-established Ryan firm had as many inactive accounts in arrears on its books as it did active accounts and, at all but the highest levels, non-performing loans outnumbered and outweighed credit advances to active accounts. Templeman, having only opened for business earlier in the decade, had yet to accumulate such a heavy load, but the clear similarity in the patterns of inactive accounts is highly suggestive.

The amounts owed by the majority of both active and inactive accounts were often small. Indeed, accounts with outstanding balances of less than ten dollars, which were not shown on the graph, constituted the majority of active accounts: 371 at Ryan's and 287 at Templeman's. More than two-thirds of the inactive accounts at Templeman's, 121 out of 170, were also for sums less than ten dollars. Although relatively fewer at Ryan's, these small non-performing loans are the largest single category: 190 of the 738 inactive accounts. From the merchant's perspective, the frequently small-scale nature of the non-performing loans did not mean they were somehow less of a problem. Rather, it underlined the ubiquitous nature of the problem, one whose scale and prevalence meant it could not be resolved merely by instituting better controls over credit.

If the situations facing Philip Templeman and James Ryan were at all indicative of the problems facing merchants in other communities, then the historiographical implications are profound indeed. It has long been argued that fishers and their families had little choice: because, tied by lines of credit to particular merchants, when they purchased goods it was at the high prices charged a captive market, and when they sold their goods it was at artificially low prices. Truck was an unequal exchange, whereby merchants systematically exploited tied producers.

The situation revealed here was qualitatively different. First and most importantly, for the overwhelming majority of clients the amount of credit outstanding at year-end was not enough to tie them to anything. Second, every year a significant number of clients who had received substantial credit advances walked away from their obligations. For the merchant this individual freedom of action, or as progressive historiography calls it "human agency," meant a substantial cost, which had to be met somehow. The most obvious and perhaps only way merchants could recover the expense of these unproductive credit advances was to factor them into the pricing policies at their stores. There were two alternatives: pay less for goods purchased or charge more for goods sold.

Graph 1

The state of all client accounts with credit advances larger than 10 dollars on 31 December 1889.

Clients of James Ryan

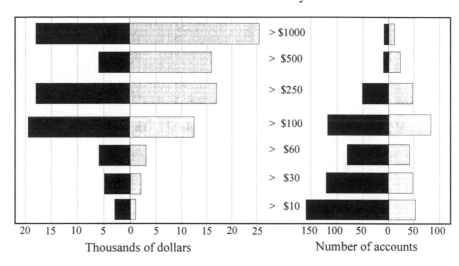

Clients of Philip Templeman

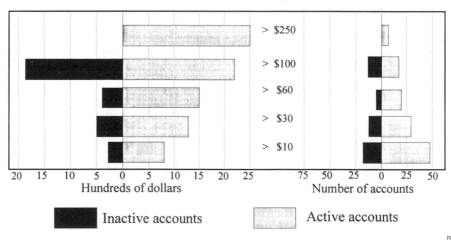

Ryan and Templeman were fish merchants. Cod fish and oil were not their only business, but it was an important aspect of both firms' activities. So, in terms of prices for goods purchased that they could control, or at least attempt to control, fish products loomed large. Relative to their fish trade, the amount of each year's new non-performing loans was also very large. They ranged from a fifth to a third of the total price paid for fish products. It is possible to establish the firms' gross operating margins in this trade by comparing the annual total amounts paid for fish products with their sale price. If these merchants were aware of the problem outlined here, then one would expect to see a pricing strategy aimed at increasing revenues in 1891 to cover the costs of 1890's large portfolio of non-productive loans. The accounts are consistent with such a strategy: Ryan's margin increased 39%, while Templeman's rose a more modest 32%. Such a reactive strategy, however, would have been insufficient to deal with a long-term problem of this scale. Although mark-ups were in some years substantial, as is clear from Table 2, they were linked to volume and that was not something the merchants could easily influence, let alone control. So, firms must also have had recourse to the alternative of charging more for goods sold at their stores. Since non-productive loans were a cost all merchants faced, competition between merchants would have had little or no impact on this strategy.

Table 2

The gross mark-up on cod, cod oil and pickled fish by Ryan & Templeman.

	1889		1890		1891	
Ryan's purchase price	$37,180		$38,103		$53,401	
Ryan's sale price	$48,704		$46,669		$65,302	
Gross mark-up	$11,524	31%	$8,567	22%	$11,901	22%
Templeman's purchase price	$10,364		$8,606		$13,063	
Templeman's sale price	$11,579		$11,503		$16,881	
Gross mark-up	$1,216	12%	$2,900	34%	$3,818	29%

Seen in this light, the unequal exchange of truck was not the result of merchants' systematic exploitation of tied producers. The problem was not systematic, it was systemic: the product of sharply differentiated social relations of production and exchange within the fishery. Truck prices were onerous precisely because the ties did not bind. Individuals could and did get out. This individual exercise of human agency was, however, seriously circumscribed by age and gender. Unquestionably, young, unmarried men were the most likely to walk. Their choices left the community to pay the bill. Families, with their heavier reliance on store-bought staples, would have borne the brunt of these costs. The situation was not, however, as linear as this might suggest. There were overlapping dialectical tensions. On the one hand, individuals were pitted against community and, on the other, producers were pitted against merchants. Neither dynamic was new nor unique to the inshore fishery, but it was their interaction that is important here. Unequal exchange justified, indeed for some necessitated, these individual actions which, however, weakened any community-based response to the merchants' strategies. These dynamics were politically and culturally significant and the knowledge that the exercise of individual freedom has high social costs has remained an important ethical dimension in the evolution of fishing communities.

Beyond Suppliers and Clients

Ryan's principal suppliers were Bowring Brothers of St John's and their affiliate, C.T. Bowring and Co. of Liverpool, England. Bowring also handled most of Ryan's fish and, due to the importance of fish shipped to them late in the season, at year-end Ryan was owed substantial sums by these suppliers. The ledger opens on 1 January 1889 with an outstanding balance owed Ryan by Bowring of $19,549. If by year-end it was down to $8,552 it would nevertheless stand at $6,781 and $8,168 the following years. In addition to Bowring, Ryan purchased supplies from 16 firms and one individual, none of whom provided at year-end any substantial advances of credit. The norm was to balance to the cent. The account of the Dana Hardware Co. of Boston was balanced to the cent every three months.

Ryan was the leading firm in Bonavista and other local merchants carried on substantial business with it. Indeed, Philip Templeman was Ryan's second largest supplier of salted cod, maintaining in two of the three years a small positive balance. I have classified these local firms as clients. In addition, James Ryan had an affiliated firm, James Ryan and Co., located up the coast at King's Cove. This affiliate did a very large business with the parent firm. Over the three years, its annual volume of purchases increased from $7,162 to $13,197, paid for in part by the quite exceptional positive balance of $15,700 the affiliate had built up on the books of James Ryan by the end of 1889.

The much smaller and more recently established firm of Templeman presents a different image. Over the three years studied, Templeman's total indebtedness to his suppliers increased steadily from $4,605 to $6,989. While his principal supplier, Goodfellow and Co., accounted for more than 90% of the advances, seven smaller suppliers almost invariably ended the year being owed funds by Templeman. The sole exception was a local firm, Baine, Johnston and Co., whose large and active account at year-end owed Templeman quite small sums in two of the three years.

On a yearly basis, Ryan and Templeman had, therefore, quite different relations with their suppliers. Ryan was not dependent on credit advances and, coupled with its pre-eminence in the region, this allowed it a certain degree of autonomy, perhaps most evident in its direct relations with a Boston-based supplier. In contrast, Templeman was increasingly bound by a chain of credit. In a context where systemic costs limited price competition, these firms' differing credit situations led to differing credit policies towards their clients. These differing strategies had paradoxical effects on the problem posed by the substantial movement in their clienteles.

There are several different ways of looking at this question of movement. The first is simply in terms of active accounts. Table 3 shows the total number of client accounts over the three years, the number active in any single year, and the relatively small number of "core" accounts, a quarter at Templeman's and closer to a fifth at Ryan's, which were active in all three years. Clearly the more or less constant movement suggested by the non-performing loan data reflected a larger process of change, for not just debtors but all clients could and did move around a great deal. As important as this movement was for the firms involved, it was potentially less important than changes within their core accounts, fluctuations in the volume of their retail businesses, or the evolution in the relationship between debits and credits of clients active in a given year.

Table 3

Number of active client accounts each year.

	Total	In 1889	In 1890	In 1891	1889-91
James Ryan	1450	673	740	954	319
Philip Templeman	671	434	428	453	169

Graph 2 plots the principal elements needed to assess these more complex elements of change. Ryan's retail business was roughly five times the size of Templeman's; therefore, to allow for a clear comparison of the importance of change, it was necessary to construct this graph using two different scales. As is readily apparent, both in terms of the relative importance of core clients and in terms of the volume of retail business, the two firms were quite similar. It was in terms of the annual balances in retail business that the differences were most apparent. Over the three years Ryan improved his situation by a dramatic $48,395, while Templeman's business remained consistently dependent on debt financing. This remarkable difference in fortunes was directly related to the firms' differing credit strategies.

Both Table 3 and Graph 2 highlight a significant increase in the number of active accounts at Ryan's in 1891. This growth resulted almost solely from a dramatic increase in accounts active on the credit side of the ledger. Now 1891 was a relatively good year in the fishery and the value of fish delivered to Ryan was up 40% over the previous year. It was, however, a change in credit policy, rather than the improved fishery, that most contributed to this dramatic improvement in the balance sheet. In 1891, close to 30% of Ryan's clients came into his premises and paid in cash or in kind a total of $32,397 without purchasing a single item. In all, 278 clients had credits, but no debits, posted against their accounts. I suspect Ryan had lowered the boom on many of these clients, forbidding them further purchases at his store until they cleared up their outstanding debts. It was an understandable policy, inasmuch as these accounts, many inactive for years, represented debts totalling $35,288.

It was fortuitous for Ryan that he chose a relatively good year in the fishery to attend to the problem of outstanding debt. Significantly, however, only $123 of the credits were for fish products. Having cut them off at his store, perhaps he did not expect these errant clients to make their remittances in fish. Nevertheless, the paltry nature of this sum does speak to the complexity of the social relations in the fishery where substantial payments could be made without any mention of the thousands of quintals of salted cod on which they were in large measure based. Paradoxically then, Ryan's clientele increased because in a relatively good year in the fishery he was sufficiently autonomous from his suppliers to sacrifice short-term sales—note the differences in debits between the two firms in 1891—in order to get his own house's credit in order.

No similar pattern was discernible at Templeman's, nor, given Templeman's increasing dependency on credit from his suppliers, would one have expected otherwise. His goods had to be moved because payments had to be made. Instead, as Templeman's credit facilities with his suppliers expanded, so too did his extensions of credit to retail clients. In 1889, credit from suppliers accounted for 60% of Templeman's $7,614 in credit advances. By 1891, credit from suppliers constituted 66% of a sum, that at $10,630, had grown by 40%.

Graph 2
The annual movement in active client accounts.

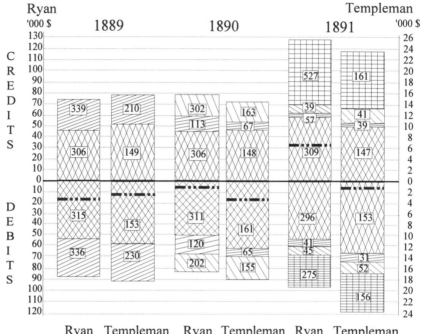

Accounts	Ryan	Templeman	Ryan	Templeman	Ryan	Templeman
with credits	645	359	721	378	945	388
Accounts with debits	651	383	633	381	676	392
Total number	673	434	740	428	954	453

Active in 1889 In 1890 In 1891

From 1889 to 1891 ▬••• Balance

rchs

These differing strategies were developed in a context where much that was fundamental about the fishery was beyond the control of merchants. As intermediaries, merchants aimed at maximising their volume of business which necessarily meant, in a context where people moved about with such apparent ease and yet hierarchies were significant, developing strategies focused on particular clients. Assessing the success of these mercantile strategies is difficult, particularly when carried out at our present level of abstraction. Year-end figures do not permit a detailed examination of the relationship between merchants and their most important clients, but they can allow for a preliminary assessment. Any such examination, however, depends a great deal on the analytical categories used to organise the data. Three questions need to be addressed: How does one define a prominent client? Are there ways to test the reliability of the proposed definitions? What should be the basis for a comparative presentation of the data?

What constitutes a prominent client is a problematic question on two levels. The general problem is how to define the categories so they are not simply arbitrary boxes, but have some meaning in terms of the processes being studied. The more specific problem is how to ensure that the differences in the scale of operations between these two firms do not introduce serious distortions in the images of the relations each firm maintained with its most prominent clients. Particular clients could be considered important for various reasons. I defined prominent clients in two ways: by the amount of outstanding credit they had at year-end and by the value of fish products sold.

On Graph 1 it was evident at both firms that a distinct minority of active accounts, those over $100, accounted for the greatest part of outstanding credit. Ryan had made, however, substantially greater advances to active clients than Templeman, ten times more to a clientele that was only double the size. So in relative terms, a $100 credit advance at Templeman's would be the equivalent of $500 at Ryan's, the level at which Ryan's active accounts began to outweigh and outnumber inactive accounts, suggesting that Ryan too may have considered it significant.

Defining prominent clients in terms of fish sold poses slightly different problems of scale. The clientele at Templeman's, a much smaller and newer firm, included fewer traders and dealers than Ryan's. The bulk of fish sold to him came directly from the producers themselves. Based on my knowledge of the records gained in the data entry stage of our research, I felt that a reasonable definition of prominent fishers would be those with sales to Templeman of more than $100. Fish sales to Ryan were between three and four times those to Templeman. If, however, I defined a prominent fish supplier to Ryan as one who sold $350 worth of fish, it would seriously distort the comparative image, because it would effectively eliminate trap crews. At prevailing prices, $350 meant 90 quintals of merchantable quality salted cod. A single crew would have been unlikely to catch and, more importantly, their shore crew would certainly have had great difficulty in curing to this high level 45,000 pounds of cod a season. In short, I would be comparing successful fishers at Templeman's with traders, dealers, and Labrador schooner masters at Ryan's. My choice of $250 as the cut-off for Ryan represents, therefore, an arbitrary compromise.

Identifying ways to verify the reliability of these definitions was the next problem. Clearly, over such a short period, one would not expect there to be dramatic movement; if there was such movement, then the problem would probably lie with my definitions of who was a prominent client. The question then became what benchmarks should be used to measure movement? The intermediary role of the merchants and the remarkable similarities between the two firms on Graph 2 prompted me to select the value of all retail transactions, both debits and credits, as the principal benchmark figure. In conjunction with the proportion of accounts selected, these figures could then be used to evaluate the reliability of the definitions. By both measures the results were remarkably stable. At both firms roughly a tenth of active accounts were responsible for half of all retail transactions and substantially greater amounts of fish. Indeed, the similarities in the patterns suggest that these admittedly clumsy constructions correspond, albeit imperfectly, to a structural reality underlying relations between merchants and clients.

The most effective way of presenting the data is not simply a technical question, for it raises analytical and conceptual issues. Normally in economic and business history, a dia-chronic approach is used, wherein the size of the circles in the Venn diagrams used in Graph 3 would be based on the value of each variable and they would change independ-ently each year. So, for example, in 1889 the circle representing all fish sold to Templeman would be larger than that for credit, because sales were larger than advances. In 1890, the circle representing fish sales would decrease by 17%, but the one for credit would increase by 29%; while in 1891 the circle for fish would increase 52%, but the circle for credit by only 8%. This method applies a disciplinary consensus that accords analytical primacy to certain variables, conceived as elements of a model capable of explaining broader histori-cal change.

There are problems with this consensual approach. The diachronic method abstracts variables from their historical context to analyse in isolation their change over time. So, for example, the differing sizes of the diagrams in 1891 would be quite unrelated to the situa-tion in that year and could only be understood by referring back to the arbitrary base year of 1889. Like the social sciences from whence it came, diachronic analysis is fundamentally a-historical. Instead of testing an explanation of change against the complexity of the past, it radically transforms and simplifies the past by imposing upon it an explanatory model developed in the present. When the evolution of variables selected to fit the model is then examined, it is always outside their historical context. For any attempt to understand the historical dynamics of the inshore fishery, this method would be particularly inappropriate, because the differing models used are all based on economic theories developed to manage advanced capitalist economies. Whatever one may think of late-19th-century outport New-foundland, it certainly was not an advanced capitalist economy.

An alternative method of organisation would be to respect the historical logic of the source. The information in the ledgers was maintained in accounts and the merchants' decisions to advance or refuse credit were, presumably, made on an account-by-account basis. So, the organisation of Graph 3 gives analytical primacy not to economic variables, but to the number of active accounts at the firm each year. The base circles represent these accounts and are all drawn to scale, from Ryan's 954 in 1891 down to Templeman's 428 in 1890. These circles are then used as the basis for the diagrams detailing economic factors, because these factors only make historical sense in light of who chose to deal with each firm each year. The number of accounts, however, is not the same as the number of clients, because people could choose to have more than one account at a firm. So, this analytical method of presentation visually subordinates changes in retail transactions, fish sales, and credit advances by according primacy to the choices people made. It draws one's attention to the relative positions of the accounts of differing types of prominent clients and their positions vis-à-vis other clients' accounts. It underlines the choices merchants made in extending credit and fishers made in deciding to whom to sell their fish.

Graph 3 allows a further test of the validity of the widespread assumption that the lines of credit preceded and dictated the contours of the relationship between fisher and mer-chant. If this assumption were valid, then the composition of the two groups of prominent

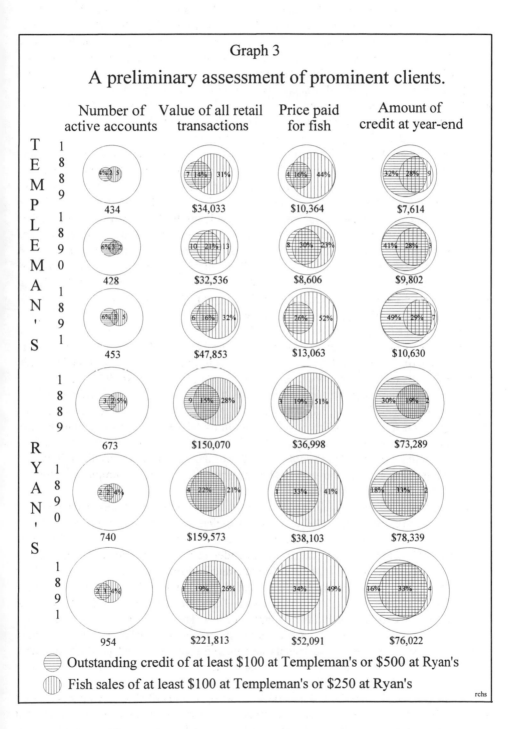

Graph 3

A preliminary assessment of prominent clients.

clients would be very similar, because those making the largest remittances in fish would be those who had received the largest credit advances. At neither firm did this prove to be the case. Although affected by the quality of the fishing season, most notably in 1890 at Templeman's, accounts responsible for roughly half of all the fish sold and a quarter of all retail transactions were accorded only limited advances of credit. Or put another way, in 1889 at both firms, accounts receiving close to a third of the credit supplied only 3% to 4% of the fish. At neither firm was this relationship static. By year-end 1891, at Templeman's half the advances were credited to accounts that had supplied only 1% of the fish, while at Ryan's the same category supplied even less fish, but received only 16% of the credit. These differing patterns are consistent with the contrasting credit strategies discerned earlier and confirm the differing directions implicit in these strategies.

Graph 3 highlights the relative importance of certain categories of accounts. Clearly particular clients would have been significant in mercantile strategies that aimed at maximising the volume of business. As Graph 3 confirmed, credit strategies differed at the two firms. At Ryan's new large extensions of credit were few in number. So the 28 "core" large debtors accounted for an increasing share of available credit; by 1891 they had absorbed 43% of all credit extended to active accounts. These were for the most part established clients who were allowed to maintain large negative current accounts year after year. A minority of these debtor accounts were not credited with any significant sales of fish; however, these people were usually partners in other "fish" accounts.[5] In short, at Ryan's significant extensions of credit increasingly privileged certain large clients who figured prominently among the firm's suppliers of fish products.

In contrast, at Templeman's only four of the 13 "core" large debtors regularly supplied significant amounts of fish. Templeman's more liberal credit policy meant that these core accounts were of declining relative importance, accounting for 26% of outstanding credit by the end of 1891, though many of the new recipients of his largesse did not prove to be large suppliers either. Over the three years, aside from "core" debtors, 39 clients ended a year owing more than $100. In 1891 only 16 sold any fish at all to Templeman. Operating under greater constraints Templeman had fewer options, but his more accommodating credit policy did not generate a greater supply of fish. When a select group of clients receive 53% of the outstanding credit and yet supply only 22% of the fish there is a problem. It was, of course, a familiar one. By year-end 1891 non-performing loans at Templeman stood at $4,665, a 95% increase in two years.

Perhaps the most remarkable aspect of Graph 3 was the extraordinary dependency of both firms upon relatively few clients for their supplies of fish products. In 1891, 83% of Ryan's fish came from only 7% of the active accounts, while at Templeman's 78% was supplied by 8% of his accounts. These figures underscore the importance for the firms of maintaining a core group of large fish suppliers and building on that core over the three years. Table 4 presents the changing numbers of clients and the relative importance of the fish sold by 81 clients at Ryan's and 54 clients at Templeman's, who in at least one of the three years met my definition of a large supplier of fish. There was quite substantial movement in and out of these categories, but the minority of consistently large suppliers figured prominently. At Ryan these "core" clients accounted for half of the fish products supplied, against a third at Templeman which was far more reliant on its many small suppliers. Only once, however, did a large supplier cross over.[6]

Finally, the last line of Table 4 may be the most significant of all, for it renders explicit what is only implicit on Graph 3. At both firms, the majority of clients sold their merchant no fish products at all. At Ryan's, these clients accounted for 39% of the total retail business in 1889, a share which rose to 44% by 1891. At Templeman's, these clients' share rose steadily, from 21% to 34%. I believe these non-fish accounts provide the key to understanding the often contradictory credit strategies of outport merchants.

Table 4

The changing significance of large suppliers of fish products.

| | Ryan's | | | | | | Templeman's | | | | | |
| | 1889 | | 1890 | | 1891 | | 1889 | | 1890 | | 1891 | |
	#	%	#	%	#	%	#	%	#	%	#	%
Core prominent clients	27	47	27	52	27	50	13	29	13	34	13	35
Others prominent in 1889	20	24	14	13	8	4	16	31	10	9	4	3
Others prominent in 1890	13	12	18	23	12	13	8	11	11	19	5	12
Others prominent in 1891	19	9	26	19	36	33	14	7	13	12	25	43
All prominent clients	63	77	65	83	68	84	43	64	37	60	41	78
Total of all accounts	232	100	196	100	205	100	236	100	200	100	181	100

Conclusion

This analysis of merchant accounts from a late-19th-century Newfoundland outport has revealed a complex and unexpected situation. Contrary to the historiographical consensus, fishers were not all the same. There was significant social differentiation among petty commodity producers. Furthermore, although merchant credit did play a prominent role, credit was much more selective and substantially less binding than previously thought. So, in accounting for change in these communities, how are we to understand merchant credit strategies? Admittedly, it is tempting to explain these strategies solely in terms of the political economy of the fishery. After all, catching and processing cod were the principal economic activities linking outport Newfoundland to the broader north Atlantic economy and beyond. Tempting but a-historical, because the salt-cod trade was not, nor had it ever been, the only economic activity within these communities; however, the economic organisation of most productive activities—from housing construction and small boat building to agriculture, food processing, and garment making—was local and informal in nature.

With a dismissive arrogance, born of the division between mental and manual labour, academics and bureaucrats have all too often categorised these essential and widely diverse forms of productive labour as simply subsistence activities; however, the importance and variety of these activities meant they were much more than that. They constituted an informal economy.

This informal economy was in continuous and complex interaction with the formal economy. They had a symbiotic relationship, but understanding the former is difficult because no reliable ways of establishing its size exist. It rarely involved cash payments or waged labour, nor did it respond to the same market pressures as the formalised cod and sealing industries. Indeed, its pricing and exchange mechanisms depended on a completely different cultural and ethical matrix.[7] Despite these important differences, the informal economy of outport Newfoundland operated within and was an integral part of one of the oldest capitalist societies in the Americas. So, just as in the formal economy, value created by productive labour in this informal economy could be and was appropriated.

The appropriation of value within the informal economy operated on a variety of different levels. First and foremost, it was central to the political economy of patriarchy within the family. This level paralleled the accounting for women's labour in the formal economy, where despite their importance in the creation of value—proper processing of cod added as much as 40% to the price of the finished product—extremely few women had accounts of their own. Second, the labour of young people was not accorded its full value. Here too, the practice was visible in the accounts when young women working as domestics in the households of the more successful fishers were credited only $14 to $18 for their entire year's work. Undoubtedly, this discriminatory practice was a contributing factor to the high levels of geographic mobility among young men already noted. Third, certain types of labour involving higher, socially defined, skill levels, such as boat building, were accorded greater exchange values. Fourth, ownership of capital goods, such as horses, commanded a greater share of the value produced. Appropriation at each of these levels shared two characteristics: they were within the realm of production and they reinforced existing forms of inequality within the community. These social relations of production were not unique to Newfoundland. In their broad outlines, as sketched here, they were to be found operating within the informal, moral economies throughout Europe and the Americas.

Value created in the informal economy could also be appropriated by those who, like merchants, were outside the realm of production. The most obvious form of mercantile appropriation was the unequal exchange which acted like a tax on the imports—flour, cloth, nails, twine—necessary for the functioning of the informal economy. A more insidious form of mercantile appropriation depended upon a firm's success in inserting itself as a "clearing house" for the settlement of debts within the local economy. Again an element of the formal economy illustrates how this process worked. Generally shares and wages in the fishery were not paid directly to the people involved; instead they were posted on the merchants' books as credits to the accounts of crew members. Most of these credited funds would then be spent on purchases at the merchant's store, which was one of the reasons why merchants cultivated relations with the larger traders and dealers. It was not just fish they brought to the premises, they generated associated business as well.

As intermediaries in the multi-faceted and hierarchical socio-economic structures of the outports, merchants like James Ryan and Philip Templeman aimed at maximising their total volume of business, not just, nor even primarily, their trade in cod. The key to success lay in capturing as much as possible of the informal economy. Ryan and Templeman used credit to ensure that the maximum amount of the total economic activity in the community passed through their books. For both firms this meant extending their "clearing house" business to include all the myriad debts, obligations and exchanges of the informal economy.

Evaluating these firms' success in applying this strategy is difficult, particularly from the abstract perspective of year-end figures. Nevertheless, two perplexing aspects of these firms' businesses can now be seen in a different light. First, those apparently excessive extensions of credit, which could not be justified in terms of the fishery, now begin to make sense. They were loss leaders; a way of ensuring that these prominent community members' non-fishing activities would generate business for the merchant. Both Ryan's core of prominent clients and Templeman's more liberal credit policy were important here. Second, the relative importance of those accounts which sold no fish whatsoever can be seen as a partial indicator of the extent to which these firms were able to draw within their grasp significant aspects of the informal economy. Over the three years, the value of Ryan's non-fish producing clients' retail business rose from 158% to 185% of the value of his trade in fish. Philip Templeman oversaw a spectacular growth in his share of this business. It rose from 64% of his fish trade in 1889, to 99% in 1890 and reached 124% by 1891. Clearly, Philip Templeman, if he had not already known, quickly realised that there was more than cod around the bay.

Twenty years ago, the late David Alexander wrote what has come to be considered the definitive article on the traditional economy of Newfoundland.[8] A key element of this oft-cited and frequently reproduced piece has not received the attention it merited. Alexander suggested that the informal economy contributed as much to Newfoundland's gross national product as the fishery. If my analysis of these ledgers is correct, his may well have been an overly conservative estimate. It was, nevertheless, an extremely important insight— one, I suspect, shared by Philip Templeman and James Ryan. Sadly, it went largely unheeded by historians and social scientists, because we failed to recognise the historic centrality and complexity of work in outport life. When the final accounting for the changes which have led to the commercial extinction of the cod is drawn, our failure will loom large on the debit side of the balance sheet.

Notes

1.   This article is based on research funded by the Social Sciences and Humanities Research Council of Canada, whose support is gratefully acknowledged. The grant was awarded to Rosemary E. Ommer as principal researcher and myself as co-investigator. These funds allowed us to hire and benefit from the counsel of David Bradley and Robert Hong, who were responsible for the data entry of Ryan's and Templeman's records respectively. Rosemary and I collaborated on earlier attempts to make sense of this data in papers presented to the Canadian Studies Association of New Zealand in 1992 and the Canadian Sociology and Anthropology Association in 1993. We would like to thank Doug McCalla for his astute criticisms of that work. The present article was written, however, by myself while on sabbatical in Montréal. The support of the Faculty of Arts of Memorial University, which accorded me this leave of absence from my normal teaching duties is gratefully acknowledged.

2.   The image and the crises are not unrelated. See my "The Costs of Modernity: The role of the human sciences in the Newfoundland crisis" in *Culture, Technology and Change in the Americas, Volume 3*. Edited by Ward Stavig and Sonia Ramirez Wohlmuth. Center for Caribbean and Latin American Studies, University of South Florida, 1996, pp 318-31.

3.   The Ryan records are at the Maritime History Archives of Memorial University and the Templeman records are at the Provincial Archives of Newfoundland and Labrador.

4.   Robert C.H. Sweeny, with the collaboration of David Bradley and Robert Hong. "Movement, Options and Costs: Indexes as historical evidence, a Newfoundland example." *Acadiensis*, 22, 1 (Fall, 1992), pp. 111-21.

5.   For example, Abraham Skiffington and Company's account only had fish credited to it and debits balanced exactly the amount supplied. These debits were primarily transfers to the personal accounts of Abraham and his son, which had negative balances every year and no fish sales posted.

6.   William of William Tilly left Templeman for Ryan in 1890, leaving behind a debt of $470 upon which he had made only derisory payments by the end of 1891. Presumably Ryan was aware of Tilly's situation, nevertheless he chose to extend Tilly credit in both 1890 and 1891.

7.   Following the path-breaking work of George Rudé and E.P. Thompson in the 1960's, historians refer to this matrix as a moral economy. Moral economies were common throughout Europe and the Americas in the early modern period, when informal economies predominated. They provided flexible systems which were roughly equitable, but with serious gender inequities, locally defined, but widely practised. They used a labour-based evaluation of production costs to establish prices for purposes of consumption and exchange.

8. David Alexander, "Newfoundland's Traditional Economy and Development to 1934" *Acadiensis*, 5, 2 (Spring, 1976), pp. 56-78.

Twentieth-Century Fishery

Credit: William W. Warner

chapter
ten

"RECURRING VISITATIONS OF PAUPERISM"
CHANGE AND CONTINUITY IN THE NEWFOUNDLAND FISHERY

James E. Candow

Introduction

The Newfoundland fishery, as pursued by Europeans and their descendants, has exhibited certain recurring features. Although numerous species were exploited, cod has been the mainstay until the present decade. Despite the disappearance of the British migratory fleet in the 19th century, Newfoundland's fishing grounds have remained attractive to other nations, often to the chagrin of Newfoundland residents. During the course of its history, the fishery has also been characterized by periods of resource scarcity and market instability. Since 1832 these crises have prompted government financial support to the industry. Fishers responded by broadening the resource base, migrating to less populous areas of Newfoundland and Labrador, and adopting new fish-catching techniques. Twentieth-century adaptations include a shift from saltfish to fresh and frozen fish, and the application of scientific knowledge, economic development theory, and government regulation. All have failed to check resource depletion, which since World War II has occurred at unprecedented levels because of the impact of modern technology.

The decline of the British migratory fishery during the 19th century was largely a result of the growth of the resident population and the emergence of St. John's as the emporium for the entire island. These trends were facilitated by the expansion of the seal hunt and the Labrador fishery. By 1844 seal products accounted for over one-third of the value of all Newfoundland exports, second only to the venerable cod. Capital investment in sealing during the era of sail peaked in 1857 when 370 vessels and 13,600 sealers went to the ice.[1] France had acquired treaty rights to Newfoundland's northeast and west coasts during the 18th century, and when the French returned to the *petit nord* after the Napoleonic Wars, the resident fishery that had taken root there shifted to Labrador. By 1850, roughly 700 Newfoundland schooners were fishing there.[2] This fishery peaked in the mid-1880s when 9.1% of Newfoundland's total population participated in the annual migration "down to the Labrador."[3] By contrast, the Newfoundland-based bank fishery of the Napoleonic Wars did not last much past 1815, although there was a short-lived revival in the 1840s.[4]

For Newfoundlanders, the growth of the seal hunt and the Labrador fishery was fortuitous. The island's population of 193,124 in 1884 was almost four times what it had been in 1816. This astounding increase coincided with declining average catches in the inshore fishery brought on by overpopulation, especially in Conception Bay. Average catch per fisher in the 1880s was only about 25% of the early-19th-century average, despite the invention of the cod trap in 1866 and its proliferation during the following decades.[5] In the seal hunt, falling profits in the 1850s led to the introduction of steam-powered ships in

1862. Because of their greater strength and mobility in the ice, steamers were more efficient than sailing vessels. As a result, the industry needed fewer sealers: some 8,000 were displaced between 1860 and 1881.[6] Also, the high costs of steamers and new seal oil refining facilities forced many old firms out of the industry and led to its concentration in St. John's by century's end. Still, if it had not been for the employment that sealing and the Labrador fishery provided, Newfoundland's economic day of reckoning would have occurred much sooner.

The 1860s offered a taste of Newfoundland's future. The inshore fishery was characterized by poor catches between 1861 and 1864, while at Labrador catches were depressed for most of the decade. Newfoundlanders responded in the now familiar pattern of moving into undeveloped or underdeveloped areas. Whereas before 1860 few white people had fished north of Cape Harrison, Labrador, by 1870 Moravian missionaries counted over 500 Newfoundland schooners between it and Cape Chidley.[7] On the island there was population movement into the west coast, where Newfoundlanders insisted they had concurrent rights to the fishery with France. But the chief manifestation of the crisis was the dependence of one-third of the able-bodied population on government relief payments.

This development was the result of a fundamental change in the colony's economic structure. During the transition from a migratory fishery to a resident one, British and Irish merchants who formerly hired migratory servants on a cash basis to work in the fishery instead became traders who supplied residents with fishing gear, food, and other necessities in exchange for cash or fish, especially the latter. This practice continued under the resident merchant class, and by the early-19th century credit prevailed throughout the fishery, except in the seal hunt, which in some areas at least operated on a cash basis since 1832.[8] Merchants advanced supplies to fishers in the spring before the fishing season began and took their fish as payment in the fall after the season ended. If a fisher had a surplus after squaring the account, the merchant issued a voucher for the appropriate amount, which the fisher drew on to purchase supplies during the winter, after which the cycle began anew. If, however, a fisher failed to catch enough fish to cover the spring advance, the merchant gave the winter "outfit" on credit, and the fisher was even deeper in debt when the next fishing season rolled around. The system reduced many fishers virtually to a state of bondage, but the resultant debt load could and did ruin many a merchant.

After 1832, when Newfoundland achieved representative government, the credit system added a new wrinkle. When catches were low, most merchants opted out of the responsibility to carry fishers through the winter months, and instead offered credit for the actual fishing season only. The Newfoundland government stepped into the breach and assumed the obligation of winter support in the form of relief payments, which became a permanent feature of Newfoundland life.

Although prosperity returned with a successful fishery in 1869, the crisis prompted business leaders and politicians to examine other ways of improving the economy. The adoption of the steamship in the seal hunt and the cod trap in the inshore and Labrador fisheries was indicative of one response: raising the technological ante. On the French Shore, Newfoundland and Canadian entrepreneurs established a thriving lobster fishery that by 1887 provided employment for 1,000 Newfoundlanders, many of them women and girls working in canning factories.[9]

The government, perhaps inspired by the examples of New England and Nova Scotian fishers, who by the 1870s had joined the French on the Grand Banks, sought to expand the fishery by reviving the defunct Newfoundland bank fishery.[10] New England schooners and dories had replaced French greenfishing ships and chaloupes, but the bulter system, known as dory trawling, remained the preferred method of fishing. Thus the Newfoundland government introduced a bounty system to encourge local schooner construction. Four bankers were outfitted in 1876, and by 1880 there were 32, carrying 432 crewmembers. The bank fishery reached a peak of 330 vessels and 4,401 crewmembers in 1889, accounting for 20% of all exports of dried cod.[11] During the 1890s it averaged only 99 vessels and 1,324 crewmembers per year, hardly compensation for the thousands displaced in sealing and shipbuilding.

Despite technological advances, the expansion of the fishery onto the Grand Banks, and the emergence of the lobster fishery, the fishery as a whole continued to decline. Seal oil prices fell after 1880 as mineral oils became more widely available, and because of over-harvesting, catch levels were only half of what they had been at mid-century. By the 1880s seal oil accounted for only one-eighth of the value of Newfoundland's exports, and by the end of the century less than one-tenth. The seal hunt had been crucial to the economic viability of the Labrador fishery, and as steam replaced sail in the seal hunt, it had negative repercussions on the Labrador fishery. Schooners remained idle during what used to be their most profitable period—the spring seal hunt. Sealing steamers also usurped the role of schooners in carrying passengers to Labrador, and with this trend the St. John's steamer owners gained a bigger share of the supply trade, further extending the capital's economic dominance.

Beginning in the mid-1880s the market price for Newfoundland and Labrador salt cod fell steeply because of inferior quality, marketing chaos, and increased international competition from Iceland, Norway, and France.[12] Many of the quality problems originated in Labrador. After the adoption of the cod trap, which permitted the capture of huge concentrations of fish during the four-to-six weeks cod run, fish were processed only to the saltbulk phase during trap fishing, thus reducing the time available for air drying.[13] The entry of steamships into the carrying trade in the mid-1880s likewise contributed to a decline in quality, and created its own marketing problems.[14] Because the first fish to reach market fetched the best price, and because steamers dramatically reduced shipping time, exporters sacrificed the cure in order to get their fish to market as soon as possible. On the other hand, the arrival of several steamers at market around the same time created gluts that invariably drove prices down. These trends became more general in the mid-1890s with the inauguration of regular steamship service between St. John's, New York, Liverpool, and Glasgow.

The confluence of these various trends spelled big trouble. Employment in the fishery shrunk from its historic peak of 60,419 in 1884 to 53,502 a mere seven years later.[15] The island's traditional economy had reached its limits, and, for many, emigration was the only option. Net emigration in the period 1884-1901 ranged from an estimated 1,500 to 2,500 people per annum, and major centres such as Harbour Grace suffered population decline.[16] Looking for a scapegoat, the government prohibited bait sales to French fishers after 1887 in an attempt to cripple the French bank fleet. There was much chest-thumping in St. John's, but the French found alternative bait supplies, and the advent of French steam-trawlers in the first decade of the 20th century did away with the need for bait altogether.[17] Thus there was small loss in 1904 when France relinquished all legal claims to the French Shore.

The crisis provoked a range of other responses. In 1887 the Newfoundland government appointed a commission to investigate the operations of fisheries departments in other countries and to make recommendations on whether the Newfoundland government itself should have such a department.[18] The Fisheries Commission recommended the appointment of a full-time superintendent and emphasized the importance of "at once introducing the proper means for the artificial propagation of codfish, with the view of restocking our exhausted bays and fishing grounds near the coast."[19] In 1889 the government appointed Norwegian Adolph Nielsen to the superintendent's position; four years later it created a Department of Fisheries, renamed the Department of Marine and Fisheries in 1898. Until the department's creation, Nielsen and the Fisheries Commission functioned in a quasi-departmental capacity.

By the late-19th century, many of the central problems of modern fisheries, and their proposed solutions, were taking shape. This was evident in the reports of the Newfoundland Fisheries Commission/Department of Marine and Fisheries. Superintendent Nielsen strongly advocated artificial propagation (or aquaculture) as a means of re-stocking Newfoundland's inshore waters.[20] His experiments with cod and lobster hatcheries in the 1890s were inconclusive, as were similar experiments in Scotland, Norway, and the United States. Enthusiasm for artificial propagation was complemented by concern for "reckless and injurious methods of fishing."[21] Accordingly, the Department of Marine and Fisheries advocated stricter enforcement of regulations on mesh size in cod traps to enable smaller fish to escape, thereby ensuring recruitment to the stock.

Early Development Schemes

As a result of the crisis, the government sought new approaches to economic development that might free Newfoundland from its historic dependence on the fishery. The search began in earnest in the 1880s when the government tried to stimulate the economy through public works and the encouragement of a local manufacturing sector. The main public works were the new St. John's drydock (completed in 1884) and the trans-Newfoundland railway, which was to provide access to the largely imaginary agricultural and mineral resources of the interior and the west coast. The railway, built between 1881 and 1898, and subsequent branch-line construction, diverted some labour from the fishery, but its main legacy was a crippling debt load that plagued successive governments.[22]

This development strategy, patterned after the Canadian and American examples, gave way to an approach that favoured the use of foreign capital to develop land-based natural resources. The Bell Island iron ore mine opened in 1895, followed by the Grand Falls newsprint mill in 1909, the Corner Brook newsprint mill in 1925, the Buchans lead, zinc, and copper mine in 1928, and the St. Lawrence fluorspar mine in 1933. Whereas fishery products accounted for 90% of the value of Newfoundland exports in the period 1891-95, by 1931-35 they had fallen to 30%.[23]

Meanwhile, the fishery's woes continued. Newfoundland's saltfish exports grew by 3.27% between 1898 and 1913, but in the same period Canada's grew by 3.45%, Norway's by 4.75%, and Iceland's by 10.58%.[24] The fishery rebounded dramatically during World War I, when the enemy naval presence in the North Sea precluded fishing by Newfoundland's competitors. During the war the average price for saltfish rose by 84%.[25] Accordingly, Newfoundland's fish exporters turned a blind eye to the industry's problems.

A general trade depression set in after the war, compounded in Newfoundland's case by the resumption of fishing in the North Sea and by rising prices for fishers' provisions, most of which were imported from North America. By 1921 fish prices were only half of what they had been in 1918. In 1920 the Minister of Marine and Fisheries, William Ford Coaker, introduced legislation to standardize the cure of saltfish and to regulate prices through a coordinated marketing effort among licensed exporters.[26] However, chaos in the Italian market following rapid devaluation of the lira, and the restriction of credit by Canadian banks in St. John's, led to panic and caused some merchants to sell below the set prices. Coaker's scheme was soon in a shambles, and in the spring of 1921 the "Coaker Regulations" were withdrawn.

International saltfish production continued to rise, reaching an all-time high of 316,874 metric tonnes in 1928.[27] Most of this was attributable to increased efforts by Norway and Iceland, especially the latter, which in 1928 overtook Newfoundland as the world's leading saltfish exporter. During the inter-war period Spain and Portugal revived their distant water fisheries, joining France on the Grand Banks. This, combined with Scandinavian competition, undermined the position of Newfoundland saltfish in its traditional Mediterranean markets, forcing a reorientation to Brazil and the Caribbean. After 1929 Newfoundland's share of the Brazilian market shrunk steadily in the face of Norwegian and Icelandic competition, fostering even greater dependence on the Caribbean, which accounted for 39.3% of Newfoundland's saltfish exports in the 1930s.[28]

The Saltfish Conundrum

Despite its overall decline as a source of export earnings, in terms of employment the fishery still had no rival, accounting for 36,886 jobs in 1935 out of a total labour force of 88,710.[29] The forestry sector provided 8,049 jobs, and the mining sector 1,821. However, pay in the fishery was low, and the main consumers were poor agricultural workers in countries that depended on one or two key staples, notably sugar and coffee. The economies of such countries were therefore susceptible to "violent changes in prosperity," which in turn were reflected in the prices that consumers were able to pay for saltfish.[30]

Given the low prices for salt cod, it stood to reason that unless the average annual production of 40-50 quintals per fisher could be significantly increased, the industry was destined to remain on a treadmill.[31] The development of the railway and improvements in mechanical refrigeration offered hope for new products—fresh and frozen fish—that could bring higher prices. Neighbouring Nova Scotia had been sending fresh fish, preserved in ice, by train to Quebec, Ontario, and New England since the late-19th century, and by refrigerated train cars after 1907.[32] Small amounts of frozen herring, packed in straw, had been exported from Newfoundland since the 1850s, primarily for use as bait by the New England bank fleet.[33] The opening of the trans-Newfoundland railway in 1898 and inauguration of a ferry service between Newfoundland and Nova Scotia prompted Newfoundland's first tentative forays into the fresh/frozen trade.[34]

Although the early history of Newfoundland's fresh/frozen trade is still little known, it appears to have been modest until World War I, when disruption of the North Sea fishery prompted British demand for frozen fish, which previously its own fleet had provided.[35] The Reid Newfoundland Company, owners of the railway and coastal steamer service, undertook the most ambitious program, erecting a central cold storage unit at St. John's and a number of smaller depots along the coast.[36] Its coastal steamers collected fish from these depots and brought it back to St. John's, where steamers transported it to England. British demand ceased abruptly when the North Sea fishery resumed after the war, and the Reids' operation came to an end.

Throughout the 1920s there appears to have been slow but steady activity in the new sector, centred on the west and southwest coasts, and producing mainly chilled salmon for British and American markets.[37] In 1928 Job Brothers, in partnership with the Hudson's Bay Company, equipped the S. S. *Blue Peter* with a brine freezer and began collecting salmon and blueberries from cold storage plants along the northeast coast and Labrador for export to Britain.[38] The British trawler fleet continued to meet the domestic market's requirements for fresh and frozen fish other than salmon.

The trend towards fresh/frozen production gained further momentum in 1930 when the Newfoundland government and the British Empire Marketing Board arranged for Harold Thompson, a Scottish government marine biologist, to conduct a survey of the Newfoundland fishery and to make recommendations for a marine research program.[39] In a far-sighted report, Thompson correctly predicted that the spread of cold storage facilities in Newfoundland must ultimately lead to a corresponding decline in saltfish production.[40] While recognizing that saltfish would remain the dominant product of the Newfoundland fishery "for many years to come," Thompson identified the need to effect "a gradual increase in the proportion of the amount of fish to be marketed... in a form—frozen or canned—which will make a wider appeal and on the whole fetch improved prices." He forecast eventual increased demand for frozen fish in the United States as fish stocks declined off American coasts and refrigeration facilities spread across the country. He further argued that to take advantage of these developments, and to maintain a viable frozen fish sector, Newfoundland would have to develop an offshore trawler fleet to ensure a year-round supply of fish that the highly seasonal inshore fishery could not provide.

Not surprisingly, given his profession, Thompson also stressed the importance of scientific research in a reorganized fishing industry. He recommended that the Newfoundland government establish a fisheries research bureau, and confidently proclaimed that scientists would be able "to anticipate to some extent the local prospects of success in the fishery of the ensuing season." In this, he was to be sadly mistaken. The only immediate result of Thompson's report was an agreement between the Newfoundland government and the British Empire Marketing Board to fund a five-year research program into the fishery. Thus in 1931 the Newfoundland Fishery Research Station was established at Bay Bulls.

Conditions in the fishery worsened during the Great Depression, and by the spring of 1933 a quarter of Newfoundland's 285,000 residents were receiving government relief.[41] In 1934 the British government rescued Newfoundland from fiscal oblivion by suspending democracy on the island and substituting an appointed Commission of Government, as recommended by the 1933 Newfoundland Royal Commission (named the "Amulree Commission" after its chair, Lord Amulree).

In its early approach to fishery management the Commission of Government largely followed the advice contained in the Amulree Commission report. It had emphasized that "recurring visitations of pauperism necessitating large payments for relief have always been a feature of the island's economic history and must continue to be so, while the fishery is conducted on its present basis and remains the mainstay of the country."[42] To break the cycle, two things were needed:

> first that the fishery should so far as possible be conducted on a rational and scientific basis and, secondly, that provision should be made either by individual fishermen, or by their employers, in good years to tide them over the bad seasons which sooner or later overtake them.[43]

The report singled out the credit system as the root cause of most of the industry's problems, and concluded that the fishery "can never prosper, or indeed be fully developed by Newfoundlanders, while the present system lasts."[44] Accordingly, the report recommended that the credit system should be dismantled gradually in order ultimately to place the fishery on a cash basis. Unlike Thompson, the commissioners felt that it made more sense to revive the fishery by improving the production and marketing of saltfish, with which Newfoundlanders were most familiar. They downplayed the potential of the fresh/frozen sector because of lack of freezing plants, and transportation problems inherent in Newfoundland's scattered population and the island's distance from the United Kingdom, where, they noted, "fresh cod is a cheap fish."[45]

Under Commission of Government, the former Departments of Marine and Fisheries and Agriculture and Mines were combined into a single Department of Natural Resources. Within the new department, the Newfoundland Fisheries Board was created in 1936 to oversee all aspects of the fishery, including marketing and scientific investigation. During its first years in office the commission stimulated shipbuilding by making funds available for schooner construction and became directly involved in the fishery by making gear available to "reputable" fishers on a hire-purchase basis.[46] It conducted a vigorous education program to improve the cure of saltfish and established bait depots at selected locations to ensure a steady bait supply. More importantly, it launched an attack on the credit system by encouraging the formation of co-operatives. Progress in this area was painfully slow, although there were some successes, notably the Bonne Bay Fisheries Co-operative Society, which by 1940 was the largest lobster co-operative in the world.[47] Finally, the Commission, through the Department of Rural Reconstruction, implemented a land settlement scheme that it hoped would draw thousands away from the fishery and give rise to a thriving agricultural sector.[48]

The land settlement scheme proved a failure, and the fishery, already reeling from the Great Depression, experienced more problems. The 1937 saltfish catch was significantly lower than in previous years; the Spanish Civil War closed off a still important market; and Brazil raised its import duties on fish.[49] By April 1939 there were nearly 85,000 people on government relief, the highest number since March 1934.

The Emergence of the Fresh/Frozen Trade

It was within this context that the government began to encourage greater development of the fresh/frozen sector. Since 1935, experimental voyages by government scientists aboard the steam trawler *Cape Agulhas*, and concurrent efforts by Crosbie and Company's steam trawler *Imperialist*, had confirmed the seemingly limitless potential of the Grand Banks for trawling.[50] Meanwhile, the electrification of rural Newfoundland had proceeded apace, making it possible to erect fish processing plants at more outport locations.[51] The government began to provide funding to companies wishing to establish plants, such as Bonavista Cold Storage Co. Ltd., incorporated in 1939.[52] But its boldest move was a 1939 agreement with the New York-based General Seafoods Corporation, the world's largest frozen seafood producer.[53] In exchange for a 22-year monopoly, General Seafoods agreed to erect a processing plant at La Poile and to purchase fresh fish from Newfoundland fishers. As part of the agreement, the corporation had received a commitment from the United States Customs Commission that it would not assess import duties on fish processed at the La Poile plant. However, this aroused so much opposition in New England that the United States Treasury removed the exemption, thus killing the deal.

The aborted agreement with General Seafoods was indicative of rapid growth in the fresh/frozen sector and of a shift in its market orientation from the United Kingdom to the United States. Exports of fresh and frozen cod fillets increased dramatically from 39,335 lb. in 1936 to 781,958 lb. in 1938, of which 96% went to the United States.[54] However, the outbreak of World War II disrupted the North Sea fishery once again, creating a strong demand within the United Kingdom for Newfoundland frozen fish. It quickly supplanted the United States as the main market, and held this position until the war ended. Fresh and frozen cod exports ballooned from 2,294,300 lb. worth $34,414 in 1939 to 35,176,430 lb. worth $5,864,038 in 1945. By the latter date there were 44 filleting plants and 18 freezing plants on the island.[55]

Saltfish production actually declined from 1,045,236 quintals in 1939 to a wartime low of 703,456 quintals in 1943, before rebounding to 955,217 quintals in 1945. Any shortfalls were more than offset by the highest prices since 1919; by the orderly wartime marketing of all saltfish through the Washington-based Combined Food Board, established in 1943; and by reduced numbers of fishers as thousands flocked to construction jobs on Canadian and American military bases. The number of males employed in the inshore fishery fell from 18,622 in 1939 to 13,724 in 1941. By 1945, however, employment was above pre-war levels, having reached 19,650. The wartime construction boom was over, and labour as always had returned to the employer of last resort.

In 1944 the Commission of Government announced its post-war economic rehabilitation program, one that "anticipated much that would happen in Newfoundland during the next two decades,"[56] especially in the fishery. Emboldened by the performance of the fresh/frozen sector, the government abandoned its promotion of fisheries co-operatives in favour of an approach that emphasized frozen fish production as a means of improving fishers' incomes and eliminating the dreaded credit system.[57] On a per-pound basis, prices were higher for frozen fish than for saltfish, and payments to fishers in the fresh/frozen sector were in cash.[58] The plan called for utilization of species other than cod; development of an offshore trawler fleet; centralized processing plants; cash payments to inshore fishers, and a combined wage/bonus system for trawler crews; fisheries-related vocational education; and enhanced scientific research.

The National Convention, which met from 1946 to 1948 to consider Newfoundland's constitutional future, fully endorsed the commission's vision of fishery development. After Newfoundland became Canada's tenth province in 1949, that vision merged with the Canadian model for east coast fisheries development that sought growth in the fresh/frozen sector based on sales to affluent urban markets in Canada and especially the United States.[59] The British North America Act assigned responsibility for "sea coast" fisheries to the federal government, and, except for opposition from the saltfish sector, Newfoundland's leaders embraced the new approach. This entailed neglect of Newfoundland's traditional fishery sectors, and abandonment of its established trading partners, the saltfish-consuming countries of the Mediterranean, the Caribbean, and South America. The Newfoundland government pumped $13 million into fish plant construction, and trawlers quickly displaced schooners in the bank fishery, with the last of the bank schooners sailing in 1953.[60] In 1955 the export value of frozen fish surpassed that of saltfish for the first time.[61] Also, in 1950 the number of Newfoundlanders making the annual migration to Labrador dipped below 1,000 for the first time in nearly 150 years.[62] Fewer than a dozen Newfoundland motor vessels participated in the seal hunt in any given year during the 1950s, as Norway and transplanted Norwegians operating from Nova Scotia took control of the industry. Though a few diehards persisted in both sealing and the Labrador fishery after the 1950s, their days were effectively over. In such a climate the old saltfish firms either converted to the fresh/frozen fishery or, more typically, withdrew from the fishery altogether to focus on retailing. Saltfish, which accounted for 87% of all Newfoundland groundfish landings in 1948, made up only 8% by 1972.[63]

The switch from saltfish to fresh/frozen entailed commercial exploitation of species other than cod, including redfish, plaice, flounder, haddock, and halibut, most of which were processed into fish sticks for the American market. Before the advent of mechanical refrigeration, cod dominated because its non-oily flesh was ideally suited to salting and drying, and it kept for months in the hottest climates. Fisheries scientists, now employed by the Fisheries Research Board of Canada, played a key role in locating these "under-utilized" stocks and in pioneering new vessel types to provide access to them. Thus longlining experiments at Bonavista in the period 1950-55 proved the existence of cod stocks 25 miles offshore in depths of 120-160 fathoms, where they were inaccessible to traditional inshore boats.[64] Longliners were decked vessels, 35-65 feet long, powered by diesel engines. They were so-named because they employed long lines with baited hooks, spread along the ocean floor, to catch fish. It was not unlike the old dory trawling technique, except that longliner crews employed mechanical haulers and shooting gear. By 1967 there were 174 longliners in Newfoundland, many of which had abandoned longlines in favour of more efficient gill nets.[65]

The new technology and exploitation of under-utilized species helped to bring about a long-awaited but modest increase in fishers' incomes. Average annual earnings per person rose from $744 in 1953 to $954 in 1963.[66] Fishers' families also benefitted from access to Canadian social programs after confederation, which assumed a role similar to that of credit and government relief in earlier times. The average family allowance in 1949 was $200 per year, a significant portion of family income.[67] However, the key social program was unemployment insurance, which initially was not extended to fishers. They finally became eligible to receive it in 1957. Total payments to fishers in the fiscal year 1957-58 amounted to $1,759,000; two years later that figure had almost doubled, and "UI" was well on its way to becoming a mainstay of the Newfoundland economy.[68]

Ironically, improved incomes and Canadian welfare state largesse acted as a magnet to pull people back into the fishery, the very opposite of what government was trying to achieve. Mechanization of the forest industry and the failure of the Newfoundland government's industrialization program accentuated the trend. The number of fishers increased from 16,820 in 1953 to 21,407 in 1963.[69] The wealth, however, was not evenly spread. In 1965 there were some 700 persons in the offshore fleet, but they accounted for about one-third of all fish landings.[70] In 1963 the Newfoundland Fisheries Commission blamed the persistence of low earnings in the inshore sector on outmoded boats, gear, and fishing techniques.[71] Salvation lay in reducing the number of fishers and improving the available technology.

Modernization

The Fishery Commission's assessment was symptomatic of the post-war approach to solving "the economic problem of the Newfoundland fishery."[72] Modernization became the battle-cry of federal social scientists—mainly economists—who began to wrestle with the problem. They felt that the fishery should be treated like any other Canadian resource industry, i. e. its primary objectives should be increased productivity and maximum profits. They claimed that this could only be achieved by industrializing the fishery and reducing the number of fishers.[73] The approach was predicated on widespread acceptance of the "tragedy of the commons" theory. In an influential article written for the Canadian Department of Fisheries in 1954, economist H. Scott Gordon argued that the fishery, as a common property resource (one in which there was complete freedom of access) would tend towards resource depletion as unrestricted fishing effort increased.[74] And, as the resource declined, so too would fishers' incomes. To prevent these twin evils, government would have to restrict access to the fishery.

Senior bureaucrats gradually embraced Gordon's ideas, which dovetailed nicely with modernization theory. Within the Department of Fisheries, this was reflected in the growing influence of economists at the expense of biologists and other fisheries scientists. Gordon's article foretold this struggle in its condescending tone towards biologists, and its depiction of them as well-intentioned but ill-equipped to deal with economic problems. Ironically, Gordon's theory was based on assumptions about human behaviour that were more properly the domain of anthropologists, sociologists, or psychologists.[75] Undaunted, the economic experts who subsequently descended upon outport Newfoundland produced some truly chilling social engineering documents. In a 1966 study of the St. Barbe coast fishery, W. A. Black recommended that "radio or television programs should be beamed to the coast in order to stimulate a desire for goods and services."[76] This should, he continued, "result in a more aggressive attitude towards prosecuting the fishery in order to obtain the desired goods through increased production and income." Clearly, no stone was to be left unturned.

On the policy level the new ideology took several forms, of which the Newfoundland Fisheries Household Resettlement Program of 1965 was the most extreme.[77] A joint federal-provincial initiative, it replaced the Newfoundland government's 1953 resettlement program, which provided financial assistance to individuals who wished to move to larger communities. The 1965 program greatly increased the amount of economic assistance available to those wishing to move (thereby making relocation more attractive), and steadily lowered the level of community consent that was needed before a move could take place.[78] It incorporated the federal government's industrial strategy of "growth pole development" for Atlantic Canada, which was directly tied to expansion of the offshore fleet and the processing sector. Borrowing from the work of economist François Perroux, government economists and planners envisioned several thriving communities (growth poles) that would stimulate economic activity in surrounding areas.[79] Between 1953 and 1974 an estimated 250 communities and 30,000-40,000 people were moved, constituting "one of the largest government-initiated mass population movements in Canadian history."[80] Some managed to find employment in fish plants or on offshore vessels, but for the most part the anticipated spin-off jobs never materialized, and unemployment in the so-called growth centres hovered around 20%.

International Competition and Regulation

Between 1965 and 1972, the heyday of the federal-provincial resettlement program, the number of fishers in Newfoundland declined from 21,701 to 15,961.[81] No doubt this was partly the result of resettlement, which stripped many inshore fishers of their ability to continue in the industry. But other factors contributed to unemployment in the fishery. The American market, though indeed capable of paying higher prices than the old saltfish markets, was prone to price reverses during periods of over-supply, or when competing products such as meat and poultry became cheaper. Both conditions applied in the early seventies. But the main problem in this period was the drastic decline of fish stocks because of overfishing by technologically-advanced international fleets.

Except during wartime, there had of course been an international presence on Newfoundland's fishing grounds since the commencement of European exploitation. After World War II, fishers from France, Spain, Portugal, the United States, and Canada returned to the

Grand Banks. They did so in greater numbers than ever before, and with new technology.[82] Gradually these traditional visitors were joined by fishing fleets from other countries, including the Soviet Union, Denmark (Faroes), Federal Republic of Germany, Iceland, Great Britain, Norway, Italy, and Poland. Factory freezer trawlers and mother ships appeared, enabling fish to be processed and even packaged at sea. By 1963 non-Canadian vessels accounted for over half of the total catch in Newfoundland waters.[83]

As early as 1951 Newfoundland trawler fishers attributed a scarcity of haddock on the Grand Bank to Spanish over-fishing.[84] Before the decade ended, haddock had disappeared as a commercially viable species on the Grand Bank. In 1959 scientists discovered new cod spawning grounds off the Labrador coast, and in no time international fleets added Labrador to their itinerary. The pressure was not confined to the offshore. Until 1964, when Canada declared a nine-mile contiguous zone adjacent to its three-mile territorial limit, international vessels fished up to three miles from the Newfoundland coast. Canada continued to recognize France's right, because of historical usage, to fish in the contiguous zone. Canada's declaration of a 12-mile territorial sea in 1970 led to the 1972 Canada-France Fisheries Agreement. France agreed to withdraw its factory freezer trawlers from the Gulf of St. Lawrence after 15 years. A limited number of trawlers from St. Pierre and Miquelon could continue to fish along the coasts of Newfoundland, Nova Scotia, and in the Gulf of St. Lawrence, in return for reciprocal rights for Canadian fishers along the coasts of the two French islands.

Largely as a result of intense fishing by international fleets, the total catch of cod in the northwest Atlantic rose from approximately 900,000 metric tonnes in the 1950s to a historic peak of nearly 2,000,000 metric tonnes in 1965.[85] This kind of effort could not be sustained, and catches declined steadily during the early 1970s, falling to less than 500,000 metric tonnes in 1977. Newfoundland fishery statistics reflect these trends. The province's catch reached what was then an all-time high of 484,743 metric tonnes in 1969, then fell almost without interruption to 249,605 metric tonnes in 1974. Between 1974 and 1976 the federal government paid out more than $140 million in subsidies (mainly inventory and income support) to prop up the east coast fishery.[86]

The decline of northwest Atlantic fish stocks in the early 1970s was proof that the existing regulatory regime had failed. The International Commission for the Northwest Atlantic Fisheries (ICNAF) had been founded in 1949 to promote scientific investigation into and preservation of the region's fish stocks, and was comprised of member states whose vessels fished in the region. Its attempts to conserve stocks through regulations on fishing gear gave way in 1970 to a quota regime for major fish stocks. ICNAF set a Total Allowable Catch (TAC) for individual species based on the concept of maximum sustainable yield (MSY)—the maximum level at which a stock could be fished and still retain the ability to replenish itself.[87] By the mid-seventies it was obvious that the TACs were too liberal, but even so, member states routinely ignored them.

Meanwhile the third Law of the Sea conference (LOS III), which began in 1973, was moving towards recognition of the right of coastal states to manage marine resources up to 200 miles from their coasts. In 1976, LOS III failed to produce a general agreement, prompting the Canadian government to announce that it would unilaterally implement a 200-mile fishery zone in 1977. The Canadian zone took effect on 1 January 1977, and with it came a new regulatory regime. Inside the 200-mile limit the Canadian Atlantic Fisheries Scientific Advisory Committee (CAFSAC) had quota jurisdiction, including responsibility for under-utilized species that Canada continued to make available to other countries. Since parts of the continental shelf (Flemish Cap and southern Grand Bank) lay outside the 200-mile limit, there was still a need for international co-operation in these areas. Thus in 1979 ICNAF was replaced by the Northwest Atlantic Fisheries Organization (NAFO). As part of the new regime, both CAFSAC and NAFO abandoned the MSY concept as the basis for setting TACs, opting instead for a concept known as $F_{0.1}$, a measure of fishing mortality designed to identify "approximately 20% of the exploitable biomass for harvest each year," thereby ensuring "a healthy and steadily growing stock and a TAC that increased proportionally in successive years."[88]

The 200-mile limit and the new regulatory regime were intended to conserve fish stocks and restore the economic health of communities dependent upon the fishery. In the years immediately following 1977 these goals appeared to be realized. Fish stocks recovered, prices and consumption rose, and both the inshore and offshore sectors expanded. By 1981 there were 28,587 registered fishers in Newfoundland, a 41% increase over 1977. There were also 9,415 fish plant workers (33% increase); 17,135 registered fishing vessels (23% increase); and the total catch was 498,721 metric tonnes (27% increase). Despite these numbers, the industry found itself in a crisis because of a combination of high debt loads and high interest rates, soft markets, and huge inventories of unsold product. Fishers' incomes were on their way back down, and the two largest processors—Fishery Products Ltd. and the Lake Group—were in serious financial difficulty. The situation was much the same in the other Atlantic provinces, so in January 1982 the federal government appointed a Task Force on Atlantic Fisheries.

The task force, chaired by Michael Kirby, was asked to recommend "how to achieve and maintain a viable Atlantic fishing industry, with due consideration for the overall economic and social development of the Atlantic provinces."[89] The mandate neatly encapsulated the essential problem that had plagued governments since before confederation, namely how to strike a balance between the fishery's role as a major employer and the need of workers in the industry to earn an adequate income. Put another way, was the fishery to be a "social fishery" propped up by government, or an "economic fishery" able to stand on its own?

The task force ascribed the crisis to the tragedy of the commons, claiming that too many people were competing for a diminishing resource. It was a simplistic analysis that ignored market problems and the catching capacities of post-World War II fishing vessels. Nevertheless, licensing, introduced in 1975 ostensibly to restrict entry to the cod fishery, had functioned merely as a form of registration. The provincial government, chastened by criticism of the resettlement program, and mindful of the voting clout of inshore fishers, had championed their cause and kept access as free as possible. The situation began to change only in 1981 when licensees were divided into full-time (bona fide) and part-time fishers. Part-timers were restricted to catching groundfish, were not to use large vessels and certain gear types, and could not qualify for most government subsidies. Still, the tradegy of the commons theory failed to recognize that there would still be competition for the resource, and the capital-intensive nature of that competition might be more significant than mere numbers of fishers.

The Kirby task force made three main recommendations: (1) that the industry "should be economically viable on an ongoing basis"; (2) that workers should receive "a reasonable income as a result of fishery-related activities, including fishery-related income transfer payments"; and (3) that Canadian fishers should have priority access to fish stocks within the 200-mile limit.[90] The first recommendation was intended to release the federal government from the burden of "periodic bailouts" that were endemic to the industry. Because the task force assigned top priority to this recommendation, its report has since been portrayed as having tipped the scales in favour of the capital-intensive offshore fishery. Hearkening back to Harold Thompson more than half a century earlier, the report foresaw a continuing role for trawlers in the industry because this was the only means of ensuring a year-round supply of fish for processing, and also because they were the only vessels capable of reaching certain offshore stocks. The second recommendation entailed a rejection of the so-called "rural-romantic school of social thought," which allegedly held that employment levels should be maintained regardless of income. But the task force also acknowledged that various federal social benefits constituted 19% of the average fisher's income, the lion's share of it unemployment insurance payments.

In the aftermath of the Kirby report, the federal and provincial governments, some of the chartered banks, and the major processors worked out a deal that resulted in the creation of two super companies—National Sea Products Ltd. in Nova Scotia and Fishery Products International Ltd. in Newfoundland. It was an inauspicious beginning for an industry that was supposed to become economically viable. Nevertheless, both companies soon became profitable because of healthy catches and price recovery, and the east coast offshore fleet continued its unbridled expansion, essentially replacing the pre-1977 capacity of the international fleet.[91] The Newfoundland fish catch reached an historic peak of 515,464 metric tonnes in 1986. Although the catch dropped slightly to 499,086 metric tonnes in 1987, the high prices obtained that year made it the best ever in terms of landed value ($292,697,000). This was reflected in employment levels, which reached 29,022 for fishers and 11,206 for plant workers.

Collapse

The turnaround in the industry's fortunes masked, temporarily at least, a fundamental problem. In the early 1980s Spain and Portugal began intensive fishing on two areas of the Grand Bank that lay outside the 200-mile limit. These were known as the nose and tail of the bank, and together comprised only 12,000 square nautical miles out of a total area of over 100,000 square nautical miles.[92] But they were crucial spawning and nursery areas for cod, redfish, caplin, turbot, and several flatfish species. Management responsibility for these so-called "straddling stocks" (because they straddled the 200-mile limit) fell to NAFO. In 1986 NAFO set a quota of 25,665 metric tonnes (all species) for European Union (EU) vessels on the nose and tail of the Grand Bank, yet the total reported EU catch that year was over 110,000 metric tonnes. These figures did not include unreported catches, catches of non-commercial species that were dumped at sea, and catches by non-NAFO fleets that were simply fronts for EU vessels operating under flags of convenience. For the remainder of the decade there continued to be wide discrepancies between NAFO quotas and reported catches. Since, as Pierre Trudeau once remarked, "fish swim," overfishing of straddling stocks had a negative effect on Canadian catches inside the 200-mile limit. Combined with over-expansion of the Canadian offshore sector in the same decade, and destructive fishing practices by all countries, the end result was that cod stocks off Newfoundland were under extreme pressure.

As early as 1983 inshore fishers in Newfoundland began to complain that the offshore fleets were catching too many fish, which they gauged by declining inshore catches and reduced fish size, the latter a sign that fish were not growing to maturity.[93] In 1987 the Newfoundland Inshore Fisheries Association (NIFA) commissioned a study of Department of Fisheries and Oceans (DFO) stock assessment techniques. Three Memorial University of Newfoundland bioligists concluded that DFO's estimates of stock size were too generous and that stocks were being overfished.[94] Two years later DFO scientists, using new modelling techniques and improved data, reached the same conclusion, admitting that their estimate of fishing mortality was too low, from which it followed that their predictions of future stock growth were overly optimistic. Accordingly, CAFSAC recommended a TAC of 125,000 metric tonnes for northern cod (the cod stocks extending from the northern half of the Grand Bank to mid-Labrador) in 1989, down from 266,000 metric tonnes annually from 1984 to 1988.

Notwithstanding this scientific advice, the Minister of Fisheries and Oceans, Thomas Siddon, set a TAC of 235,000 metric tonnes for the 1989 season. An outraged NIFA launched an unsuccessful court challenge in an attempt to lower the TAC. Politically, it was still unthinkable to reduce the TAC to the recommended level because it would have caused massive unemployment in Newfoundland fishing communities. However, the minister did create an independent panel to review scientific advice provided to the government since 1977 on norther cod and "to make recommendations regarding stock assessment methods and means with a view to better forecasting the size, growth potential and behavior [sic] of the stock in future."[95]

The panel, chaired by Dr. Leslie Harris, President and Vice-Chancellor of Memorial University of Newfoundland, submitted its report in February 1990. Broadly, its recommendations were that (1) the accepted level of fishing mortality should be lowered still further; (2) Canada should seek international permission to manage straddling stocks, and that failing such permission it should act unilaterally to manage them; (3) DFO should review its management structures in order to achieve a "more focused and coordinated approach to the management of the northern cod stocks"; (4) DFO scientists should take a more ecological approach in their research, instead of studying cod in isolation; (5) DFO scientists should examine the relationship between inshore and offshore cod populations; and (6) the federal government "should carefully reexamine its biological, ecological, and socio-economic goals in respect of the fisheries to ensure that they are clearly defined, internally consistent, and attainable." Although vaguely worded, this last point seemed to open the door to a reduction in the number of fishers.

Since the release of the Harris report there have been dramatic developments. In January 1990 the Minister of Fisheries and Oceans, Bernard Valcourt, announced a further reduction in the northern cod TAC to 197,000 metric tonnes, despite a CAFSAC recommendation of 125,000. As the Harris report had noted, a reduction to CAFSAC's recommended level "would precipitate social and economic repercussions of a particularly drastic nature."[96] Nevertheless, because there was so much over-capacity in plants, vessels, and the labour force, the minister's compromise triggered plant closures. On 7 May the minister announced a five-year $584 million Atlantic Fisheries Adjustment Program to assist those affected. He also declared a moratorium on new fishing licenses and prohibited some part-time fishers from catching groundfish. Although the northern cod TAC for 1991 was set at 185,000 metric tonnes, fishers managed to catch only 127,000, lending further credence to scientists' claims that they had underestimated fishing mortality in the 1980s. The 1992 TAC was also 185,000 metric tonnes, but ridiculously low catches early in the year confirmed that something was very wrong. On 24 February 1992 the Minister of Fisheries and Oceans, John C. Crosbie (who succeeded Valcourt in April 1991), lowered the TAC to 120,000 metric tonnes and shut down the offshore fishery.

With the industry reeling from reduced TACs and declining catches, the presence of EU fishing vessels on the nose and tail of the Grand Bank became more and more galling to Canadians, especially Newfoundlanders. The Canadian government, spearheaded by Crosbie, stepped up its efforts to halt international overfishing of straddling stocks. In a speech on 10 January 1992 to the Royal Institute of International Affairs Conference on International Boundaries, Crosbie drew attention to an increase in Spanish fishing vessels in the northwest Atlantic since 1990, when Namibia expelled them from its 200-mile zone because of overfishing.[97] He also noted the growth in the number of vessels operating under flags of convenience on the nose and tail of the Grand Bank. Canada, he continued, had been working through diplomatic channels since 1990 to achieve a resolution of the problem. Increasingly, however, it appeared that changes to international law were required in order to give greater force to the rights of coastal nations to control fishing of straddling stocks. The alternative, he warned, was unilateral action by coastal states. In June, Crosbie attended the United Nations Conference on Environment and Development where he continued to press the case for more effective conservation of high seas stocks.

Canada achieved some successes on the diplomatic front during 1992. On 28 May, Spain announced that it would withdraw 44 pair-trawlers from the nose and tail of the Grand Bank effective 1 June. This, however, still left a fleet of 34 Spanish pair-trawlers in the region. On 10 June an ad hoc Court of Arbitration handed down its decision on the Newfoundland-St. Pierre and Miquelon boundary dispute (caused by overlapping 200-mile limits). It gave France fishing rights in a corridor, 200 nautical miles wide, extending south from St. Pierre to the edge of its proposed 200-mile limit, and established a 24-mile boundary around the coast of St. Pierre and Miquelon. These allotments amounted to 18 % of the total area that France had sought in the arbitration. The decision was a major victory for Newfoundland after nearly 300 years of fisheries negotiations with France. In earlier times, Newfoundland diplomats who achieved similar victories were accorded hero status. Such was the gloom in the fishing industry during the summer of 1992 that Crosbie never received due credit for the accomplishment.

That gloom became even more pervasive on 2 July, a day that will go down as one of the darkest in Newfoundland's chequered history. At a press conference in a St. John's hotel, Crosbie announced what many had feared but none had dared to speak: northern cod stocks had fallen to their lowest recorded level, and to continue fishing might lead to their extinction. Accordingly there was to be a two-year moratorium on the catching of northern cod, throwing some 19,000 people out of work. Fishery workers would be retrained for jobs in other sectors of the economy, and as many as 4,500 older fishers would be offered early retirement packages, indications that the federal government intended to use the crisis to reduce permanently the number of inshore fishers, something governments had talked about for decades.

Fishers and plant workers received interim financial assistance of $225 per week until a more detailed, long-term aid package could be worked out. The $587-million Northern Cod Adjustment and Recovery Program (NCARP) increased the amount of financial assistance to a maximum of $406 per week for the duration of the moratorium (to 15 May 1994).[98] There was more to come. On 18 December at a Halifax press conference, Crosbie announced a reduction of 69,000 metric tonnes in the 1993 groundfish TAC for the Gulf of St. Lawrence, Sydney Bight (waters between Cape Breton and Newfoundland), and Scotian Shelf. Crosbie warned that the Atlantic Canadian fishery was embarking on a new course, and that it would employ "thousands less" in the processing sector, as well as fewer fishers.[99] The $191-million Atlantic Groundfish Adjustment Program (AGAP) followed in April 1993 to assist the 7,600 fishers and plant workers (mainly Nova Scotians, but some from southwestern Newfoundland) laid off as a result of the 18 December quota cuts. On 20 December 1993, new Fisheries and Oceans Minister Brian Tobin announced the 1994 Atlantic Groundfish Management Plan that completely shut down cod fisheries off the south coast of Newfoundland, in Sydney Bight, and the southern Gulf of St. Lawrence, and reduced flatfish quotas on the Scotian Shelf from 14,000 metric tonnes to 10,000. The minister estimated that these measures would bring to 35,000 the number of unemployed fishers and plant workers in Atlantic Canada in 1994.

When NCARP and AGAP expired on 15 May 1994, they were replaced by The Atlantic Groundfish Strategy (TAGS). TAGS was a $1.9 billion, five-year program designed like its predecessors to provide income assistance and "labour adjustment" to affected fishery workers in Quebec and the Atlantic provinces, the majority of them Newfoundlanders. It reduced the previous income assistance by 6%, lowering the weekly average from $300 to $282.[100] In order to receive assistance, individuals initially had to take educational upgrading or participate in community development projects, many of which were "green" projects in aquaculture and silviculture. Such retraining was supposed to help people "achieve labour market self-sufficiency through meaningful employment," or, in lay terms, to prepare them for jobs outside the fishery. Specifically, TAGS aimed for a 50% reduction in processing capacity after five years, which, together with other measures such as early retirement and buying back fishers' licenses, would eliminate an estimated 15,000 jobs.

Federal government projections that 26,500 eastern Canadian fishery workers would qualify for assistance under TAGS proved well off the mark: in July 1996 there were 37,000 recipients, of whom 25,500 were Newfoundlanders. While the fishery accounted for less than 4% of Newfoundland's gross domestic product, 25,500 was still a significant number in a provincial workforce of some 245,000. By now TAGS was nearly $500 million over budget, and the government responded by eliminating the program's job-training component. Meanwhile, the government had announced its vision of a "core fishery" that would provide employment for 13,250 professional fishers in Atlantic Canada. Its subsequent decisions to reduce unemployment benefits and to charge fees for wharf access and other services have caused unease among many inshore fishers, who see these changes as measures designed to turn the industry over to the offshore sector.

Conclusion

A mere 500 years after John Cabot's historic voyage, numerous marine species that once belonged to the northwest Atlantic ecosystem have become extinct.[101] Northern cod may or may not be the latest member of this dubious club. Certainly, there have been periods in the past when cod was scarce, possibly because of over-exploitation, or perhaps because of environmental factors such as lower water temperatures. However, there can be no denying that the collapse of nothern cod stocks is a local manifestation of the post-World War II technological onslaught against the world's natural resources.

Technology and greed, along with the failure of science, economics, politics, and regulation, have brought us to the brink of the ecological abyss. Today, turbot (also known as Greenland halibut) is the only commercially viable fish stock remaining on the Grand Bank. Canada's seizure of the Spanish trawler *Estai* in March 1995 for overfishing turbot just outside the 200-mile limit was in large part responsible for a subsequent United Nations agreement on management of straddling and migratory fish stocks. Ratification of the agreement is, however, a long way off. The "turbot war" and the continuing presence of international fishing vessels in and around Canadian waters are testament to the enduring international character of the northwest Atlantic fishery.

In the end, important questions remain. Will northern cod return? Will we learn anything from its collapse? And even if we do, will it be possible to apply those lessons?

The answers to these questions will largely determine the future shape of Newfoundland society, for cod appears set to rule even from beyond the grave.

Notes

1.   James E. Candow, *Of Men and Seals: A History of the Newfoundland Seal Hunt* (Ottawa: Environment Canada, 1989), p. 30.

2.   W. A. Black, "The Labrador Floater Codfishery," *Annals of the Association of American Geographers*, Vol. 50 (September 1960), p. 268.

3.   Michael Staveley, "Population Dynamics in Newfoundland: The Regional Patterns," in *The Peopling of Newfoundland: Essays in Historical Geography*, ed. John J. Mannion (St. John's: Memorial University of Newfoundland, 1977), p. 70.

4.   *Newfoundlander* (St. John's), 11 July 1844, p. 2.

5.   John J. Mannion, "Introduction," in *The Peopling of Newfoundland*, p. 12.

6.   Candow, *Of Men and Seals*, p. 43.

7.   W. G. Gosling, *Labrador: Its Discovery, Exploration, and Development* (London: Alston Rivers, 1910), p. 413.

8.   For contrasting views on the credit system, see *Merchant Credit and Labour Strategies in Historical Perspective*, ed. Rosemary Ommer (Fredericton: Acadiensis Press, 1990).

9.   Captain Charles Campbell, R. N., "The Lobster Factories on the West Coast," enclosure no. 8 in Campbell to Hamond, 10 October 1888, in D. W. Prowse, *A History of Newfoundland from the English, Colonial, and Foreign Records* (London: Macmillan, 1895), p. 556.

10.   Ruth Fulton Grant, *The Canadian Atlantic Fishery* (Toronto: Ryerson, 1934), pp. 9-10.

11.   Newfoundland. *Journal of the House of Assembly, 1890*, Appendix, pp. 94-95.

12.   Shannon Ryan, *Fish Out of Water: The Newfoundland Saltfish Trade 1814-1914* (St. John's: Breakwater, 1986), p. 40.

13.   John Proskie, *Report on the Newfoundland Fisheries: An Appraisal of the Labrador Fishing Industry* (Ottawa: Department of Fisheries, 1951), p. 15.

14.   Ian D. H. McDonald, *"To Each His Own": William Coaker and the Fishermen's Protective Union in Newfoundland Politics, 1908-1925*, ed. J. K. Hiller (St. John's: Memorial University of Newfoundland, Institute of Social and Economic Research, 1987), pp. 8-9.

15.   See *Census of Newfoundland and Labrador* for the respective years.

16.   David Alexander, "Newfoundland's Traditional Economy and Development to 1934," *Acadiensis*, Vol. 5, No. 2 (Spring 1976), p. 63.

17.   James Hiller, "The Railway and Local Politics in Newfoundland, 1870-1901," in *Newfoundland in the Nineteenth and Twentieth Centuries: essays in interpretation,* ed. James Hiller and Peter Neary (Toronto: University of Toronto Press, 1980), p. 134.

18.   *Encyclopedia of Newfoundland and Labrador* (St. John's: Newfoundland Book Publishers (1967) Ltd., 1985), s. v. "Fisheries Research"; M. Baker et al., "Adolph Nielson: Norwegian Influences on Newfoundland Fisheries in the Late 19th-Early 20th Century," *Newfoundland Quarterly*, Vol. 87, No. 2 (Spring 1992), p. 25; Keith W. Hewitt, "The Newfoundland Fishery and State Intervention in the Nineteenth Century: The Fisheries Commission, 1888-1893," *Newfoundland Studies*, Vol. 9, No. 1 (Spring 1993), pp. 58-80.

19.   *Journal of the House of Assembly, 1889*, Sessional Papers, p. 614.

20.   *Journal of the House of Assembly, 1889*, Sessional Papers, pp. 620-21.

21.   Newfoundland. *Annual Report of the Newfoundland Department of Fisheries, for the Year 1893* (St. John's: Department of Fisheries, 1893), p. 3.

22.   David Alexander, "The Economic History of a Country and a Province," *Canadian Forum*, Vol. 53, No. 638 (March 1974), p. 12.

23.   *Newfoundland: Economic, Diplomatic, and Strategic Studies*, ed. R. A. MacKay (Toronto: Oxford University Press, 1946, AMS Press edition, 1979), Appendix 3.

24. David Alexander, "Development and Dependence in Newfoundland 1880-1970," *Acadiensis*, Vol. 4, No. 1 (Autumn 1974), p. 14.

25. McDonald, *"To Each His Own,"* p. 73.

26. David Alexander, *The Decay of Trade: An Economic History of the Newfoundland Saltfish Trade, 1935-1965* (St. John's: Memorial University of Newfoundland, Institute of Social and Economic Research, 1977), p. 22.

27. G. M. Gerhardsen and L. P. D. Gertenbach, *Salted Cod and Related Species* (Washington: Food and Agriculture Organization, 1949), p. 120.

28. Gerhardsen and Gertenbach, *Salted Cod and Related Species*, p. 31.

29. MacKay, *Newfoundland*, Appendix 1.

30. Gerhardsen and Gertenbach, *Salted Cod and Related Species*, p. 10.

31. The average catch figure is from H. B. Mayo, "The Economic Problem of the Newfoundland Fisheries," *Canadian Journal of Economics and Political Science*, Vol. 17, No. 4 (November 1951), p. 489.

32. Grant, *The Canadian Atlantic Fishery*, p. 83.

33. *Encyclopedia of Newfoundland and Labrador,* Vol. 2, s. v. "Fish Plants,"; Charles L. Cutting, *Fish Saving: A History of Fish Processing from Ancient to Modern Times* (New York: Philosophical Library, 1956), p. 295.

34. W. G. Reeves, "Alexander's Conundrum Reconsidered: The American Dimension in Newfoundland Resource Development, 1898-1910," *Newfoundland Studies,* Vol. 5, No. 1 (Spring 1989), pp. 14-15.

35. *Canadian Fisherman* (Gardenvale, Québec), Vol. 3, No. 1 (January 1916), p. 25.

36. *Canadian Fisherman*, Vol. 5, No. 2 (February 1918), p. 596.

37. See *Western Star* (Corner Brook), 7 July 1926, p. 3, and 8 September 1926, p. 1..

38. Robert Brown Job, *John Job's Family* (St. John's: Telegram Printing Co., Ltd., 1954), pp. 63-64.

39. *Encyclopedia of Newfoundland and Labrador*, Vol. 2, s. v. "Fisheries Research."

40. Harold Thompson, "A Survey of the Fisheries of Newfoundland and Recommendations for a Scheme of Research," *Reports of the Newfoundland Fishery Research Commission*, Vol. 1, No. 1 (St. John's: Newfoundland Fishery Research Commission, 1930), p. 18.

41. Great Britain. *Newfoundland Royal Commission 1933: Report* (London: His Majesty's Stationery Office, 1934), p. 29.

42. *Newfoundland Royal Commission 1933*, p. 50.

43. *Newfoundland Royal Commission 1933*, p. 79.

44. *Newfoundland Royal Commission 1933*, p. 103.

45. *Newfoundland Royal Commission 1933*, pp. 119-20.

46. Newfoundland. *Department of Natural Resources 1934-1935 Annual Report* (St. John's: Department of Natural Resources, 1935), p. 2.

47. James E. Candow, "'They Done Alright': A History of the Mudge Family Fishery at Broom Point, Newfoundland, 1941-1975," *Microfiche Report Series No. 427* (Ottawa: Parks Canada, 1990), p. 29.

48. Peter Neary, *Newfoundland in the North Atlantic World, 1929-1949* (Montreal and Kingston: McGill-Queen's University Press, 1988), p. 106.

49. Neary, *Newfoundland in the North Atlantic World*, p. 85.

50. Harold Thompson, "A Biological and Economic Study of Cod (Gadus callarius, L.) in the Newfoundland Area including Labrador," *Research Bulletin No. 14 (Fisheries)* (St. John's: Department of Natural Resources, 1943), p. 113.

51. On the electrification of rural Newfoundland, see Melvin Baker, Robert D. Pitt, and Janet Miller Pitt, *The Illustrated History of Newfoundland Light & Power* (St. John's: Creative Publishers, 1990), p. 173 ff.

52. *Fishermen's Advocate* (St. John's), 21 June 1947, p. 17.

53. *Canadian Fisherman*, Vol. 26, No. 12 (December 1939), p. 20; and Vol. 27, No. 8 (August 1940), p. 22.

54. Newfoundland. *Report of the Newfoundland Fisheries Board and General Review of the Fisheries for the Years 1937 and 1938 with Statistical Survey for the Period 1930-1938* (St. John's: Department of Natural Resources, 1940), pp. 46-51.

55. Newfoundland. *Report of the Newfoundland Fisheries Board and General Review of the Fisheries for the Year 1945 with Statistical Survey* (St. John's: Department of Natural Resources, 1947), p. 14.
56. Neary, *Newfoundland in the North Atlantic World,* pp. 225-26.
57. Neary, *Newfoundland in the North Atlantic World*, pp. 250-51.
58. Newfoundland. *Report of the Newfoundland Fisheries Board and General Review of the Fisheries for the Years 1939 and 1940 with Statistical Survey* (St. John's: Department of Natural Resources, 1941), pp. 24-25.
59. Alexander, *The Decay of Trade,* pp. 4-16.
60. *Fishermen's Advocate,* 16 April 1954, p. 1.
61. E. Pazdzior, "The Fishing Industry," in *Newfoundland and Labrador: The First Fifteen Years of Confederation,* ed. Ian McAllister (St. John's: Dicks, 1966), p. 125.
62. A. Prince Dyke, "Population Distribution and Movement in Coastal Labrador, 1950-1966," M. A. thesis, McGill University, Montreal, 1968, p. 120.
63. Alexander, *The Decay of Trade*, p. 149.
64. Wilfred Templeman, "The Bonavista Longlining Experiment, 1950-1953," *Bulletin No. 109* (Ottawa: Fisheries Research Board of Canada, 1956); and, by the same author, "Marine Resources of Newfoundland," *Bulletin No. 154* (Ottawa: Fisheries Research Board of Canada, 1966), p. 2.
65. Eric Benedict Dunne, "Biological and Economic Aspects of the Newfoundland Cod Fisheries," M. A. thesis, Memorial University of Newfoundland, 1970, p. 86.
66. Pazdzior, "The Fishing Industry," p. 118.
67. Richard Gwynn, *Smallwood: The Unlikely Revolutionary* (Toronto: McClelland and Stewart, 1968), p. 24.
68. Pazdzior, "The Fishing Industry," p. 131.
69. Pazdzior, "The Fishing Industry," p. 118.
70. Templeman, "Marine Resources of Newfoundland," p. 7.
71. Newfoundland. *Newfoundland Fisheries Commission: Report No. 2* (St. John's: Government of Newfoundland and Labrador, 1963), p. 1.
72. The phrase is H. B. Mayo's. See Mayo, "The Economic Problem of the Newfoundland Fisheries," p. 482.
73. See J. Douglas House, "Canadian Fisheries Policies and Troubled Newfoundland Communities," in *A Question of Survival: The Fisheries and Newfoundland Society,* ed. Peter R. Sinclair (St. John's: Memorial University of Newfoundland, Institute of Social and Economic Research, 1988), p. 179.
74. H. Scott Gordon, "The Economic Theory of a Common Property Resource: The Fishery," *Journal of Political Economy,* Vol. 62 (1954), p. 131.
75. Ralph Matthews, "Federal Licencing Policies for the Atlantic Inshore Fishery and their Implementation in Newfoundland, 1973-1981," *Acadiensis,* Vol. 17, No. 2 (Spring 1988), p. 87.
76. W. A. Black, "Fishery Utilization St. Barbe Coast," *Agricultural Rehabilitation and Development Administration Study 1043* (Ottawa: Department of Mines and Technical Surveys, Geographical Branch, 1966), pp. 29-30.
77. Matthews, "Federal Licencing Policies," p. 93.
78. Noel Iverson and D. Ralph Matthews, *Communities in Decline: An Examination of Household Resettlement in Newfoundland* (St. John's: Memorial University of Newfoundland, 1968), p. 3.
79. Miriam Carol Wright, "'The Smile of Modernity': The State and the Modernization of the Canadian Atlantic Fishery 1945-1970," M. A. thesis, Queen's University, Kingston, 1990, p. 9.
80. Ralph Matthews, "The Outport Breakup," *Horizon Canada,* Vo. 9, No. 102 (April 1987), p. 2438.
81. Canada. *Annual Statistical Review of Canadian Fisheries 1955-1976* (Ottawa: Fisheries and Environment Canada, 1977), Vol. 9, p. 123. Unless otherwise indicated, all remaining statistics are from the *Annual Statistical Review of Canadian Fisheries* for the respective years.
82. *Fishermen's Advocate,* 1 June 1951, p. 1.
83. *Newfoundland Fisheries Commission: Report No. 2,* unpaginated (but see section entitled "Fish Resources").
84. *Fishermen's Advocate,* 11 January 1952, p. 1.

85. W. H. Lear, "Atlantic Cod," *Underwater World* (Ottawa: Department of Fisheries and Oceans, 1989), p. 6.

86. Matthews, "Federal Licencing Policies," p. 92.

87. R. D. S. Macdonald, "Canadian Fisheries Policy and the Development of Atlantic Coast Groundfisheries Management," in *Atlantic Fisheries and Coastal Communities: Fisheries Decision Making Cost Studies*, ed. Cynthia Lamson and Arthur J. Hanson (Halifax: Dalhousie Ocean Studies Programme, 1984), p. 28.

88. Northern Cod Review Panel. *Independent Review of the State of the Northern Cod Stock* (Ottawa: Department of Fisheries and Oceans, 1990), p. 9.

89. Canada. *Navigating Troubled Waters: A New Policy for the Atlantic Fisheries: Highlights and Recommendations* (Ottawa: Department of Supply and Services, 1982), p. 3.

90. *Navigating Troubled Waters*, p. 60.

91. Atlantic Provinces Economic Council, "The Atlantic Fishery in the 1990s: Background to Crisis," *Atlantic Report 25* (July 1990), p. 6.

92. "Notes for an Address by the Honourable John C. Crosbie Minister of Fisheries and Oceans and Minister for the Atlantic Canada Opportunities Agency to the Royal Institute of International Affairs Conference on International Boundaries Chatham House, London," 10 January 1992 (Ottawa: Department of Fisheries and Oceans, 1992). Press release.

93. Cabot Martin, *No Fish and Our Lives: Some Survival Notes for Newfoundland* (St. John's: Creative Publishers, 1992), p. 137.

94. Martin, *No Fish and Our Lives*, p. 162.

95. *Independent Review of the State of the Northern Cod Stock*, p. 11.

96. *Independent Review of the State of the Northern Cod Stock*, p. 136.

97. "Notes for an Address by the Honourable John C. Crosbie," p. 9.

98. *Globe and Mail* (Toronto), 20 April 1994, p. A4. In April the Transitional Fisheries Adjustment Program (TFAP) provided an additional $191 million to unemployed fishers and plant workers in eastern Canada, bringing total federal aid to nearly $800 million.

99. *Daily News* (Halifax), 19 December 1992, p. 4.

100. "The Atlantic Groundfish Strategy," *News Release/Communiqué NR-HQ-94-26E*, 19 April 1994.

101. For a full account, see Farley Mowat, *Sea of Slaughter* (New York: The Atlantic Monthly Press, 1984).

chapter
eleven

THE NEWFOUNDLAND FISHERY
RESEARCH COMMISSION, 1930-1934

Melvin Baker and Shannon Ryan

The story of Newfoundland's past efforts at scientific fishery research is little known, and existing literature has concentrated on fishery research associated with the crisis of the early 1990s. This literature has tended to focus on how federal fishery scientists have been unable to make more reasonable forecasts on the amount of codfish available for commercial harvesting, and on the social and economic impacts the cod moratorium has had on local fishers.[1] Thus, the impression has been created that Newfoundland ignored scientific fishery research in the past and was, therefore, a backward participant in the international codfish trade. This paper intends to correct this view by examining the research efforts that originated in the colony in the 1880s and culminated with the formation of the Newfoundland Fishery Research Commission in 1930.

Fishery Research prior to 1930

The first efforts to place fishery research in Newfoundland on a more formal, scientific basis occurred in the late 1880s. Because of poor catches earlier in the decade, the government appointed a legislative committee to examine the state of its fisheries and to ascertain what scientific work was being carried out by its British, Canadian, and Norwegian counterparts. The result was the appointment of a Fisheries Commission in 1888 and the hiring of Adolph Nielsen, a Norwegian fisheries official,[2] to run the commission. Nielsen established a codfish hatchery at Dildo, Trinity Bay, introduced a patented Norwegian barrel for preserving bait herring, published instructions for curing codfish and herring and for the manufacture of cod liver oil, and established lobster hatcheries. In the late 1890s the Newfoundland government lost interest in his efforts and in 1898 Nielsen became involved in developing modern shore station whaling in Newfoundland with Norwegian and local business persons; the work of the cod hatchery was allowed to lapse, and the building and its equipment were sold.[3]

In 1911, the government of Prime Minister Edward Morris, acting upon a suggestion by Harbour Grace merchant William Munn, hired an expert knowledgeable in the manufacture and grading of cod liver oil. Again the government turned to Newfoundland's main competitor, Norway, and appointed Mico Siemunsen to advise both merchants and fishers on preparing cod liver oil for export. Until 1914 when ill-health forced his return to Norway, Siemunsen also provided the government with extensive information on the operations of the Norwegian fisheries department and stressed the need to establish a fishery school in St. John's to instruct fishers in modern fishing techniques.[4] In 1914 the government hired Walter Duff, an inspector with the Scottish Board of Fisheries, to examine and report on local fisheries. Duff called for the hiring of skilled fishery officers to instruct fishers in curing and packing herring for export, recommended that fishers be provided with motor boats to travel further out into the bays to catch fish, and made suggestions for the catching of under-utilized fish such as turbot, haddock, smelts, hake, flat fish, and plaice.[5] Later that same year Dr. Johan Hjort, Director of the Norwegian Fisheries Board, visited St. John's and gave an illustrated lecture showing how Norway's fisheries had benefited from scientific research, and how their research vessel had located fishing banks off Norway's coast.

Also, in 1914 the government appointed a commission to examine local fisheries. The commission reported in 1915 and recommended further scientific investigation and more government regulation by the creation of a fish inspection board. The commissioners wrote that "some attempt should long ago have been made to investigate in an intelligent, comprehensive, and scientific way, the waters and fishing grounds contiguous" to Newfoundland and called on the government to provide the necessary financial and other arrangements for this work. The report noted that "we have practically no detailed knowledge of the ocean bottom round our coast, nor has there ever been any intelligent attempt to locate new fishing areas which unquestionably exist." The legislators echoed a popular view among the public that Newfoundland was not taking full advantage of the fisheries potential along its coasts, and that "not one-half of the fish producing capacity... has been reached." The report called for a greater interventionist approach by the Newfoundland government since "further extended development cannot safely be left to private initiative alone."

In 1916 the Newfoundland Board of Trade called for a reorganization of the Department of Marine and Fisheries that would see a "much larger share of the public monies... expended in scientific investigation and practical experiment." It recommended the establishment of a fishery school and the acquisition of a vessel "properly equipped for experimental fishing and for scientific research around our coasts." The board also wanted the government to compile compulsory fishery statistics on catches as was done in Norway.[6] Despite all recommendations, immediate government action on fishery research initiatives was delayed by the demands to meet Newfoundland's war commitments.

Post-war fishery reform centred on the efforts of William Coaker and his Fishermen's Protective Union. In the 1919 general election Coaker's Union Party in coalition with the Liberals won, and Coaker was appointed Minister of Marine and Fisheries. The new government issued a series of marketing regulations to govern the fisheries: minimum prices were set for each major market, and exporters were threatened with the loss of their licenses if they breached regulations. In 1920 the legislature passed several bills containing the various fisheries reforms the FPU had advocated. These were known collectively as the Coaker Regulations, which among other things regulated prices, shipping of fish to market, all aspects of catching, processing, culling, warehousing, and transportation of fish.[7] The object was to improve the quality of Newfoundland fish and thereby to eliminate buyers' complaints about its poor quality. The government also intended to create a bureau to undertake scientific research, to be financed by an export tax on fish.[8] The Coaker Regulations failed because some opposition politicians, who were also merchant-exporters, broke ranks rather than submit to government regulation.

The proposed scientific research bureau also fell victim to the failure of the Coaker Regulations. In 1921 Newfoundland sent James Davies, a government analyst,[9] to the inaugural meeting of the International Committee on Deep Sea Fisheries Investigations, a joint creation of the Canadian, American, and Newfoundland governments. The committee's purpose was to form a "permanent means of co-operation between these countries in investigations between these countries, both those that are in progress and also those that may be undertaken in the future."[10] In the department's annual report for that year, Coaker observed that it was "humiliating that Mr. Davies representing the oldest fisheries in the New World was not possessed of any information of a scientific or hydrographic nature.... and that all the recommendations concerning these matters which have been put forward during the past 10 years have been ignored."[11] As Newfoundland's public debt grew during the 1920s, fuelled in part by efforts to repay public loans raised to finance the war effort, and in part to pay for the annual operating deficits associated with nationalization of the Newfoundland railway in 1923, local politicians were reluctant to make the necessary expenditures for fishery research. In fact, of the $50,000,000 the government raised in loans between 1918 and 1933, only about $1,000,000 was spent on the fisheries.[12]

Despite Coaker's prediction in 1921 that "Newfoundland will now be more in evidence generally as far as the main fishery problems are concerned,"[13] the colony remained on the outside of research by the International Committe on Deep Sea Fisheries Investigations Committee and did not send a representative to its meetings until 1926.[14]

The need for scientific research was emphasized in 1923 when Newfoundland participated in a Canadian research visit to its coastal waters. Archibald Huntsman[15] of the Biological Board of Canada conducted research on Newfoundland's west and southwest coasts for the International Committee on Deep Sea Fisheries Investigations Committee, examining the tides, temperatures, plankton, and other conditions affecting cod and other fish. Newfoundland's representative on the research vessel was Alan Gardiner of the British Ministry of Agriculture and Fisheries. In September, Huntsman and Gardiner addressed the Board of Trade stressing the importance of knowing where and when cod could be caught and urging Newfoundland to commence the scientific study of all cod stocks.[16] The problem remained a financial one, and Newfoundland's only contribution to the international committee was an attempt to maintain better annual codfish catch statistics. As the British Trade Commissioner to Canada and Newfoundland observed in his annual report on Newfoundland for 1925,

> the need for the employment of an expert to follow scientific developments in fishery matters abroad and to apply suitable experience to Newfoundland has frequently been suggested by ministers of the government and others. The funds for such an appointment and for the establishment of the necessary biological laboratory had not as yet been provided.[17]

The founding of Memorial University College at St. John's in 1925 was a major influence in stimulating public interest in scientific fishery research. College president John Lewis Paton sought to promote greater public awareness and study of Newfoundland's natural resources in general and fish in particular. Soon after assuming the presidency, he outlined Newfoundland's approach to fishery research:

> No pisciculture—no stations for marine biology. No understanding of life-story of the cod. Troubled with bait—no study of the question. No guidance. All rule of thumb. As it was in the beginning so it is now. Same as to drying and curing of fish—they tell me Norwegians are far ahead of us, but we go on in the same old rut.[18]

Although biology was not part of the college's curriculum in its first year, Paton made inquiries among English academics for advice on the nature of the program that should be established. He asked F. E. Weiss of the University of Manchester for help in recruiting a suitable candidate to teach biology. While Weiss was making arrangements to interview candidates,[19] another applicant appeared. George Sleggs, who had been seeking a position in Canada, wrote McGill University on 2 April 1926 for a position there or elsewhere. William Blackall, the Church of England Superintendent of Education and a member of the Board of Trustees of Memorial University College, was in Montreal in early April and, after being notified of Sleggs' availability, immediately wrote him. Since 1925 Sleggs had been working on a doctoral degree at the Scripps Institution of Oceanography of the University of California. He had previously conducted fishery research for the British government, including work in 1920 on the distribution of plaice and sole in the Irish Sea. On 10 May 1926, Sleggs accepted Paton's offer of appointment; Paton also arranged for Sleggs to be seconded during the summers to the Newfoundland Department of Marine and Fisheries to conduct local fishery research and secured its support to help finance a biology laboratory at the college.[20]

During July 1926, the North American Committee on Fishery Investigations (formerly the International Committee on Deep Sea Fisheries Investigations) met in St. John's and expressed its satisfaction with the recent work undertaken by Newfoundland in the collection of codfish catch statistics. The committee's presence prompted the Newfoundland government to take some action and in August 1926 placed a vessel at Sleggs' disposal. The vessel examined temperature distribution and carried out drift bottle experiments along the island's eastern coast from St. John's to Bonavista Bay.[21] Besides studying the migratory patterns of cod in Newfoundland waters, Sleggs gave public lectures on the preservation of fish by salting and smoking.[22] In a lecture to the Rotary Club later in the year, he outlined the research work that needed to be undertaken. Noting that the "fishing industry... has not been in a very flourishing condition," Sleggs called for improved methods of preparing fish for markets.[23]

The Report of the Imperial Economic Committee on trade between Great Britain and its overseas dominions and colonies was issued in 1927. The committee's report on fish stressed Britain's need for more fresh fish, since its own industry could not meet market demands.[24] However, the Newfoundland fish trade was dominated by salt codfish exports, although some companies had made efforts to diversify. One, Job Brothers and Company, was active in exporting fresh-frozen salmon to Britain.[25] The report also noted that Newfoundland's cod liver oil was priced too high compared with its Norwegian counterpart and that greater care was necessary to ensure that only the "clear colourless oil which is drawn off at the beginning of the steaming is placed on the market as medicinal oil." The committee recommended that the Empire Marketing Board conduct investigations into the preparation and medicinal value of cod liver oil in Newfoundland.[26]

After the report's release, Sir Halford McKinder, president of the Imperial Economic Committee for the British government, visited St. John's. Addressing the Board of Trade on 9 September 1927, McKinder said that the British government was making one million pounds available annually for the marketing of foodstuffs produced in its overseas dominions and sold in the United Kingdom. He emphasized that there was no future for dried and salted codfish, except in the poorer economies of the world, and that Britain and other wealthy countries wanted fresh fish. Concerning the need for local fishery research, he suggested that Newfoundland make arrangements with Canada for co-operative research that could be undertaken in a fishery research station that the Empire Marketing Board was prepared to assist Canada to establish at one of its eastern ports. McKinder told President Paton of Memorial University College that any biological work done in Newfoundland should be done under the general direction of A.G. Huntsman and other Canadian scientists.[27]

In early 1929, President Paton prepared a proposal that would further enhance the study of marine biology at Memorial University College. Having failed to secure financial assistance of $100,000 from the Rockefeller Foundation to endow a chair in biology at the college, and believing that Newfoundland could not afford its own independent fishery research laboratory, Paton prepared an application to the Empire Marketing Board and wrote Huntsman on 6 March 1929 for his support. The proposed funding would also be used to set up a research laboratory that could carry out work assigned to the college by Huntsman and to demonstrate to local people techniques in the curing and canning of fish. In the draft proposal he prepared for the board, Paton was very critical of the conservatism inherent in the Newfoundland fishery.

> The plain facts of the case are as follows: Those engaged in the fish business of Newfoundland—merchant, planter and fisherman alike—are, in their conservatism, persisting in putting new wine into old bottles. With little (and indeed frequently with no) modification they are following the methods of drying, curing and marketing the fish that have been inherited as it were from past centuries. Tastes change, economical and geographical conditions change, science throws her light on many things that were to our forefathers darkness, this world of to-day cannot work and live as did the world of three or four hundred years ago.... Greater knowledge of her fisheries is necessary, new methods, new markets are imperative. There can be no doubt that from the harvest that lies at her feet, Newfoundland could be prosperous; the potential value of her fisheries is surely immense in food products of many kinds, medicines, cattle feeds, fertilizers, oils, fats, skins, glues, etc., but until research comes to the rescue much or must run to waste or remain latent.[28]

Paton also expressed his frustration with local politicians who paid only token attention to improving local fisheries, and with the general public's scepticism towards scientific research. He wrote Huntsman that politicians could not be depended upon to provide and sustain the funding required: "Directly 'Brother Ass' gets up in the House of Legislature, and asks—'What's the good of paying a man to be chucking bottles into the sea?'—their assistance collapses," a reference to Sleggs' 1926 drift bottle work. In 1928 Paton had attempted to convince the fisheries department to convene a conference of business representatives interested in lobster canning on the west coast. As he told Huntsman, he "offered any help that the college might be able to provide through its biological and chemistry staff. But I have heard no more. This is the sort of dead end that we are up against if we look to the government to take the initiative."[29]

The college proposal included the rental of a research vessel each summer, oceanographic equipment, research laboratory equipment, the stocking of a library with standard works in marine biology and fisheries, and the erection of a fisheries research building attached to the college.[30] As Paton envisaged it, the college's research program would be in cooperation with and under the general direction of the new Canadian fishery research station at Halifax, and would employ recent college graduates who had continued their studies at Acadia and Dalhousie Universities. A joint research station, as McKinder had suggested in 1927 and which Paton strongly supported, was not to be; Newfoundland proved reluctant to commit itself financially and, in any case, the Canadians were evidently not prepared to devote part of their resources to Newfoundland's research needs.[31]

In July 1929, biochemist Dr. Jack Drummond of the University College of London, visited Newfoundland on behalf of the Empire Marketing Board to examine local oil refineries and their products. In an address to the St. John's Rotary Club, he stressed the importance of science to industry and said that research was what Newfoundland needed for its fishing industry. He observed that the oil he had examined was of good quality, but that there was considerable room for improvement.[32] During his 1929 visit, Drummond apparently discussed with Prime Minister Sir Richard Squires the necessity for greater fishery research, and gave him a proposal from the Empire Marketing Board for this purpose. In late 1929, Squires attended the Imperial Conference in London where agreement was reached on a proposal that would see both the board and the Newfoundland government provide an annual sum of £5,000 to establish and maintain a fishery research program for a five-year period. It would also provide a maximum of £5,000 in capital for the joint research scheme. The board also agreed to pay half the cost for a marine biologist to visit Newfoundland to investigate how a research scheme could be carried out, and to suggest estimates for it.[33]

On the recommendation of Jack Drummond, the Empire Marketing Board selected Dr. Harold Thompson, a biologist with the Scottish Fishery Board, to undertake the proposed survey. A First World War veteran, the Scottish-born Thompson had received his bachelor of science degree in zoology and chemistry from Aberdeen University in 1920 and his doctoral degree in 1925. From 1922 he had been a marine biologist with the Fisheries Board of Scotland and was a specialist in North Sea haddock.[34] Thompson carried out his Newfoundland survey between 15 July and 25 October 1930.

The Commission and the Bay Bulls Research Laboratory

On 21 August 1930 the Squires government appointed a Fishery Research Commission consisting of Clyde Lake (Minister of Marine and Fisheries), Prime Minister Squires, John Paton, James Davies (Newfoundland's Acting High Commissioner in London, who was appointed secretary of the commission), Sir William Coaker, Frederick Alderdice (Leader of the Opposition), and Leonard Outerbridge, of Harvey and Company. The Commission held its first meeting that same day and met with Thompson who submitted an interim progress report. Any successful research program, he wrote, required a well-equipped laboratory with research materials to be acquired from two sources: the first from a research vessel, and the second from fishers who would be trained by fishery officers. The laboratory should be close to government facilities and services in St. John's, so he suggested a fish plant at Bay Bulls. The plant was owned by Harvey and Company which promised to make part of the facility available for a nominal rent and to share their fishmeal, smoke house, and cold storage departments. This plant was situated "on a deep and sheltered waterway, open all the year round, and capable of developing into a busy modern fishing centre." As for a vessel, a trawler would be needed and the only one available in Newfoundland belonged to Harvey and Company. The firm agreed to make this vessel available to the laboratory for six weeks each summer. Thompson stressed that Newfoundland's research should "dovetail into and be correlated with similar work in Canada and the United States."[35]

Before returning to Britain, Thompson met with the Fishery Research Commission again on 20 October 1930, to present estimates of the capital and annual maintenance accounts required for the research scheme, which had to be approved by the Empire Marketing Board. For capital expenditures of approximately $50,000, about $31,750 would be required to equip the laboratory, while $3,700 would be needed for the installation of scientific gear on the trawler. He recommended that the balance be retained for emergencies and further installations. For salaries, rental space, and the trawler, another $50,000 would be needed. The commissioners told Thompson to inform the Empire Marketing Board of their wish that the proposed research scheme should begin as soon as possible and no later than 1 May 1931.[36] In 1931, the board accepted Thompson's final report, which contained the main recommendations he had made earlier to the Fishery Research Commission. There were also suggestions to improve the cure of salt codfish, to market a larger portion of fishery products in the canned, smoked, or frozen state, and calls for investigations of the life-histories of cod, salmon, herring, and squid.[37]

Thompson accepted the directorship of the Newfoundland program, agreeing to a five-year contract, the life of the agreement between Newfoundland and the Empire Marketing Board.[38] Faced with the decision by Harvey and Company to sell their vessel the *S.S. Cape Agulhas*, the commission recommended that the government purchase the vessel and lease it to the laboratory. The government was initially reluctant to do so, but, when Thompson suggested that the work of the laboratory would otherwise be hampered for 1931, the government relented.[39]

The fishery research program established by Thompson in 1931 was practical in nature and had three main objectives. The first was to survey Newfoundland's fishery resources that were being actively developed or capable of being exploited. This work included studying the life-histories of the principal fishes in local waters by examining fluctuations in their numbers and movements. It also involved the compilation of statistics demonstrating the "maximum, minimum and normal densities of numbers of the stocks, these to serve in future as a guide to the existing trade and to possible new enterprises, and as part-basis for future protective or other legislation."[40] The second objective was to examine existing methods of fish processing and to suggest improvements. The third was to find new ways of utilizing waste products of the fishery.[41] Specific research carried out by the commission included the influence of currents upon the two different types of cod (that "born in and 'acclimatised' to cold and relatively warm-water conditions respectively") off Newfoundland, improved methods to secure more oil from cod livers, the canning of fish products, and general public educational programs.[42] Thompson and his staff actively participated in the North American Council (formerly Committee) on Fishery Investigations. At a meeting in Ottawa in 1931, Thompson explained the work of the commission, and subsequently became a regular participant at meetings and in its general research programs.[43] The laboratory took on the task of collating all hydrographic data collected by Canada, the United States, Newfoundland, and France in the northwest Atlantic. Thompson's work in the 1930s was supplemented by research work carried out by French marine scientists and by Sleggs of Memorial University College, who in 1933 under the auspices of the Fishery Commission published a study of the caplin in Newfoundland waters.[44]

The establishment of a fishery research program in the midst of a world depression, which saw demand for Newfoundland's fish exports drop drastically, came at a time when Newfoundland's financial problems were reaching crisis proportions.[45] Since 1920, the government had operated its annual budget on a deficit basis. In 1931, the Canadian banks refused to do any further business with the country because of its financial condition. In 1931 the national debt stood at approximately $100 million, with interest payments consuming 65% of current revenues. Prime Minister Squires appealed to the Canadian banks to reverse their decision, because he needed a loan to help the government meet its debt interest payments due on 30 June 1931. Eventually, Squires' personal intervention with Canadian Prime Minister R. B. Bennett resulted in the latter persuading the banks to lend Newfoundland an additional $2 million to meet its June 30 deadline. The loan, however, came with several preconditions that Squires had no alternative but to accept. The government agreed to a policy of tariff revision and retrenchment in public expenditures. It would also ask the British government to appoint a financial advisor to advise on the reorganization and coordination of the various public services and to make recommendations with a view to strengthen the dominion's finances. In August 1931, the Squires' government made arrangements with the British government for the appointment of Sir Percy Thompson, deputy chair of the Board of Inland Revenue, as financial advisor.

Further retrenchment proved publicly unpopular for Squires, who lost a general election in 1932 to Frederick Alderdice of the United Newfoundland Party. With the threat of Newfoundland's default, Great Britain offered further financial assistance in return for Alderdice's accepting the appointment of a Royal Commission of enquiry into Newfoundland's financial condition and its future financial prospects. This commission's main recommendation was that parliamentary democracy be suspended until Newfoundland was financially self-supporting; in the meantime, it should be governed by a commission of six appointees with equal representation from Britain and Newfoundland. The Commission of Government assumed office on 16 February 1934 with Alderdice serving as one of the Newfoundland representatives.[46]

The Fishery Research Commission operated against this background. On 19 September 1931, the commission had received a telegram from the Empire Marketing Board asking it to reduce expenditures on the joint research scheme. The commission was told not to approve any capital expenditures, to reduce staff salaries by 10%, and not to use the trawler for the 1932 summer research period. The commissioners decided upon a maximum $5,000 reduction in capital expenditures, and to make no additional staff appointments. They refused to cancel the trawler research program for 1932, but informed the board that they would seek a reduction of the vessel's rental charge. Regarding the staff salaries, the commission decided to allow Thompson to make a representation to the board.[47] Thompson warned that "any considerable whittling down of the scale of operations (such as would occur were the use of the trawler withdrawn) would result in the almost complete cessation of the existing public interest in the scheme." The British members of his staff all had five-year contracts and they would have to agree to it, which he believed would be doubtful.

At the 9 January 1932 meeting of the Fishery Research Commission, the future of the research scheme was debated with Sir Percy Thompson present in an advisory role on Newfoundland finances. He stated that the research station was a "vital public service and one which must be budgeted for in the annual vote to the Department of Marine and Fisheries." However, to keep the scheme alive, the commission decided to guarantee the Marketing Board that a 10% cut would be made on its annual maintenance account.[48] This cut was achieved by the staff agreeing to reductions in their salaries.[49] For the springs of both 1933 and 1934 the government also used the *Cape Agulhas* for commercial trawling experiments and for employment purposes in general.[50]

Financial considerations continued to trouble the commission, for the following year the British government abolished the Empire Marketing Board effective 1 October 1933. The British government wanted to reduce its financial commitment to the research program, but members of the Fishery Commission were determined to hold Britain to its financial obligations until 31 March 1936.[51] The research program itself received favourable comments by witnesses who testified in camera before hearings of the Royal Commission on Newfoundland's political future. Raymond Gushue of the Newfoundland Board of Trade informed the commissioners that the laboratory did "very good work,"[52] a view endorsed by the commissioners themselves in their report to the British and Newfoundland governments. The laboratory, they wrote,

> has already succeeded in doing admirable work and is recognized both in Canada and the United States as a leading authority on the deep sea fishery of the Western North Atlantic. Its potential importance to the industry can hardly be exaggerated. Scientific investigation cannot, however, give full results so long as the administrative services of the government are inefficient and the industry itself remains unorganized.[53]

Government inefficiency was also a major aspect of Harold Thompson's evidence in camera before the commissioners on 3 April 1933. He considered the Department of Marine and Fisheries to be inefficient and more concerned with "job farming" instead of encouraging local fisheries. The change of government in 1932, he noted, was a case in point, where the captain and crew of the *Cape Agulhas* were dismissed by the new government despite their familiarity with the vessel and the work of the scientists. The Fishery Research Commission also came in for scathing criticism. He told them that "we have a Research Commission to make appointments, for making the major decisions, such as expenditures. It is very difficult to get them to agree. They contend with leaders of all parties—Squires and Alderdice—who do not meet at the same table."[54] In 1934 the Commission of Government abolished the Fishery Research Commission and placed the Bay Bulls laboratory under the management of the newly-established Department of Natural Resources.

Conclusion

On the expiry of the five-year lease with Harvey and Company in 1934, the government purchased the premises containing the laboratory for $25,000.[55] In 1935 the future work of the Bay Bulls laboratory was placed in the hands of a government commission under the auspices of Supreme Court Judge James Kent appointed to examine local fisheries in general. The Commissioner for Natural Resources, Sir John Hope Simpson, asked the Kent Commission for a recommendation on whether the Bay Bulls laboratory should be continued, and if so, whether it should be moved to St. John's. Kent's reply on 10 October 1936 was a strong endorsement of the laboratory and its research, and he saw no advantage in moving it to St. John's. He recommended that it be vigorously supported by the government.[56]

When Thompson's five-year contract expired in 1936, the government offered him only a one-year extension. Thompson took this as an indication that it no longer wished to retain his services, and he started a job search that led to his appointment as Director of Fisheries Research in Australia,[57] where he subsequently wrote two studies on the biology of haddock and codfish in Newfoundland waters. In his last annual report for the laboratory in 1935, Thompson reflected on his tenure in Newfoundland. He noted that on the practical research side over the five years of operation, "more fishery developmental projects broke down owing to lack of provision in the matter of ensuring the necessary supplies of raw material, than owing to imperfections of processing technique." It was necessary to emphasize this point, he wrote, because there was a "popular tendency, not shared by those who have real experience of the fish trade or of fisheries' science, to consider that great forward steps could be made by the expenditure of large capital sums and the application of what are vaguely called formulae."[58]

On 19 April 1937 the Bay Bulls facility was destroyed by fire, resulting in the loss of a considerable amount of scientific records, equipment, and library.[60] Despite this setback, the government decided to continue the scientific work by combining fishery laboratory work with the services provided for public health and the government analyst. In 1940 fishery research was continued as part of the Newfoundland Government Laboratory in a new building opened at St. John's.[61]

By 1940 marine fishery research had become strongly entrenched and Newfoundland thereafter relied on its own well-trained people to fill the positions necessary for this work. Its establishment owed much to the determination of President Paton of Memorial University College. The college continued to prepare students for marine biology and these students became important figures in fishery research in post-confederation Newfoundland. Paton and his fellow college professors and students strongly believed, as George Whiteley wrote in 1932, that the fishery laboratory may "suggest changes in different cures, perfect new ones, or advise that certain things should or should not be done." However, "if a better article is to be offered to world markets," he continued, "the burden and responsibility rests on the trade and on every fisherman engaged in it."[62] How Newfoundland has adapted to technological and scientific change in the fishery during the 20th century is still little known and worthy of further study.

The Fishery Research Commission was a major attempt by the Newfoundland government to develop an institution to study the fisheries. It certainly made a gallant effort in this direction and through the Bay Bulls laboratory brought knowledge of the fisheries to the forefront both locally and internationally.[63] However, the Fishery Research Commission was established during a time of local and international chaos and any potential long-term benefits of its work could not counteract the short-term problem of selling fish. Nevertheless, the Fishery Research Commission was a watershed in Newfoundland's efforts to understand its marine resources.

Notes

1. See, for instance, Alan Christopher Finlayson, *Fishing for Truth: A Sociological Analysis of Northern Cod Stock Assessments from 1977-1990* (St. John's: Institute of Social and Economic Research, 1994); Manuel Do Carmo Gomes, *Predictions Under Uncertainty: Fish Assemblages and Food Webs on the Grand Banks of Newfoundland* (St. John's: Institute of Social and Economic Research, 1994); David Ralph Matthews, *Controlling Common Property: Regulating Canada's East Coast Fishery* (Toronto: University of Toronto Press, 1993); and R. Andersen, "The Need for Human Sciences Research in Atlantic Coast Fisheries," *Journal of the Fisheries Board of Canada*, vol. 35, no. 7 (1978), pp. 1031-49.

2. Melvin Baker and Shannon Ryan, "Adolph Nielsen," *Dictionary of Canadian Biography*, vol. 13 (Toronto: University of Toronto Press, 1994), pp. 768-69.

3. M. Baker, A.B. Dickinson, and C.W. Sanger, "Adolph Nielsen: Norwegian Influence on Newfoundland Fisheries in the Late 19th-Early 20th Century," *Newfoundland Quarterly*, vol. 87, no. 2 (Spring 1992), pp. 25-32, 35. See also Keith Hewitt, "The Newfoundland Fishery and State Intervention in the Nineteenth Century: The Fisheries Commission, 1888-1893," *Newfoundland Studies*, vol. 9, no. 1 (Spring 1993), pp. 58-80; Provincial Archives of Newfoundland and Labrador (PANL), GN2/5A, Special File of the Colonial Secretary's Office, file 32-k, W.M. Macfarlane to A.W. Piccott, 29 April 1911.

4.   Newfoundland, *Annual Report of the Department of Marine and Fisheries for 1911*, pp. xxxiii-xliv, and for *1912*, pp. 55-64. His view was also supported by the Royal Dominions' Commission, which visited St. John's in July 1914 as part of its survey of the natural resources of the British Empire. See Newfoundland, *Annual Report of the Department of Marine and Fisheries for the Year 1916*, p. 28; and paper on cod liver oil read by W.A. Munn before the Royal Dominions Commission, July 1914, published in the *Daily News*, 20, 21, 23 November 1914.

5.   PANL, GN2/5A, file 251, Governor W.E. Davidson to J.R. Bennett, 24 July 1914; and Walter Duff, *The Fisheries of Newfoundland: Lecture delivered in St. John's, Newfoundland* by Mr. Walter Duff of the Fishery Board for Scotland (1914).

6.   Cited in the *Daily News*, 7 April 1932. See also *Seventh Annual Report of the Newfoundland Board of Trade for 1915*, p. 4.

7.   David Alexander, *The Decay of Trade: An Economic History of the Newfoundland Saltfish Trade, 1935–1965* (St. John's: Institute of Social and Economic Research, 1977), pp. 23–24.

8.   Ian D.H. McDonald, *"To Each His Own": William Coaker and the Fishermen's Protective Union in Newfoundland Politics, 1908-1925*, edited by J. K. Hiller (St. John's: Institute of Social and Economic Research, 1987), p. 96.

9.   The Welsh-born Davies had been appointed to this position in 1914. The government analyst was responsible, for example, for examining the quality of food such as milk and butter. See PANL, GN2/5A, file 27-6 "Report of the Government Analyst for 1915."

10.   Canada, *Fifty-Sixth Annual Report of the Fisheries Branch, Department of Marine and Fisheries for the Year 1922* (Ottawa: F.A. Acland, 1923), p. 23. Canada and the United States each had three representatives on the committee, while Newfoundland was given one. France received a place in 1922 on the committee because of its fisheries off St. Pierre and Miquelon and carried out fishery research off Newfoundland. See, for instance, "Preliminary Report on the Fishing Campaign in Newfoundland and Greenland in 1930, by Commander Beauge, in charge of an investigation undertaken by the Office Scientifique et Technique des Peches Maritimes, December 1930" in PANL, GN34/2, Department of Marine and Fisheries, file "Fishery: Research Commission." The committee was a predecessor of the International Commission for the Northwest Atlantic Fisheries (ICNAF). See Kenneth Johnstone, *The Aquatic Explorers: A History of the Fisheries Research Board of Canada* (Toronto: University of Toronto Press, 1977), p. 119.

11.   Newfoundland, *Annual Report of the Department of Marine and Fisheries for the Year 1921*, p. 18. The extent of scientific research in Newfoundland was the work in 1918 by Davies who had "examined microscopically some hundreds of codfish scales.... to determine approximately the ages of the codfish caught for that year." See *Fifteenth Annual Report of the Council of the Newfoundland Board of Trade, 1923*, p. 19.

12.   McDonald, *To Each His Own*, pp. 132-33.

13.   Newfoundland, *Annual Report of the Department of Marine and Fisheries for the Year 1921*, p. 18.

14.   Newfoundland, *Annual Report of the Department of Marine and Fisheries for the Year 1926*, p. 17.

15.   Huntsman (1883-1973) was curator of the St. Andrews Biological Station from 1915 to 1919 and its director from 1919 to 1934. From 1924 to 1928 he was also director of the Fisheries Experimental Station at Halifax. See Jennifer Hubbard, "Home Sweet Home: A.G. Huntsman and the Homing Behaviour of Canadian Atlantic Salmon," *Acadiensis*, vol. 19, no. 2 (Spring 1990), pp. 40-71; and W.E. Ricker, *The Fisheries Research Board of Canada: Seventy-Five Years of Achievements*, Report No. 2, August 1975.

16.   *Daily News*, 19 September 1923.

17.   F.W. Field, *Report on the Trade, Industries and Resources of Newfoundland for 1925*. (London: 1926), p. 13.

18.   MUN, President's Office, Box PO-9, file "Biology, JLP," Paton to George Russell, nd.

19.   MUN, President's Office, Paton Papers, Box 1, file "January 1926," F.E. Weiss to Paton, 18 January 1926; file "March 1926" Weiss to Paton, 2 March 1926; and Box 2, file "May 1926" Weiss to Paton, 13, 26 May 1926. See also Box PO-9, file "Fishery Research, 2," Paton to James Nelson Gowanloch, 5 February 1930.

171

20. MUN, President's Office, Paton Papers, Box 2, file "April 1926" Blackall to Paton, 26 April 1926; Box 1, file "Enclosures: Letters sent to Paton but not written to him," Sleggs to the President, McGill University, 2 April 1926; Box 2, file "May 1926" Sleggs to Paton, 10, 25, 30 May 1926. While he taught biology at Dalhousie University during 1921-22, he was also involved in drift bottle experiments in the Cabot Strait. During 1924-25 he was an instructor in biology at the University of Saskatchewan where he incidentally had sat in on a public lecture that Paton had given as part of a public speaking tour of Canada prior to his appointment as president of Memorial University College. During his 1921 drift bottle experiments, he had spent "two pleasant days" in St. John's. Sleggs resigned from Memorial University College in 1933.

21. *Daily News*, 10 July 1926; Newfoundland, *Annual Report of the Department of Marine and Fisheries for the Year 1926* (St. John's: 1927), p. 14; and *Twenty-First Annual Report of the Council of the Newfoundland Board of Trade, 1929*, pp. 33-46. Sleggs' observations on these meetings were published in the annual reports of the Newfoundland Department of Marine and Fisheries.

22. *Annual Report of the Department of Marine and Fisheries for the year 1926* (St. John's: 1927), pp. 19-20.

23. *Daily News,* 26 October 1926.

24. The theme of Newfoundland concentrating more on fresh fisheries had long been advocated by former premier Edward Morris who, following his resignation from politics in 1917 and retirement to Britain, lobbied British industrialists and businessmen to invest in the local fresh fish industry. See issues of the St. John's *Colonial Commerce* for January and April 1918, for articles on Morris's activities in Britain and his "Cold Storage in its Application to the Newfoundland Fisheries," *Newfoundland Quarterly*, vol. 22, no. 2 (Autumn 1922), pp. 30-31, and "Newfoundland's Contribution to the Empire's Larder," in *Ibid.*, vol. 30, no. 3 (1930), pp. 10-11.

25. Melvin Baker, "Hazen Algar Russell," *Newfoundland Quarterly*, vol. 89, no. 3 (Spring/Summer 1995), pp. 35-37.

26. "Report of the Imperial Economic Committee on Fish," in *Daily News,* 9 September 1927.

27. *Nineteenth Annual Report of the Council of the Newfoundland Board of Trade, 1927*, p. 30; *Daily News*, 10 September 1927; and MUN, President's Office, Box PO-9, Paton to Arthur Barnes, 1 March 1929.

28. MUN, President's Office, Box PO-9, file "Fishery Research, 2," Paton to Arthur Barnes 1 March 1929, and to Huntsman, 8 March 1929. See also file "Biology, JLP," Paton to Pearce, 13 October 1926.

29. MUN, President's Office, Box PO-9, file "Fishery Research, 2," Paton to Huntsman, 8 March 1929.

30. *Ibid.*, "Scientific Research in connection with the Fisheries of Newfoundland."

31. See comments by Leonard Outerbridge in the *Daily News*, 7 April 1932.

32. *Daily News*, 9 August 1929.

33. PANL, GN8/2, Richard A. Squires Papers, file 31.ii, Passfield to Governor Sir John Middleton, 2 December 1929.

34. A copy of Thompson's resume is in PANL, GN8/6, Frederick Alderdice Papers, file 15.i.

35. PANL, GN34/2, Box: Fisheries and Lifesaving, file "Fishery Research, 1930-31," Progress Report by Harold Thompson, 23 August 1930. Minutes of meetings of the Fishery Research Commission for 21 August, October, 30 December 1930, 8 January, 21 February, and 6 March 1931, are located in this file. See also Arthur Barnes to H.B. Clyde Lake, 21 August 1930, for the appointment of the commissioners. On the question of a site for the laboratory, Thompson wrote that it "might even be situated in the City itself, but for the facts that the suitable sites on the waterside are occupied and that a pure sea-water supply for fish tanks and other purposes is necessary. There is much to do with our limited capital, and the least expenditure will be incurred if we can rent and adapt suitable existing premises."

36. PANL, GN34/2, file "Fishery Research, 1930-31," Minutes of the 2nd meeting of the Fishery Research Commission, 20 October 1930.

37. Harold Thompson, *Reports of the Newfoundland Fishery Research Commission, Vol. 1, No. 1: A Survey of the Fisheries of Newfoundland and Recommendations for a Scheme of Research* (St. John's: Newfoundland Fishery Research Commission, 1931), pp. 9-12, 18-28.

38. PANL, GN8/2, file 31.ii, James Davies to H.B. Clyde Lake, 26 January 1931, enclosing a copy of the contract terms offered to Thompson.

39. *Ibid.*. See also minutes of the meeting of the commission for 21 February 1931.

40. Harold Thompson, "Fishery Research in Newfoundland," in J.R. Smallwood, ed., *The Book of Newfoundland*, vol. 2 (St. John's: Newfoundland Book Publishers, 1937), p. 211. The work of biologist Whiteley is described in his article, "Marine Research and the Scientific Method," *Newfoundland Quarterly*, vol. 31, no. 4 (Spring 1932), pp. 28-31, and his autobiographical *Northern Seas, Hardy Sailors* (New York: W.W. Norton & Company, 1982), pp. 83-110.

41. Thompson, "Fishery Research in Newfoundland," p. 211.

42. *Ibid.*, pp. 212-13.

43. PANL, GN34/2, Box "Fisheries and Lifesaving," file "Fisheries: International and Scientific Investigations," minutes of the 18th meeting of the North American Council on Fishery Investigations. See also Thompson, "Fishery Research in Newfoundland," p. 212.

44. See *North American Council on Fishery Investigations. Proceedings 1934-1936* (Ottawa: 1939), pp. 23-26; and George F. Sleggs, *Observations upon the economic biology of the caplin* (St. John's: Fishery Research Commission, 1933), pp. 7-8.

45. For a discussion of Newfoundland's economic problems in the early 1930s, see Peter Neary, *Newfoundland in the North Atlantic World, 1929-1949* (Montreal and Kingston: McGill-Queen's University Press, 1988), pp. 12-43.

46. *Ibid.* See also his "'With great regret and after the most anxious consideration': Newfoundland's 1932 Plan to Reschedule Interest Payments," *Newfoundland Studies*, vol. 10, no. 2 (1994), pp. 250-59.

47. PANL, GN8/6, file 15.1, file "Fishery Research, 1929-32," minutes of an informal meeting of the Fishery Research Commission, 23 September 1931.

48. *Ibid.*, Minutes of the meeting of the Fishery Research Commission, 9 January 1932.

49. PANL, GN8/6, file 15.1, file "Fishery Research, 1929-32," minutes of the meeting of the Fishery Research Commission, 24 November 1932.

50. *North American Council on Fishery Investigations. Proceedings 1931-1933, no. 2* (Ottawa: 1935), p. 26; and H.F. Gurney, *Economic Conditions in Newfoundland, March 1935* (London: 1935), p. 13.

51. MUN, President's Office, Box PO-9, file "Fisheries Research," minutes of the meeting of the Fishery Research Commission for 4 October 1933.

52. PANL, MG 300, Charles A. Magrath Papers, Microfilm Reel 2, Evidence of Raymond Gushue, 30 March 1933.

53. Great Britain, *Newfoundland Royal Commission 1933 Report* (London: His Majesty's Stationery Office, 1933), p. 114.

54. PANL, MG 300, Microfilm Reel 2, "Evidence of Harold Thompson, 3 April 1933."

55. *North American Council on Fishery Investigations. Proceedings 1931-1933, no. 2* (Ottawa: 1935), p. 26.

56. *Report of the Commission of Enquiry investigating the Sealfisheries of Newfoundland and Labrador other than the Sealfishery, 1937* (St. John's: 1937), pp. 155-60.

57. See Peter Pownall, *Fisheries of Australia* (Farnham, Surrey, England: Fishing News Books Ltd., 1979), p. 106.

58. Harold Thompson, *Annual Report of Fishery Research Laboratory 1935* (St. John's: 1936), p. 7.

59. Malcolm MacLeod, "Prophet with Honour: Dr. William F. Hampton, 1908-1968, Newfoundland Scientist," *Newfoundland Quarterly*, vol. 81, no. 1 (Summer 1985), pp. 29-36.

60. *Daily News*, 20, 21 April 1937.

61. Wilfred Templeman, "Fisheries Research," in *Encyclopedia of Newfoundland and Labrador*, vol. 2 (St. John's: Newfoundland Book Publishers, 1984), p. 170.

62. Whiteley, "Marine Research and the Scientific Method," p. 31. As Newfoundland-born mathematician and Paton's successor in 1933 as President of the College, Albert Hatcher, told the Rotary Club in 1929, "Newfoundland needs science.... If our country is to advance into prosperity, all trades, professions, business and science must move forward together." See *Daily News*, 12 October 1929.

63. See Shannon Ryan, *Fish Out of Water: The Newfoundland Saltfish Trade, 1814-1914* (St. John's: Breakwater Books, 1986) for many examples of Britain's support for the Newfoundland saltfish trade.

chapter
twelve

"COME ON ALL THE CROWD, ON THE BEACH!": THE WORKING LIVES OF BEACHWOMEN IN GRAND BANK, NEWFOUNDLAND, 1900-1940[1]

Cynthia Boyd

Within living memory, people from the community of Grand Bank on the south coast of Newfoundland can vividly recall their working lives during the bank fishery of the saltfish era. Grand Bank merchants financed the building of banking schooners and outfitted the boats with fishing gear and bait; dory fishers caught and processed the fish; and beachworkers, primarily women, cured the fish for market.

This paper focuses on the working lives of the women whose occupation was that of "beachwoman."[2] Beachwomen, like their sisters on the flakes, worked as makers of fish. "Making fish" for women in most Newfoundland outport communities is defined as splitting, gutting, and salting fish in addition to spreading and drying it.[3] But because of the long-held tradition of women confined to sun-drying fish in Grand Bank, making fish is here defined by this activity only.

Making Fish

From the early-19th century through World War II, the people of Newfoundland and Labrador were employed within a household-based coastal fishery (also referred to as an inshore fishery) in which fish were sold to local merchants.[4] Fish were "made" every day except Sunday during the spring, summer, and fall. Fish caught by male fishing crews was brought to the stage where women (and some older children) prepared, salted, and spread the fish to dry on the flakes. While women's wages were not significant, their production of light-salted codfish increased the earnings of their husbands and their fishing crews.[5]

Flakes consisted of open, elevated wooden platforms resting on stakes or "longers" that were driven into the ground for support.[7] Frequently women placed a layer of spruce boughs over the flakes before laying the fish there to dry. In Labrador, fish was spread to dry on "bawns" (small rocky beaches) as well as flakes.[8] In the town of Grand Bank, the natural topography of the land allowed for the fish to be spread on beaches. The task of drying fish on beaches was accomplished by a "crowd" or crew of "beachwomen".

To the Banks

During the late 1870s, the Newfoundland government provided bounties for the construction of vessels in excess of 25 tons, to promote an offshore bank fishery.[9] The banks included those off the south, southwest, and southeast coasts of Newfoundland, as well as those off Nova Scotia and Labrador. Over a period of several years, bounties increased as more and more shipowners moved into the bank fishery. Grand Bank merchants rallied to the cause by building schooners at a frantic pace. While the bank fishery decreased in other areas of Newfoundland in the 1880s, the south coast communities of Burin and Grand Bank steadfastly maintained an impressive banking fleet.[10] This essay is a collection of oral anecdotes and historical material of the bank fishery covering the period from 1900 to 1940.

Each spring the captains and their crews readied their schooners for the first voyage to the western banks. Herring was the prime bait at this time. By June, banking schooners were outfitted with a different bait—caplin—for fishing on the Grand Banks. In September, banking schooners loaded up with squid bait and headed for the coast of Labrador for the last trip of the season.[11] The entire season lasted from April to November.

After the heavily salted catch was brought ashore, fishers washed it in pounds—wooden crates (10-12 sq. ft.) with openings along the bottom.[12] The wet fish was left to drain in large piles referred to as "waterhorses" or "waterhorse fish" on the cobblestone beaches.[13] In most cases, the women employed to dry the fish were relatives of dory fishers already working aboard the schooners.

Initially, a "boss woman" was employed by the merchant, and she, in turn, handpicked her crowd of 8, 10, or 12 beachwomen. A typical beach crowd could consist of young, middle-aged, or elderly women. While no records exist to indicate exact numbers of different aged women, a typical "beach crowd" might consist of at least two or three young girls of 13 years and up, two to four women in their mid 30s or 40s, and two or three women in their 50s or 60s.[14] Usually the boss women were older, having had substantial beachwork experience and recognized for their ability to "know the fish."[15]

Beachwomen were paid by the merchant in the form of handwritten credit notes. With few exceptions, payment was "taken-up" at the merchant's store in the fall.[16] This was much like the "truck" form of payment between fishers and merchants in which there was a cashless barter of fish for goods.[17] In such an arrangement, women traded hours of beachwork for foodstuffs and merchandise. Despite non-monetary payment, beachwork carried a level of autonomy for these women. In the words of one beachwoman, Olive Pope, "You weren't making very much but you knew you were making something."[18]

Except when the merchant came to inspect beachwork, beachwomen were under the supervision of the boss woman. Boss women and their crowds often met on the bridge between Riverside East and Riverside West in Grand Bank to discuss the skies, deciding whether they threatened cloud, fog, or rain. If fish got wet or received too much sun, it was "dun" fish, unsuitable for the export market.[19] Although the boss woman ultimately decided if they would spread fish, all beachwomen were keen observers of weather conditions as the welfare of the fish was everyone's concern. In the words of one fishmaker, Dulcie Grandy of Grand Bank and Fortune, fish was paramount to butter:

> What they would do if the fog would come in and the rain, they'd ring the bell, the fish bell, you see...and when you hear the bell ring, you'd have to drop everything and go, no matter what you were doing. Everybody be going...you'd see them going down, I mean they wouldn't walk, they'd fairly run. They figured they'd get there to the beach to take up their fish because that was just as precious to them as if they owned it, you know. That was their bread and butter. You wouldn't take your time and say, well now, I've got to finish...I got bread in the oven...you'd just go on. Drop everything and go![20]

The issue of "owning" fish as expressed by Dulcie Grandy sheds light on an underlying fact: while women (and men) laboured extensively making and drying fish, they did not own the fish, the merchant did. Regardless, they had to pour all their efforts into making a "good, white" fish for market because their wages depended on it.[21]

A typical day at the "beachracket"[22]

Inevitably, the amount of fish caught during the season determined the amount of beach labour required. The glut of the fishing season occurred in July and August. Drying fish on beaches was precise work. In sun and wind, fish got "sunburnt" very quickly; sunburnt fish was usually so dry it broke apart in one's hands.[23] Fish dried on flakes allowed for air to circulate underneath it, drying the fish quickly. But fish dried on beach rocks often got crispy on one side only. For each side of the fish to receive equal amounts of air and sunshine, and not too much of either, careful timing was essential.

Beachwork commenced by women taking "yaffles" (small armloads of fish) and spreading them on the beach.[24] The boss woman saw that the fish was placed head to tail and "face-up" (flesh side up) on the beach. Sometimes the fish was so big that it would take two women to lift one fish. This continued until noon or 1:00 p.m., at which time the women would break for lunch.[25] If they lived close by, beachwomen could make it home for a quick bite, but in most cases they brought food and water and had their lunch on the beach. At 2:00 pm, the women returned to work and continued to spread the fish and/or turn the fish "back-up" (tails exposed) face-down.

By early evening, sufficiently dried fish was taken up from the beach and placed in small piles referred to as "faggots."[26] These piles were increased each day, eventually taking the shape of beehive-like mounds. Piling the fish in this manner required the most highly skilled beachwomen who likened the piling to "clapboarding a house" or "pooking" a haystack.[27] French fishers actually described these piles of fish as haystacks (*meulons*).[28] Any or all of these piles were covered with tarpaulin or old sail canvas, especially if the weather was at all menacing, or if the sky was "loury" (cloudy).[29] If it was windy, rocks were placed on top of the canvas to stop it from blowing away and exposing the fish.

The drying process could take upwards of 10-15 days. Sometimes beachwomen worked until 6-7 p.m., and if the catch was a particularly good one, they could be on the beach well after dark: "you couldn't get off the beach until the fish come up off the beach."[30] On other occasions the weather changed so drastically that fish had to be gathered at once to save it from being ruined. Becky Rose recalled an unusual event on the beach in which the skies darkened from a solar eclipse:

> I remember one time we had a pile of fish and the sun went into an eclipse and it got right dark...and everybody was hurrying up trying to get the fish up and I don't say there was one [fish] any longer than ten inches. It was all little tiny fish and when that pile was weighed I believe there was 150 quintals into it. What a pile of fish![31]

When several hundred quintals of fish were ready to be graded, a male employee of the merchant firm would come down to the beach with a horse and cart. The boss woman and some beachwomen followed the load of fish to the merchant's fish room or store[32] where the culler graded the fish. On some occasions, this took place right on the beach. Because several boss women were presented during the grading they watched to ensure that their piles of fish were not inadvertently mixed together.[33] Once the fish had been culled, the number of quintals was tallied in a tally or fish book. The tally book indicated how many quintals of fish were "merchantable" and which were "cullage," along with the name of the boss woman whose crowd had made the fish.[34] At the end of the season, the merchant and his clerks would determine the beachworkers' wages based on the amount of fish they had cured.

Wages

Beachwomen who worked from 1920 to 1940 never made more than $100 in any given season. Even the boss women only made an additional $10.[35] Unfortunately, $100 was the exception rather than the rule. According to Alice Forsey, "I never made a hundred dollars a year in my life, not down there...if you made ninety dollars or close on to a hundred, you made a good summer."[36]

In the fall, beachwomen arrived at the merchant's store to settle up. Most beachwomen referred to this as "taking up" goods to feed them and their families through the winter: flour, molasses, raisins, sugar, and salt pork were staples. Most beachwomen sewed proficiently and knitted everything from sweaters to long underwear. Therefore, cloth, thread, and needles from the merchant's shop were also desirable goods. Many women found themselves unable to purchase anything in the fall, having taken up their goods during the summer. This was particularly the case for beachworkers with large families, as well as those women whose husbands had been lost at sea. Remarking on beachwork, Becky Rose put it quite frankly: "There was nobody on the beach that had plenty because they wouldn't be there if they did...when they went on the beach they had to go."[37]

Beachwomen's subsistence activities, such as maintaining small gardens and some livestock, gave them an advantage. Root vegetables could be stored in cellars, and wool from sheep could be carded and later knitted into clothing and blankets. Beachwomen often relied on their daughters to help meet their families' needs. While their mothers were at work, daughters often took their place at home: baking bread, watching younger children, cleaning house, and weeding gardens. To alleviate the financial burden of large families, some daughters worked "in service" for other families.[38] Still other daughters in their early-to-late teens stayed in Grand Bank working on the beach alongside their mothers, aunts, and grandmothers. As Olive Pope recalled, she quit school to make fish:

> I went on the beach when I was 13. I said to Mom, "no good for me to go to school because I'm going to be put back," and when you get put back you get right disgusted. When you're the oldest you got to...help your mother do all the work, [with] your father gone. He was cook on the boats, so he was gone all the time.[39]

Working Conditions

Beachwomen had to balance beachwork with family responsibilities and household chores. They rose as early as 3 or 4 a.m. in order to wash clothing and knead bread dough before going to the beach.[40] Only on rainy days could they get a break from the work.[41] One beachwoman stated: "every day you had fish...and you wouldn't go berry picking if it was raining because that's the only day you'd get off."[42] After a morning's work on the beaches, they often went into the woods outside town to gather "blaze boughs" for fires. Blaze boughs were spruce branches left over from when the men had previously cut timber.[43] They tied a bunch on their backs and returned home or to the beach to have lunch. Fred Rogers worked for the Grand Bank firm G. and A. Buffet. He vividly remembered the toil of the beachwomen:

I can see them, all those women come up from the beach and go up Bennet's Hill and get a turn of boughs. Pack them up...and put them on their back and bring them down to cook their dinner with. They'd been on the beach all morning. And some women would go in the morning to get a turn before they went on the beach.[44]

Working conditions on the beach were conducive to injury and illness. Beachwomen endured hours of back-breaking labour in hot sun, walking on uncomfortable rocks, and handling bony codfish. They often complained of sore feet, abrasions or boils on their hands and arms, as well as night blindness. Alice Forsey described her experience being night blind:

Sometimes you'd be on the beach perhaps two weeks solid sunshine, you'd get what you'd call "night blind." Soon as the sun would start to go in if you got night blind, you wouldn't be able to see nothing. It would be a long period in which you would be looking down on the rocks, the sun reflecting from the rocks you'd be blind.[45]

Louise Belbin, who worked on the beaches for 14 years, remembered that her hands often suffered damage: "Sundays when the children go to Sunday school, I used to soak me hands and pick out bits of bone because you never had nothing on your hands, handling the fish."[46] Olive Pope remembered the agony of a stiff back at the start of the fishing season: "The first few days when we started in the spring, everybody's back would be alike. You walk up the road...your back be that stiff you couldn't straighten up. After awhile, you'd get used to it."[47] Housework awaited them after their day on the beach. One chore that every woman had to do was take out the toilet pail or the "kettle," a chore incumbent to women only: "Everyday you'd have to go with that, rain or shine....Come off the beach at night-time that was the first thing you had to do before you take off your clothes, go down to the brook [and] take your toilet pail. Mom go or I'd go."[48]

Overall, beachwomen were not as concerned about working conditions so much as they were about the time spent away from their families. If the boss woman was especially understanding, she allowed the women to bring infants to the beach. Carried to the beach in baskets, or left in the fish store, infants were not a problem. Toddlers, however, required more attention, and they were usually left with older siblings at home. In some cases, children were a constant worry to their mothers working on the beaches. With husbands at sea, beachwomen had no choice but to leave their children at home to take care of each other. Olive Pope remembered a beachwoman she worked with who had a kink in her neck from constantly looking up at her house. Her worry stemmed from the fact that her six year old was at home alone taking care of a small infant.[49]

Beachwomen felt the frustration and the pain of raising children without their husbands, often because the men were absent so frequently, or because they had met a watery grave. Louise Belbin's husband was a schooner captain who was gone for months at a time:

I was mother and father of the family. That's all you could say because the man was never home. One time he was gone 11 months....When he come home, Mabel, our first one we had...when he opened the door and come in with his suitcase, she said: "Mommy, who's that?" And I said: "That's your father," and she run up into my arms and snuggled right into me and wouldn't look around to him. He said: "My dear, it's awful hard when you got to ask your mother who your father is."[50]

Working Dress

The trademark of every beachwoman was her working attire, especially her white sun bonnet. The sun bonnet, or "sunshade," was an ornately crafted hat that served to protect the woman's face and neck from sun and wind. Sun bonnets were worn by women in other parts of Newfoundland as well, particularly in Fortune, Jersey Harbour, Placentia, and Fogo.[51] In addition, similar headgear was often worn by women involved in making fish, gardening, and farming in communities of the Gaspé and the Magdalen Islands, St. Pierre and Miquelon, and the British Isles.[52] While Grand Bank beachwomen commonly called sun bonnets sunshades, other names applied to the same hat in Newfoundland were skully, scolly, slouch, thrum cap, and Dolly Varden. In Europe they were called mob caps, poke bonnets, morning caps, tunnell-shaped bonnets, hoods, mutches, and uglies.[53]

Every beachwoman had at least two sun bonnets because their constant use required frequent washing. Most were made from heavy cotton flour sacks. Despite the task of soaking flour sacks to remove the name of the flour company, they were utilized for just about everything from bloomers to tablecloths.[54] The hat usually consisted of a front rim with a metal or bamboo wire insert that would bend to fit around the face; the main body was sewn with a number of decorative tucks and folds and two tying ends; and lastly, a long flap of material attached to the bonnet at the rear covered the neck.[55] One Grand Bank resident conjured up this memory of beachwomen: "You never get that out of your eyes...20 to 30 people walking down in the morning with white bonnets on."[56]

In addition to her sun bonnet, the beachwoman wore a large white apron covering a pair of clam-digger pants (short pants at mid-calf length), a white blouse or shirt, or simply a light cotton dress. The apron was often made from potato sack material from whence it got the name "wrapper apron."[57] Beachwomen wore wool socks to buffer sensitive skin from coarse canvas shoes or leather (later rubber) boots.

Social norms were strictly adhered to in this Methodist community, especially in the case of reverence for the dead. The bank fishery was notoriously dangerous, and every time a dory crew left a schooner they were at risk. When someone died, everyone related to the deceased donned black armbands and black clothing from head to toe. In the house, blinds were drawn and all mirrors were turned inwards. Upon hearing that their father had been lost at sea, Becky Rose's family proceeded to dye all their clothes black. When her father returned to Grand Bank as alive as the day he left, the family continued to wear black clothes until they could afford something else.[58] While in mourning, beachwomen were no exception: they wore black sun bonnets, black dresses, and black shoes. In the heat of summer, black beach clothes drew heat and increased the usual discomfort. Some women continued to wear black as a statement of devotion to a loved one, prompting friends to intervene out of concern for their health.

Conclusion

Beachwork and fishmaking are all but gone, save for people's memories of a past way of life. While many recall those days with a sense of nostalgia, others take a less rosy view. Times were hard, there was no money, and people were impoverished. Toiling over beachwork, tending gardens, clothing and feeding often large families—how did they manage? In the words of Simeon Grandy, "they must have been made of iron."[59] Many, like Louise Belbin, worked 14 years on the beaches and raised families simultaneously. Perhaps her stoicism can be applied to all fishmakers of the past: "you had to be strong your whole life to work."[60]

Cod fish drying, Grand Bank, Newfoundland. 1949.
Credit: National Archives of Canada.

Notes

1.   This title was taken from an interview with Winnie Prior in 1993. Although she has lived in Grand Bank since the mid-1970s, Mrs. Prior grew up in the now-resettled community of Jersey Harbour, on the south coast. Her mother used to make fish on flakes and beaches in this community. She was the boss of the rest of the women who were the shore crew. She would sing out "Come on all the crowd...." when it was time to continue making fish after lunch. Recorded interviews are located in the Memorial University of Newfoundland Folklore and Language Archive, [MUNFLA] St. John's, NF. The *Dictionary of Newfoundland English* (2nd Ed.) by G.M. Story, W.J. Kirwin, and J.D.A. Widdowson (St.John's: Breakwater Books, 1990), pp. 124-25, refers to a crowd as an "organized, integrated group of people esp. a fishing or sealing crew." The term beachwoman can be found in Story et al, pp. 34-35.

2.   While conducting field research for the Newfoundland Museum in 1993-94, I spoke with over a dozen beachwomen who made fish in Grand Bank spanning the early 1900s to the 1940s. With kindness and hospitality, these women opened their homes to me. I would also like to thank Walter Peddle of the Newfoundland Museum, Dr. Marilyn Porter (MUN), and Dr. Garth Fizzard (MUN) for their assistance in this research. All photographs are courtesy of the Provincial Archives of Newfoundland and Labrador (PANL) or Charlie Cousens of St. John's.

3.   Story et al, p. 322.

4.   For a description of a household-based fishery, see Barbara Neis, "From 'Shipped Girls' to 'Brides of the State': The Transition from Familial to Social Patriarchy in the Newfoundland Fishing Industry," *Canadian Journal of Regional Science*, 16, 2, (Summer 1993), pp. 185-211. See also Gerald M. Sider, *Culture and Class in Anthropology and History: A Newfoundland Illustration* (Cambridge: Cambridge University Press, 1986), p. 22.

5.   In an unpublished paper, Ellen Antler described how women were not actually paid "when their labour was applied to their own 'crowds' but their effort did make a difference to the income of their

181

households." Furthermore, in their making fish on flakes, they produced the highly valuable "light salted fish" which was $5.16 a quintal over fresh fish in the 1950s. See Ellen Antler, "Women's Work in Newfoundland Fishing Families," Centre for Newfoundland Studies, Memorial University of Newfoundland, St. John's (unpublished paper), 1976, p. 4.

6.   Dona Lee Davis, *Blood and Nerves: An Ethnographic Focus on Menopause*, (St. John's: ISER, 1983), p. 73.

7.   Hilda Chaulk Murray, *More than 50%: Woman's Life in a Newfoundland Outport 1900-1950*. (St. John's: Breakwater Books, 1979), p. 14; See Story et al, pp. 187, 313 for definitions of flakes and longers.

8.   Story, et al, p. 31.

9.   Garfield Fizzard, *Unto the Sea: A History of Grand Bank*, (Grand Bank Heritage Society, 1987), p. 110.

10.   Fizzard, p. 110.

11.   Fizzard, p. 127.

12.   Story, et al, p. 389.

13.   Story, et al, p. 602.

14.   Even though merchant records, such as that of Lake and Lake, of Fortune, and Forward and Tibbo, of Grand Bank, have extensive ledgers and specific "fish" books showing the numbers of quintals of fish made by a specific woman, the details surrounding the exact make-up of each beach crowd are not present. Other scholars, particularly Dr. Marilyn Porter, have written about the dichotomy of beachwork, bosses and beachwomen within the context of a much broader subject, that of the life histories of mothers and daughters in Grand Bank. See Marilyn Porter, "Mothers and Daughters: Linking Women's Life Histories in Grand Bank, Newfoundland, Canada," *Women's Studies International Forum*, 11, 6 (1988), pp. 545-58. Porter indicates that no less than 20 women worked in a beachcrowd. While I do not dispute this, my research is based on living memories of beachwomen in the community; over a dozen women that I interviewed commonly described a beachcrowd of 8 to 12 women with the numbers and ages of women varying.

15.   Several beachwomen defined boss women as individuals whose prior experience making fish on beaches set them apart from other beachworkers. It is likely that merchants recognized this and gave them full authority when choosing their crowds. Interestingly, the expression "to know the fish" was applied to men who culled fish, or graded fish for the export market.

16.   Story, et al, p. 555.

17.   Sider, p. 18; Story, et al, p. 585.

18.   Interview with Olive Pope, Grand Bank, 1993.

19.   Story, et al, pp. 159-60.

20.   Interview with Dulcie Grandy, Grand Bank, 1993.

21.   The late Sarah Hickman of Fortune gave this literal description of a light salted fish ready for market. Interview with Sarah Hickman and her daughter, Dulcie Grandy, 1993.

22.   Story, et al, p. 34.

23.   Story, et al, p. 546.

24.   Story, et al, p. 621.

25.   Descriptions of fish lying either "face-up" or "back-up" were commonly the same among all the beachwomen I interviewed. Winnie Prior remembered that a white flag would be raised as a signal to the women in the community of Jersey Harbour when it was time to return to the flakes (and in some cases, small beaches). Interview with Winnie Prior, 1993.

26.   Story, et al, p. 167.

27.   Interview with Alice Forsey of Grand Bank, 1993. Pooking refers to piling hay around the centre stick of the stack (interview with Becky Rose, Grand Bank, 1993; Story, et al, p. 386). In an article in Grand Bank's local newspaper, *The Southern Gazette*, former beachwoman Harriet "Cis" Welsh mentioned a "piling boss." This reference is the only one in which a distinction is made between a boss woman and another type of boss. She does not distinguish the bosses' activities in her discussion. See Robert Parsons, "The Beachbonnet: Its Heritage," *The Southern Gazette*, 20 May 1987.

28. Robert de Loture, *History of the Great Fishery of Newfoundland*, (Washington, D.C., U.S. Fish and Wildlife Service; Special Scientific Report - Fisheries No. 213, April 1957), p. 20.
29. Interview, Alice Forsey, Grand Bank, 1993.
30. Interview with Olive Pope, Grand Bank, 1993. Mrs. Pope indicated that although summer nights were rarely pitch black, there were times where they would actually have to "feel" for the fish in order to pick them up. In her article for *The Southern Gazette,* "Cis" Welsh mentioned the use of flashlights when working late nights, presumably in the late 1940s. See Parsons, 20 May 1987.
31. Interview with Becky Rose, Grand Bank, 1993. She provided a similar story for a folklore student at Memorial University (see Diane Forsey, "Women's Work in Outport, NFLD in the Past," MUNFLA MS #84-157).
32. Story, et al, pp. 418, 536.
33. Interview with Fred Rogers, 1993.
34. The merchant ledgers of Forward and Tibbo are located in the Centre for Newfoundland Studies Archives, Memorial University of Newfoundland, St. John's, Collection no. 043. Definitions of merchantable fish versus cullage describe fish that is respectively, suitable for market and that which is the lowest grade fish, typically sunburnt, broken, slimy, or maggot-filled fish. Story, et al, pp. 127, 129, 327.
35. Interview with Louise Belbin, Grand Bank, 1993.
36. Interview with Alice Forsey, Grand Bank, 1993.
37. Interview with Becky Rose, Grand Bank, 1993.
38. Before she worked on beaches and raised a family, Louise Belbin worked "in service" with a merchant's family in Grand Bank. Interview with Louise Belbin, 1993. For further discussion on this type of child labour see Neis, "From 'Shipped Girls' to 'Brides of the State'," pp. 185-211.
39. Interview with Olive Pope, Grand Bank, 1993.
40. Women who worked on flakes spreading and turning fish inevitably balanced a myriad of chores as well. See Antler, 1976.
41. Interviews with the Sarah Hickman and Dulcie Grandy, Fortune, 1993. Interview with Olive Pope, Grand Bank, 1993.
42. Interview with Olive Pope, Grand Bank, 1993.
43. Story et al, 1990, p. 48. Interview with Fred Rogers, Grand Bank, 1993.
44. Interview with Fred Rogers, Grand Bank, 1993. The word "turn" here refers to the retrieval of another pile of wood sticks for burning. Beachwomen also referred to the household chore of getting a "turn" of water for the barrels kept inside the front porch of the house. See also Story, et al, p. 588.
45. Interview with Alice Forsey, Grand Bank, 1993.
46. Interview with Louise Belbin, Grand Bank, 1993. For information related to the injury and health costs of bank fishing see Raoul Andersen's excellent work "Nineteenth Century American Banks Fishing Under Sail: Its Health and Injury Costs," *Canadian Folkore Canadien* 12, 2 (1990), pp. 101-22.
47. Interview with Olive Pope, Grand Bank, 1993.
48. Interview with Olive Pope, Grand Bank, 1993. In another interview, Olive Pope recounted that men would never empty toilet pails. Aubrey Tizzard states: "It was an unwritten law that a woman and a woman only could empty the slop pail, even on stormy days. A man would be humiliated beyond recompense if he were seen with a slop pail in his hand" (1979:41). Tizzard noted that the path to the outdoor water closet in winter was usually shovelled by his father so that his mother could dump the pail (1979:41).
49. Personal communication, Olive Pope, 1993.
50. Interview with Louise Belbin, Grand Bank, 1993.
51. Black and white photographs in the Still and Moving Images Collection at the Provincial Archives of Newfoundland and Labrador (PANL) depict women in sun bonnets making fish. See Photo no.# VA 69/3-2 (Placentia). In a personal communication, Mrs. Garfield Fizzard described a sun bonnet in her possession that once belonged to her grandmother, who lived in Fogo. The woman apparently used it when she made fish and when she was in the garden. Dulcie Grandy and the late Sarah Hickman wore sun bonnets while tending fish on flakes (Interview with both in 1993). Winnie

Prior also adorned her head with a sun bonnet in Jersey Harbour while making fish on flakes and small beaches (Interview with her in 1993).

52. The Reverend Edwin Smith briefly mentions the work of Magdalen Islands women who dug clams, farmed, and gardened while their husbands were away fishing. Photographs throughout the text provide glimpses of the women's working/living costume: white sun bonnets for both themselves and their children. See Rev. Edwin Smith, "The Magdalen Islands," *Canadian Geographic Journal*, 6, 6 (June 1932), pp. 331-47. Roch Samson discusses the work of women and children in the Gaspé fishery. Similar to Newfoundland women, Gaspé women were rarely accounted for when drying fish. One photograph depicts women stacking fish in piles on beaches much like they did in Grand Bank, and they are clearly wearing white sun bonnets with wide rims to shield their faces. See Roch Samson, *Fishermen and Merchants in 19th century Gaspé: The Fishermen-Dealers of William Hyman and Sons* (Ottawa: Parks Canada, 1984), p. 51. An etching in *The Century Magazine* gives credence to women wearing sun bonnets while spreading fish in St. Pierre. See Anon, "A French-American Seaport—The Cruise of the *Alice May*," *The Century Magazine*, 28, 2 (June 1884), pp. 163-74. In a personal communication, Garth Fizzard mentioned the use of similar bonnets in different parts of Yorkshire and Dorset, England (Fizzard, 1993).

53. For different Newfoundland expressions for sun bonnets see: Story et al, pp. 145, 489, 496, 564. For European descriptive titles, see Diana de Marly, *Working Dress: A History of Occupational Clothing* (London: B.T. Batford, 1986), pp. 78, 113, 138.

54. Flour sacks were always found in the family home as a 100 lb. sack of flour was quickly consumed by a family of ten. Interview with Louise Belbin, Grand Bank, 1993.

55. Verna Mullins of Grand Bank has been making sun bonnets for many years based on a pattern given to her from several deceased beachwomen. The measurements (in inches) are as follows: cord (which consists of the tucks): 32 in.; wire: 21 in.; ties: 20 in. each; fly piece (tail down the back): 18 in.; sunbonnet top: 19 in." Former beachwomen, Becky Rose, Louise Belbin, and Olive Pope mentioned that there were dressmakers in the community who made sun bonnets for beachwomen who did not sew. Louise Belbin often sewed sun bonnets for fellow beachwomen, but she did not charge for them.

56. Interview with Simeon Grandy and his wife, Caroline, Grand Bank, 1993.

57. Interview with Becky Rose, Grand Bank, 1993. The word wrapper was used in rhyme as well: "Fortune gallies and the Grand Bank shags, All tied up in wrapper bags." See Story, et al, p. 620.

58. Interview with Becky Rose, Grand Bank, 1993.

59. Interview with Simeon Grandy and his wife Caroline, Grand Bank, 1993.

60. Interview with Louise Belbin, Grand Bank, 1993.

chapter thirteen

TECHNOLOGY REJECTED: STEAM TRAWLERS AND NOVA SCOTIA, 1897-1933

B. A. Balcom

At the turn of the 20th century, a number of technological innovations and improvements affected Nova Scotia's groundfish industry. In time, the changes revolutionized the industry from seasonal production of wet-or dry-salted fish principally for Caribbean markets to year-round production of fresh or frozen fish for North American consumption. Among the technological advances contributing to this transformation was the adoption of trawling from steam or diesel-powered vessels.[1] Like the introduction of dory longlining in the 1870s, and factory freezer trawlers in the mid-20th century, steam trawlers created controversy in the industry. Traditional producers attempted to restrict the new technology, maintaining that existing factors of production could be modified to meet the demand for fresh fish.

At the end of the 19th century, Nova Scotia's fishers caught a variety of species including herring, mackerel, and lobster. However, the industry's mainstay was groundfish. Cod accounted for the major proportion but related species such as pollack and hake were also taken. The groundfish industry's technology had changed little from that introduced to the region by European fishers in the 16th century. Fish were caught with hooks either on handlines or longlines and were preserved through wet-salting or, more commonly, through a combination of salting and air-drying. Dried saltfish transported and stored well and had a considerable shelf life. Its good preservation moderated time as a factor in production and marketing, and enabled production to be dispersed in small centres along the coast.

Open boats fishing close to shore and decked schooners on more distant fishing banks contributed to production. The low capital cost of the inshore fishery permitted easy entry but the boats' limited range made proximity to fish stocks a determinant in the location of inshore fishery communities. The inshore fishery was distributed fairly evenly along Nova Scotia's Atlantic coast but more sparsely along the shores of the Northumberland Strait and the Bay of Fundy. The greater range of fishing schooners made proximity to fish stocks a less critical factor; instead, this branch responded to the availability of capital and labour. The traditional division of vessel ownership in 64 shares enabled the participation of small investors. Similarly, the systematized payment of labour on a share basis, and even the division of some costs between owners and crew, limited losses to investors during poor seasons. Like the inshore fishery, the vessel fishery was distributed along the province's Atlantic coast but with a marked concentration along the South Shore, most notably in Lunenburg County.

Nova Scotia's dried fish production easily exceeded Canadian requirements, especially as market preferences restricted dried fish consumption to the Maritimes. Therefore, dried fish producers were forced to rely on export markets.[2] The export marketing of fish enabled some degree of coordination in the fishery. Merchant houses purchased fish from both vessel and boat fishers with prices based on current information of market conditions. These often changed suddenly, with shortages and gluts in particular markets having ripple effects throughout the industry. Caribbean markets generally accepted lower quality fish and were convenient dumping grounds for fish not wanted in other markets. By the end of the 19th century, production increases in countries such as Norway and Iceland had an unsettling effect on Nova Scotia's dried fish trade.

Like the dry fishery, Nova Scotia's fresh fishery depended on external markets. Halifax was the only significant provincial market for fresh fish but even its requirements were easily met by local inshore fishers. The United States and central Canada were the closest major fresh fish markets, but transportation time and costs were limiting factors. Unlike dried fish, the marketability of fresh fish was measured in days or even hours depending on temperature. Moreover, preservation was measured from the time the fish left the water rather than the time it was prepared and packaged. Prior to rail linkage, sailing vessels and steamers carried fresh fish to these markets. Geographic proximity made such shipments more practical to some sectors of the American market than to central Canada, but exports were vulnerable to tariff restrictions. In addition, the fish had to be shipped in ice prior to the development of artificial refrigeration. Ice preserved fish well but only for a limited time, particularly in summer. The difficulty of getting fish to central Canada in a fresh state was a major and ongoing one. As late as the early 1930s, "flat taste" and "objectionable odour" remained drawbacks for expanding fresh fish sales, particularly when there was only a slight price differential between fresh fish and meat.[3]

Initially, high transportation costs restricted fresh fish shipments to more expensive varieties such as halibut or salmon rather than cheaper ones like cod. As early as the mid-1840s, agents operated seasonally at the Gold, LaHave, and Port Medway Rivers on the South Shore, purchasing salmon, packing them in ice and despatching them to the American market. At the same time, steamer connections between Halifax and Boston ensured a small but steady trade.[4] In June 1868, large quantities of halibut and salmon packed in ice were shipped from Halifax to Boston and Montreal.[5] For the American market, Yarmouth's close proximity proved advantageous and the port acted as a shipping centre for southwestern Nova Scotia.[6] By 1900, the province's fresh fish production was worth $2 million, while that of canned and cured fish was $5.3 million.[7]

In addition to steamer connections, Nova Scotia's fresh fish trade depended on the development of rail transportation, but change here was slow. It was not until 1876 that the circuitous Intercolonial Railway linked Halifax to Montreal. In the United States, the first transcontinental shipment of fresh fish occurred in 1884, but high freight rates postponed further shipments until a rate reduction in 1890.[8] A similar situation curtailed the fresh fish trade between the Maritimes and central Canada. In 1886, freight trains proved still too slow for such distant markets, while express shipment costs from Saint John to Toronto were higher than the original value of the fish.[9] At the same time, southwestern Nova Scotia continued its steamer exports of fresh fish to the Boston market. In 1886, an American tariff on fish that was "not fresh" created additional encouragement to fresh fish exports.[10]

By 1900, the fresh fish trade had assumed significant proportions and was growing faster than the dried fish trade. Between 1901 and 1911, the value of fresh fish grew from $1.9 million to $2.8 million, while that of canned and cured fish grew from $5.3 million to

$7.2 million for respective increases of 46.1% and 36.8%.[11] For inshore fishers, the fresh fishery offered an alternative to the production of dried fish. For vessel fishers, the fresh fishery supplemented the dry fishery, since vessel crews engaged in it only after dried fish production had ended. The small size of the fresh fish market also limited the number of dry fishing vessels engaged in it. In 1908, the fresh fishing fleet out of Halifax numbered 13 vessels, while there were over 100 vessels in Lunenburg's salt banking fleet alone.[12] Even with the early steam trawlers, initial employment was in the salt fishery in summer and in the fresh fishery in winter.[13]

The growth of the fresh fish trade in the early 1900s had its basis in technological improvements undertaken during the 1890s. Gradual improvements in rail scheduling and freight refrigeration enabled the fresh fish market to establish itself as a viable alternative to dried fish. For example, fresh fish exports from Canso grew from 162 tons in 1891 to 970 tons in 1895, and to 1,450 tons in 1900.[14] The location of rail centres determined the location of fresh fishing centres. In 1905, Halifax and Canso handled the bulk of the province's fresh fish trade. Halifax had its own rail terminal while Canso utilized the terminal at Mulgrave. Halifax firms sent carloads of Atlantic groundfish to Winnipeg and even as far west as Calgary, but Ontario and Quebec remained the most important markets. Total shipments of groundfish from the province were estimated at 10 million lb. in 1904.[15]

By 1900, the Intercolonial Railway further alleviated past problems by establishing a daily through freight from Mulgrave and Halifax to Montreal. The Mulgrave train connected with the one from Halifax at Truro and arrived at Montreal in a total time span of 61 hours. Fresh fish shippers looked for further reductions to 55 or even 50 hours and for quicker turnaround times at Montreal for fish bound to Toronto.[16] At the same time, they criticized the efficiency of the railway's insulated "refrigerator" cars which utilized ice. New cars of similar design but improved efficiency did not arrive until two years later. Thereafter, fresh fish was shipped iced in ordinary cars in winter and in the new refrigerator cars in summer.[17] There were also improvements in cost and speed of service. During the industry's early days, express rates from Halifax or Mulgrave to Montreal had been $2.50 per 100 lb. gross, or approximately $4.00 per 100 lb. net weight of fish. By 1901, the express rate was reduced to $1.50 per 100 lb. net weight.[18] However, the Intercolonial's rate policy continued to discriminate against dealers shipping less than a full carload of 20,000 lb. In 1901, the freight rate per 100 lb. for full carload lots from Halifax to Montreal was $0.26; to Toronto, $0.32; and to Chicago, $0.60. The respective rates per 100 lb. not shipped in car load lots were $0.35, $0.46, and $0.85, or increases of 34.6, 37.5, and 41.7% respectively.[19] The freight rate schedules favoured larger entrepreneurs who could take advantage of the savings obtained through shipping complete carload lots.

Efforts were also made to speed the shipping of fresh fish from the producers to the railhead. Canso's growth as a fresh fish centre was attributable partly to its natural advantages, but more importantly to the establishment of daily steamer communication with the rail terminal in Mulgrave.[20] In 1901, the steamer *John L. Cann* was making a triangular trip daily between Arichat, Canso, and Mulgrave, but there was need for another steamer linking Arichat and Petit de Grat with Mulgrave during the fishing season.[21] By 1903, a double subsidy for coastal steamers collecting and freighting fresh fish increased demand for such vessels.[22]

Despite steamer connections, fresh fish dealers still had to contend with seasonal differences in their source of supply. During the summer, they made direct purchases of fish from inshore fishers. Such purchases benefitted fishers who received immediate returns for fish caught, instead of waiting until the end of the season as was often the case in the dry fishery. In addition, a higher price was realized for fresh than for dry fish.[23] In the winter, however, inshore fishers were only able to fish when weather and other conditions were favourable. Some salt bank fishing schooners turned to fresh fishing in winter, using ice to preserve their catch. In 1905, the regular Halifax winter fleet consisted of 15 vessels averaging crews of 12 men each. Using longlines to catch cod and haddock, the vessels fished at distances of 14 to 20 miles from land.[24] On occasion they did quite well: in 1903, a schooner landed 170,000 lb. after a ten day trip to the western banks.[25]

Along with improvements in transportation, there were concurrent efforts to reduce production time. Consideration was given to the adoption of steam trawlers with otter trawls as used in the British fishery since 1894.[26] In December 1897, A.N. Whitman and Son of Canso conducted the first experiments using the wooden steam trawler *Active* which they had acquired from Aberdeen, Scotland. They outfitted the vessel with otter boards and trawls they made themselves based on descriptions obtained from Grimsby, England.[27] Speaking in later years, Whitman noted the experiment failed because "the boat was too slow; she did not have the power to pull the nets along fast enough, the smart fish got away, and the most she brought up was slower moving varieties, such as sole and flounder which at that time were not marketable here in any considerable quantity."[28] The *Active* subsequently fished with dories, a practice the Whitmans had adopted earlier with the steamer *Seabird*.[29]

The huge bag-like nets called trawls that interested Whitman bore no resemblance to the dory longlines (also known as trawls, dory trawls, and trawl-lines) used by many Nova Scotia fishers. Adopted by the Lunenburg County banking fleet in the early 1870s, each longline had a single groundline fitted with numerous evenly spaced short lines with hooks. Set on weighted lines at either end, the groundline ran parallel and close to the bottom where groundfish fed. Fishers checked the longlines only periodically but still they removed hooked fish and rebaited the hooks manually, which made the method labour intensive. British fishers had practised beam trawling since the first half of the 19th century. Sailing vessels initially hauled the trawls but by the early 1880s steam trawlers came into use.[30] In the mid 1890s, otter boards replaced the beam previously used to keep the mouth of the trawl open. The otter trawl was a conical net with the ends of its elliptical mouth extended to form two wings. The otter boards were large rectangles of wood heavily weighted and shod with iron. Each wing of the trawl had an otter board connected to the vessel by a steel cable. When set and under tow, the otter boards slid vertically along the bottom. Water pressure against the boards kept the trawl open, while the weight of the boards kept the trawl on the bottom. The trawls used a "1:3 slope," that is the cable length was set at three times the depth of the water.[31] In 1903, fishers reportely used beam trawls only in depths less than 50 fathoms, but used lighter otter trawls at twice that depth.[32]

Despite the failure of the *Active*, Nova Scotia entrepreneurs maintained interest in steam trawler technology. In 1901, the *Martime Merchant* followed the introduction of the steam trawler *Magnific* on the Newfoundland banks. Its captain pronounced the trawler's initial voyages only "fairly successful" as the trawler contended with unfamiliar local conditions. The rocky bottoms initially encountered proved destructive to the trawls. Also, July was reported to be more suitable than earlier for trawler operations because the cod were swimming deeper by then.[33] Two years later, two steamers from Yarmouth obtained crews in Halifax to go steam trawling amid speculation of a potential revolution in Nova Scotia's fish business.[34] Such speculation proved premature: in 1905, the steam trawler *Harbinger*, which had fished from Yarmouth and other South Shore ports during the previous couple of years, was advertised for sale.[35] During the first years of the 20th century, several entrepreneurs used steam trawlers as longliners carrying dories but none of these experiments proved lasting.

In 1908, A.N. Whitman repeated his earlier experiment and purchased a used Grimsby steel steam trawler, the *Wren*. Whitman planned to use the *Wren* for salt fishing in summer and fresh fishing in winter.[36] Smaller than the newer British trawlers, the *Wren* still enjoyed moderate success for several years. The vessel ultimately proved unsuitable for winter fishing, and like the *Active*, was eventually converted to dory fishing.[37] Whitman tried to ensure success by hiring a Scottish crew to operate the trawler. The move led an official of the Fishermen's Union to charge that "grasping monopolists" were attempting to reduce fish prices and wages to the point where local labour could not engage in the industry.[38] The protests indicated early opposition by fishers to corporate control of the fresh fish trade.

In spite of the protests over the *Wren*, the next several years witnessed the establishment of a steam trawler fishery in the province. In 1910, the *Coquet*, another small trawler owned in Aberdeen, Scotland, began fishing out of Canso.[39] This effort was followed in 1911 by the introduction of the first modern steam trawler, the *Cambodia*, under contract to the Maritime Fish Corporation, Ltd. and operating out of Canso. It was joined in 1912 by three other modern British steam trawlers fishing out of Nova Scotia ports and selling their catch to Canadian producers. In 1913, the Lockport Cold Storage Co. broke the cycle of using British vessels on contract by purchasing the French trawler *Baleine*.[40] The acquisition of the early steam trawlers underscored the increasing capitalization of the groundfish industry, particularly in the fresh fish sector. Cold-storage equipment, packing equipment, by-products plants and, even steam trawlers required large-scale capital investment.[41]

The continuing opposition to steam trawlers found an often sympathetic ear in government. In 1905, the Halifax Board of Trade asked for federal assistance to bring out steam trawlers from Scotland. In his response, the Minister of Marine and Fisheries noted the destructiveness attributed to trawling, and concluded that while it could not be prevented outside Canada's three-mile territorial limit, he was "not disposed to favour it in territorial waters or grant assistance to firms adopting that method of fishing."[42] Soon complaints proliferated about foreign steam trawlers interfering with the operations of inshore fishers. An order in council of 9 September 1908 prohibited steam trawlers from fishing with beam, otter, or other trawls within the three-mile limit or within the bays and harbours of Canada. Fishermen from the *Wren* had initially received the bounty paid to vessel fishers since 1883. After protests from fishers, an order in council of 22 February 1911 excluded fishers on steam trawlers from receiving the bounty.[43]

As the Minster of Marine and Fisheries noted in 1905, Canada was powerless to regulate trawling outside the three-mile limit, where regulation could only be obtained through international agreement. In April 1909, the Nova Scotia Legislative Assembly unanimously adopted a resolution advising the federal government of the destructive nature of steam trawling, recommending that it be prohibited, and urging the Canadian government to seek through the imperial government an international convention suppressing it on the open seas. That same year at the Imperial Defence Conference, Canada and Newfoundland raised the issue of an international agreement on regulating the bank fishery. No agreement was forthcoming and two years later the Canadian House of Commons passed a resolution calling for an international prohibition of trawling on the spawning grounds of the Gulf of St. Lawrence and the offshore banks.[44]

The outbreak of war in 1914 presented the province's fledgling steam trawler fishery with both opportunities and constraints. Increased wartime demand for fish, together with disruptions in the ability of European competitors to fill this demand, resulted in a considerable increase in fish prices. Steam trawler entrepreneurs had only a limited ability to capitalize on the situation through fleet expansion. Wartime competition for capital, labour, and vessel construction restricted the acquisition of new vessels, and the fleet was only able to replace losses including those stemming from wartime circumstances. Naval authorities chartered the *Baleine* early in the war for use as a patrol vessel.[45] In August 1918, a German submarine captured another steam trawler, the *Triumph*, on the Scotian shelf east of Canso. Converted into a makeshift surface raider, the *Triumph* sank one French, two Nova Scotian, and three American schooners before finally being scuttled itself. One suspects that this fortune of war did little to soothe the antagonism felt by traditional fishers towards their trawler rivals. Although the number of steam trawlers did not increase, the proportion that was locally-owned did. By October 1918, Nova Scotians owned three of the four steam trawlers fishing in provincial waters.[46]

Efforts to build trawlers within the province also underscored continued local interest in the technology. In 1918, three trawlers were built at LaHave and Shelburne for use in saltfish production. They marked a departure from existing steel steam trawlers in that they were wooden-hulled and based on traditional schooner-hull design, but they employed crude oil engines. Misfortune dogged these schooner trawlers and the last of them sank in April 1920.[47] The sinkings discouraged further interest in the vessel type until 1928 when the *Geraldine S.* was launched at Lunenburg.[48] The Nova Scotia experience contrasted sharply with that of New England. There, fishers adopted schooner trawlers at about the same time, but, encouraged by the burgeoning American fresh fish trade, they numbered 198 in 1929.[49]

After World War I, Nova Scotia's fishing industry adjusted unhappily to lowered peacetime demand and increased international competition. The offshore fishery decreased with some schooners choosing precarious but apparently more profitable careers as rum runners. By the mid 1920s some of the traditional dry fishery's problems were resolved and a number of vessels returned to the fishery. Buoyed by growth in the fresh fish trade, the province's steam trawler fleet suffered less from the post-war disruptions and by 1926 had reached 11 vessels. The steam trawlers had a demonstrated superiority in per vessel productivity. In 1903, 170,000 lb. was regarded as an excellent fresh fish catch for a schooner on a ten-day voyage, but in 1923 the steam trawler *Lemberg* landed a record 375,000 lb. after just five days at sea.[50] The efficiency of trawlers led to continued demands for restrictions on their use. The demands were especially prevalent among inshore fishers who provided fresh fish for most of the year, but they also came from vessel fishers who were now using auxiliary engines in their schooners. The combination of railway service and cold storage facilities enabled a growing proportion of Lunenburg's banking fleet to participate in the fresh fishery, and fishers on these vessels joined the chorus against steam trawlers.

The Royal Commission Investigating the Fisheries of the Maritime Provinces and the Magdalen Islands in 1927 and 1928 provided a forum for the trawler controversy. The commission's majority report on steam trawlers grouped the objections against them into three categories. First, there were objections based on "the general conservation of the fisheries." These included trawling's destruction of cod and haddock spawn, its damage to the feeding grounds, and finally its impact on fish stocks through the taking of immature and unmarketable fish. The second category related to the "direct protection of the fishermen." Critcs noted that trawlers were foreign-owned and crewed, and that they destroyed other fishers' gear without restitution. The final set of criticisms centred on "the serious economic problems involved." These included the contention that the inferior quality of trawler-caught fish was detrimental to fresh fish consumption, and that the control exerted by trawler companies over the fresh fish industry hurt traditional fishers.[51]

Based on the available scientific evidence, the majority report rejected the first two environmental objections. Because scientists had determined that cod and haddock spawn floated, the commission did not consider steam trawling destructive to it. The commission found the arguments against trawling's destruction of fish habitat less conclusive. While the likelihood of damage was acknowledged, its extent and effects were unknown and it was even postulated that the disturbance might be beneficial. The majority report concluded that trawling was destructive to immature fish and foresaw that there would "ultimately be a very serious depletion of the fisheries." The majority members viewed the progressive decline of the North Sea fisheries since the introduction and proliferation of trawlers as an object lesson.[52]

On issues relating to the direct protection of fishers, the majority report drew mixed conclusions. As Canadians owned six of ten trawlers then operating and British subjects formed the majority of the crews, it rejected the complaints of foreign ownership and crews. It accepted the complaints concerning trawling's destruction of fishing gear, since evidence of such damage was well-documented. Moreover, Canadian fishers had no legal recourse when their gear was damaged. The majority report viewed the question of damages as well as the taking of immature fish as issues requiring international agreement. It recommended that appropriate international conferences be held to resolve the regulation of fishing vessels on the banks.[53]

The majority report deemed the economic problems posed by steam trawlers to be the most significant ones. Yet, even here it weighted the problems unequally. It thought the supposed lower quality of trawler-caught fish was not a serious issue, and felt it was unlikely to injure fish consumption. Instead, it considered control of the market by trawler firms as "the heart of the whole problem." Based on greater regularity of supply, trawler firms had assumed a pivotal role in supplying the market. In consequence, during times of abundant fish supplies, traditional producers received markedly lower prices or no sales at all. Unremunerative prices had forced many traditional fishers out of the industry. Canadian fresh fish consumption supplied from the Maritimes was estimated to be 90 million lb. The theoretical production of the ten trawlers then in operation matched this amount. The majority report maintained that in the bigger economic picture, it came down to a question of retaining either steam trawlers or inshore fishers. Taking the proposition that "it is the function of industry in any country to produce men as well as goods, to make livelihoods as well as profits," the report recommended the quick prohibition of steam trawlers.[54]

There also was a minority report to the Royal Commission Investigating the Fisheries covering the same objections to trawling but often reaching different conclusions. For example, it rejected many of the destructive tendencies attributed to trawler fishing. The minority report agreed that the issue of damage to fishing gear should be addressed in an international forum. On the central question of competition between trawler and inshore fishers, the minority report supported trawler fishers. It saw the decline in the number of fishers since the advent of trawlers as part of a larger and earlier trend in the fishery. It viewed the growth in the fresh fish trade as a result of the trawler fishery and hence a benefit to inshore fishers. The minority report blamed unremunerative prices on excessive competition between shippers that could be overcome by greater cooperation. However, it felt that the trawler sector had responsibilities as well as rights that could be met by limiting trawlers to a number required to meet market requirements exceeding the capability of inshore and vessel fishers.[55]

Based on the commission's recommendations, the federal government enacted new legislation requiring that all trawlers be owned, registered, and licensed in Canada; that trawlers not fish within 12 miles of shore; that the number of licenses be restricted; and that regulations affecting licenses be made by orders in council. The beginning of the Great Depression intensified demands for action, and in October 1929 a tax was placed on fish landed by trawlers. Although declared *ultra vires* in March 1931, the tax still caused the withdrawal of trawlers from their base at Canso. By 1932, Nova Scotia's trawler fleet numbered only six vessels, all working for one Halifax corporation. Further regulations required that additional trawlers had to be Canadian-built, and that any company applying for trawler licenses had to prove that sufficient fish could not be obtained from hook and line fishers to permit its normal volume of business. In March 1933, following further demands by inshore fishers and the Nova Scotia government, the federal government reduced the number of licenses from six to four.[56] Trawler technology had at last been bound.

During the period under consideration, fishers never seriously threatened the regional groundfish resource. However, Nova Scotia's saltfish industry came under increasing pressure in its international markets. Social and economic disruption followed as fishers dealt with unstable and declining incomes. Coincidental to these disruptions, improvements in transportation enabled expansion of the province's fresh fish exports to North American markets. Entrepreneurs seized on new technologies, most notably steam trawlers, to provide improved and regularized fresh fish supplies to expanding markets. Trawler companies increasingly exerted corporate control of the trade, often to the detriment of traditional suppliers. As traditional fishers came under increasing pressure, they demanded government intervention to protect their dominance in the groundfish industry, fresh as well as salt. In the late 1920s and early 1930s, government resolved the "trawler controversy" by severely restricting use of the new technology. Employment was protected but at the cost of lower incomes and a less efficient industry.

Government did not long maintain the restrictions on trawler technology recommended by the 1928 Royal Commission. Increasingly, officials came to view the international saltfish trade as antiquated and unsalvageable. In 1944, Deputy Minister of Fisheries, Stewart Bates, advocated "modernization" of the groundfish industry.[57] Bates' influential *Report on the Canadian Atlantic Sea-Fishery* rejected saltfish production for international markets in favour of fresh and frozen fish production for the North American market. After World War II, a heavily-capitalized, industrialized, and more centralized fishery would gradually replace the dispersed, labour-intensive saltfish trade.

Notes

1. The word trawls has been used to describe longlines using hooks as well as bag-like fishing nets towed by vessels. Through out this paper, trawls will be used to refer only to nets and longlines to the line technology.

2. Ruth F. Grant, *The Canadian Atlantic Fishery* (Toronto, 1934), p. 91.

3. Grant, *Canadian Atlantic Fishery*, p. 91.

4. "Report of the Committee on Fisheries," Nova Scotia. *Journals and Proceedings of the House of Assembly for Nova Scotia*, 1846, app. 87, p. 255.

5. *Halifax Evening Express*, 19 June 1868.

6. "United States Consular Report, Windsor, N.S.," (1885), in United States. U.S. Commercial Relations - Canada 49-1 House Executive Document No. 253, vol. 34, p. 635 also "Report of Commercial Agent Roberts," (1887) in ibid., 50-1, H. Ex. Doc. 402, p. 556.

7. Canada. *Census of Canada*, 1901 (Ottawa, 1904), vol. 2, pp. 412-13.

8. C.H. Stevenson, "The Preservation of Fishery Products," *Bulletin of the United States Fish Commission*, 18, 1898, p. 368.
9. *The Critic* (Halifax), 19 March 1886, p. 2.
10. "Commercial Relations—Canada," United States, House Executive Document, No. 420, 50th Congress, 1st Session, p. 556.
11. Canada, *Census of Canada*, 1901 (Ottawa, 1904), vol. 2, pp. 412-13 and *Census of Canada*, 1911 (Ottawa, 1915), vol. 5, p.x.
12. *The Maritime Merchant*, 12 November 1908, p. 26.
13. William Wallace, "Steam Trawling in Eastern Canada," *Canadian Fisherman*, October 1918, p. 1039.
14. "Report of the Committee on Fisheries, 1901," Nova Scotia, *Journals and Proceedings of the House of Assembly of Nova Scotia*, 1901, App. 19, p. 2.
15. *Maritime Merchant*, 12 January 1905, p. 62.
16. "Report of the Committee on Fisheries, 1901," pp. 3-4.
17. *Maritime Merchant*, 23 April 1903 and ibid. 12 January 1905, p. 62.
18. Ibid., April 25, p. 44.
19. "Report of the Committee on Fisheries, 1901," p. 5.
20. Ibid., p. 2.
21. *Maritime Merchant*, 24 October 1901, p. 27.
22. Ibid., 26 March 1903, p. 66.
23. "Report of the Committee on Fisheries, 1901" p. 4.
24. *Maritime Merchant*, 12 January 1905, p. 62.
25. *Ibid.*, 12 November 1908, p. 26.
26. Albert C. Jensen, *The Cod* (New York, 1972), p. 128.
27. Wallace, "Steam Trawling," p. 1039
28. *Maritime Merchant*, 9 July 1908, p. 30
29. Wallace, "Steam Trawling," p. 1039.
30. Jeremy Tunstall, *The Fishermen* (London, 1969), pp. 17-19.
31. Jensen, *The Cod*, pp. 127-29.
32. *Maritime Merchant*, 31 December 1903, p. 64.
33. *Ibid.*, 4 July 1901, p. 44.
34. *Ibid.*, 5 November 1903, p. 32.
35. *Ibid.*, 12 January 1905, p. 64.
36. *Ibid.*, 9 July 1908, p. 30.
37. Wallace, "Steam Trawling," p. 1039.
38. *Maritime Merchant*, 20 August 1908, p. 23.
39. Wallace, "Steam Trawling," p. 1039.
40. *Ibid.* pp. 1039-40.
41. Harold A. Innis, *The Cod Fisheries: The History of an International Economy*. Revised Edition (Toronto, 1954), p. 435.
42. Canada. *Report of the Royal Commission Investigating the Fisheries of the Maritime Provinces and the Magdalen Islands* (Ottawa, 1928), p. 90.
43. *Ibid.*, p. 90.
44. *Ibid.*, pp. 90-91.
45. Wallace, "Steam Trawling," p. 1040.
46. Ibid., p. 1040.
47. Grant, *Canadian Atlantic Fishery*, p. 92.
48. *Canadian Fisherman*, July 1928, p. 23 and October 1928, p. 23.
49. Innis, *The Cod Fisheries*, p. 423.
50. *Canadian Fisherman*, March 1923, p. 74.
51. *Royal Commission Investigating the Fisheries of the Maritime Provinces* [1928], p. 92.
52. Ibid., pp. 92-93.
53. Ibid., p. 93-94.

54. Ibid., pp. 94-99.
55. Ibid., pp. 102-14.
56. For a fuller discussion of the restrictions placed on steam trawlers following the 1928 Royal Commission see Grant, *Canadian Atlantic Fisheries*, pp. 92-93.
57. See Miriam Carol Wright, "Fishing in Modern Times: Stewart Bates and the Modernization of the Canadian Atlantic Fishery," in this book.

chapter
fourteen

FISHING IN MODERN TIMES: STEWART BATES AND THE
MODERNIZATION OF THE CANADIAN ATLANTIC FISHERY

Miriam Carol Wright

The "modernization" of the Canadian Atlantic fishery after World War II involved profound changes not only in technology and the social and economic structure of the industry, but also in its relationship to the Canadian state.[1] Although dried, salted cod had been the main export product of the Atlantic fishery prior to the war, wartime demands gave an initial boost to the nascent frozen fish industry. In fact, the need for fish from war-torn countries led to government intervention to ensure adequate supplies for Canada's allies. The state's wartime involvement in the fishery was not temporary, however. After the war, the federal Department of Fisheries turned its attentions towards industrial development and embarked on a fisheries "modernization" program, led by Deputy Minister of Fisheries Stewart Bates, author of the influential *The Report on the Canadian Atlantic Sea-Fishery* in 1944 (Bates Report).[2] Official support and assistance to the frozen fish industry became the focus for this development project in the decades following the war.

Examining the Bates Report closely provides some insights into why the federal government favoured the development of the frozen fish sector and paid relatively little attention to saltfish, either in Nova Scotia, or later in Newfoundland after that former Dominion joined Confederation in 1949. More importantly, we can gain some understanding of the full implications of modernization, an approach to economic development employed by western governments in the post-war years and which federal bureaucrats applied to the Atlantic fisheries.[3] Although Bates was writing before the modernization paradigm reached its height and before such economists as W.W. Rostow popularized the theory, Bates' analysis of the problems in the Atlantic fishery, particularly his comparisons of the saltfish and frozen fish industries, were consistent with the modernization world view. As with many modernization theorists, he pointed to internal factors such as attitudes of the population and lack of capital or technology as the reasons for the problems in the fishery. His dualistic world view, his use of a traditional/modern dichotomy to explain the fishery, supports his connection to the modernization paradigm and post-war ideas about society and the nature of progress. For Bates, modernization meant far more than building larger fishing boats; it implied adopting new attitudes, accepting new values, and turning away from the world of the transatlantic mercantile trade, with which saltfish had been associated, towards a North American, industrial consumer culture. His portrayal of the conditions in the industry, and in particular the people who worked in the fishery, reveals much about the ideological underpinnings of modernization and the power relations within government decision-making.

The fisheries modernization program, however, did not occur in isolation. In fact, it was part of broader changes occurring in the relationship between the state and civil society in most western nations—the rise of the post-war welfare state. Governments began looking for ways to build new, democratic, yet decidedly capitalist societies, and at the same time mitigate some of the ravages of unbridled capitalism felt during the Great Depression of the 1930s. Although the historiography of the origins of the welfare state is broad, encompassing a variety of interpretations and ideological positions, most would agree that one of the results was a more direct involvement of the state in the economy. Other significant changes in the relationship between the state and society that occurred following the war, and which had repercussions in the development of the Canadian Atlantic fishery, were the growth of the bureaucracy and the rise of the influence of the "expert," the civil servant.[4] These men (and I use the term advisedly here) were valued for their expertise, be it economic analysis, scientific knowledge, or skills in diplomatic relations. Viewed as objective, rational, and skilled, the new civil service mandarins came to play a major role in post-war government planning.

These trends in the relationship between the state and society, which were occurring at international and national levels, also appeared in the government's approach to Canada's Atlantic fisheries. At the outset of World War II, the federal Department of Fisheries was a relatively minor department, having only been created as a separate entity in 1930. At that time, three separate agencies fell under its auspices: the Department of Fisheries, proper, which was the regulatory body; the Fisheries Research Board, an independent biological research organization; and the Fisheries Prices Support Board. Following the war, however, the Department of Fisheries expanded not only in size, but also in scale and scope of responsibilities. New divisions within the realms of education, technology, and economic development in the fisheries such as the Markets and Economics Service, Industrial Development Service, Education, and Consumer Service heralded involvement in previously untouched areas.[5]

One of the people promoting these changes was Stewart Bates, Deputy Minister of Fisheries from 1947 to 1954.[6] Bates, a university-trained economist, was just one of many who moved into the Canadian civil service following World War II. Economics, no longer viewed merely as the "dismal science," had a new prominence in the post-war bureaucracy, as it seemed to offer a way to predict and rationalize the chaotic world of capitalism.[7] Born in Scotland in 1907, Bates brought an economic perspective to a department that had previously been dominated by biologists. Educated at the University of Glasgow and Harvard University, Bates came to Canada in the 1930s to work for the Nova Scotia Economic Council. In 1937, he was appointed to the Rowell-Sirois Commission on Dominion-Provincial Relations and wrote a report entitled *The Financial History of Canadian Governments*.[8] In 1938 he became a commerce professor at Dalhousie University in Halifax, but his academic career was short. In 1942 he became Assistant Deputy Minister of the Department of Fisheries under Deputy Minister D.B. Finn. It was during this time that he wrote the Bates Report. He left the Department of Fisheries briefly following the war when C.D. Howe, a leading figure in the wartime civil service, hired him as Director General of Economic Research for the Department of Reconstruction and Supply. In 1947 he was back in the Department of Fisheries, this time as Deputy Minister. Records of his tenure in this position made note of several of his achievements, including directing the departmental reorganization in the late 1940s, and serving as chair for several international fishing commissions such as the International Commission for the Northwest Atlantic Fisheries and the International North Pacific Fisheries Commission. He was known to his colleagues as a brilliant thinker who brought trained economic intelligence to the area of fisheries management and development.[9]

Bates' role in formulating the early fisheries modernization program and his vision of the modern Atlantic fishery are the focus here. The Bates Report has been described as a "blueprint" for fisheries development in the post-war period.[10] It was commissioned as part of the larger Nova Scotia Royal Commission on Reconstruction and Rehabilitation, a "call-to-arms" for the post-war economy. The work was a long-range analysis of the fishing industry in Atlantic Canada that included discussions of the fishery since World War I, effects of the Great Depression, the mini-boom during World War II, and proposals for the future. He included in his analysis the three main sectors of the industry—the saltfish industry, the new frozen fish industry, and the inshore fishery. He concluded that the Atlantic fishery, in all its branches, was undercapitalized, inefficient, and lacking in technology that could raise productivity. Although the war had acted as a catalyst, particularly in the frozen fish industry, more assistance was needed before the Atlantic fishery would be on its way to modernization. A healthy future for the industry lay in the further development of the frozen fish industry. The saltfish sector needed to loosen its dominance, both structurally and psychologically, in Atlantic Canada to make way for a fully modernized fishery. New technology, such as trawlers, cold storage facilities, and refrigerated transport, however, would only partially address the needs of the fishery. Without a new spirit of enterprise and a new vision for the future among company owners and shore fishers alike, Bates believed the Atlantic fishery could never fully join the North American economy and society.

Since its beginnings in the early-16th century, the fishery off the east coast of what became Canada was part of the international Atlantic trade network. Merchant ships transported dried codfish from the region to ports in southern Europe, the Caribbean, and South America. In return, the fish producing colonies received sugar and molasses from the southern regions, and consumer goods from Europe. By the time that Bates wrote his report, much of Atlantic Canada's economy had shifted landward, away from the old maritime economy, but the saltfish industry still had undeniable links to the older transatlantic trade. The markets, the technology, and the products of the industry were still oriented towards the Atlantic, and Bates argued that this was one of the saltfish industry's fundamental problems. The future success of the industry lay, he believed, in focusing on the North American market, and there was little chance for expansion if saltfish was the main product.

The problem with saltfish, he claimed, was that it was desired only by the "Mediterranean and negro races" who were too poor to pay higher prices for the product.[11] Nova Scotia fishers, who typically bought most of their consumer goods from North America with its higher price levels, could not continue to sell their products in these lower-priced markets. In contrast, Bates claimed that rising standards of living in North America would soon lead to a demand for more diversified food products, such as "appetizers and side dishes for which fish is peculiarly suited."[12] By implication, Bates argued that a modernized Atlantic fishery would have to join the fray of the North American food industry.

Organizing the industry to make it more adaptable to the North American market was something that Bates thought necessary for the fishery in Atlantic Canada. The structure of the Atlantic fishery was rooted in the early history of the saltfish trade and had not moved towards more industrial (and efficient) forms of organization. He found the Atlantic saltfish firms weak because of their relatively smaller size, their lack of centralization and consolidation, and the absence of organizations among either the fishers or the exporters. He acknowledged that there were historical reasons for this, mainly that saltfish was originally caught and produced by the fishers themselves. Scattered production points along the coast led to the rise of smaller, localized saltfish companies. Although some newer processing methods had arrived in the 20th century (artificial dryers), he claimed that the industry never fully adapted to more "modernized" forms of organization such as were found in the Pacific coast fishery. Bates argued that the fishery in British Columbia was much more capitalized, concentrated into larger firms with modern vessels and processing facilities, and geographically centred in a few main areas.[13] Clearly, with Bates putting a premium on centralization and consolidation in the fishing industry, the non-centralized, small-scale saltfish firms would seem "backward" and inefficient.

It was for the areas of technology and methods used in the saltfish industry that Bates reserved some of his harshest criticisms. Raising individual productivity through new technology was one of the main goals of the modernization program, and Bates complained about the "retardation" of fishing methods in Atlantic Canada. In particular, he attacked the continued use of schooners and dories which had been developed for the Atlantic bank fishery. Schooners were less efficient, he claimed, because they could not make regular landings, and they were limited during times of bad weather. In contrast, the newer trawlers could fish year-round, and could lead to a greater degree of integration between the harvesting and processing sectors. Bates noted regretfully that the Nova Scotia government restricted trawler licenses in the 1930s, a political decision brought about as a result of opposition from company owners, schooner owners, and shore fishers.[14] The restricting of trawlers, he argued, not only limited the possibilities for development in the saltfish sector, but, perhaps more significantly, hindered the growth of the nascent frozen fish industry. In fact, in notes made while writing the report, Bates virtually accused the saltfish industry of casting a pall over the frozen fish industry:

> ...the tardiness of development has allowed interests that were vested in the older forms of the fish trade (particularly saltfish) to maintain their influences, their outlook and vision over most of the fishing industry and its associated institutions. Even the technique of production in fresh fishing has tended to be confined within the horizons of those interested primarily in the older branch of the trade.[15]

The "outlook" and "vision" of the saltfish industry, in Bates' mind, was not of a forward-looking, modern, industrial, North American fishery, but of a stagnating fishery that looked to the past. Elsewhere in the report, Bates remarked that the fresh frozen industry entered the war "under the compulsion of methods" of the saltfish trade; without the restraining influence of the latter, the former would be able to develop more easily.[16] As well, he argued that the trawler controversy resulted in the "strangling of the spirit of experimentation in new fishing methods, both inshore and offshore."[17] Clearly, Bates believed that the saltfish industry had a negative influence on the development of a modern, industrial fishery in Atlantic Canada and could not be allowed to maintain its dominance.

Bates' attitudes towards the saltfish trade extended to the people who worked in the industry—both company owners and fishers. In characterizing the fishers, Bates drew on stereotypes of the seafaring man and of the nature of rural peoples. Historian Ian McKay has recently explored the construction of the "Folk," the portrayal of rural Nova Scotians as hardy, innocent, unchanging, pre-modern people by amateur folklorists such as Helen Creighton from the late 1920s to the early 1970s.[18] Far from being an historically accurate depiction of rural Nova Scotians, McKay argues:

'Folk' as a social category in Nova Scotia, and its promotion and naturalization as a commonsense concept, were the product of twentieth-century modernity (urbanization, professionalization and the rise of the positive state) and the anti-modernism that arose as a response to it.[19]

Bates, too, employed the prevailing stereotypes about rural Nova Scotians in his discussion of the fishing people, although he and Creighton were viewing the fishing people from opposite ends of the spectrum. He thought they needed to change their ways and become modern, while she sought to preserve their supposed traditional ways. He romanticized the inshore fishers, yet insisted that these very romantic qualities of the fishers were the source of their backwardness. Fishers were ruggedly individualistic and independent, but at the same time they were stubborn, ignorant of modern ways, and in need of guidance. "Fishermen," he remarked, "often value highly freedom and the romantic and heroic attributes of the sea; sometimes, like older European seafarers, they act as if 'seafaring is necessary, living is not necessary.'"[20]

Independence posed a problem, according to Bates, because fishers valued it more than economic gain. He wondered at those who were "apparently willing to stand much economic stress so long as this freedom is unimpaired."[21] In contrast, Bates noted, the British Columbia fishers had entered the work culture of post-war North America, and they had become "near the position of wage-earners" with organized labour unions.[22] The lack of organization among Atlantic coast fishers made them more vulnerable to "commercial exploitation."[23] Clearly, Bates thought a more industrialized work culture would add stability to the industry, particularly by preventing companies from offering such low prices to fishers. One of the expressed aims of modernization was to raise the incomes of fishers by encouraging them to fish more efficiently and productively. Bates assumed this goal would be achieved more easily if the fishing people were part of the industrial framework.

Like the inshore fishers, the people who owned and operated the companies were individualistic and independent to the detriment of the industry as a whole. Bates had definite ideas about the nature of competition among capitalist enterprises, and he suggested that the saltfish industry was at a lower level of capitalist organization. He characterized the different types of competition within capitalism as being either of a destructive, predatory nature or of a type conducive to economic and social progress:

> Competition may mean at least two things 1) the struggle between individuals to make reductions in costs and improvements in production by striving for innovations and adaptations of all kinds of their own firms 2) the struggle between individuals for a share of a commodity. The second kind of competition is predatory, and it does little either for consumers or producers, or for the march of economic progress. It is the former kind of competition that has social virtue. It was the latter kind that typified the salt fish trade.[24]

Bates claimed that the saltfish industry was so competitive that there were no trade associations by which the companies could come together to modernize their trade. "To some of them," Bates argued, "co-operation is apt to appear to take away the whole spirit of the business."[25] In fact, Bates claimed that "the social milieu in the industry was not the kind that encourages the search for modernization."[26] Part of the problem was that the industry had not been able to attract young, dynamic "men of ability" to inject new life into the firms, and that many of the firms tended to be one-person operations, limiting possibilities for expansion. Bates did not place all of the blame for the problems in the saltfish industry on the attitudes and lack of vision of the people who worked in it. He acknowledged that years of depression and low prices in the saltfish trade since the end of World War I were a major contributing factor to the lack of capitalization and innovation in the industry. Yet, Bates found little movement towards modernization since World War II, when prices and production improved substantially.

Bates' assessment of the fishing industry led him to conclude that it would not be worthwhile to ask people in the industry what they thought should be done to improve it. The various sectors, and the people within those sectors, were so individualistic in outlook, that they would not be capable of making rational judgements about the future of the Atlantic fishery. "Various particular and sectional interests can and do speak, but never on behalf of the whole industry or its general welfare."[27] Perhaps indirectly, Bates was justifying his own role in making assessments of the problems of the Atlantic fishery. People like Bates—university-educated "experts"—could be trusted to make objective analyses of conditions in the industry, but a saltfish company owner, or an inshore fisher, could not. This is significant because it reveals something of the power structures within government decision-making following World War II, when the advice of the "expert" was privileged over that of those actually working in the industry. McKay suggests that inherent in the romanticized portrayal of rural people is "a systematic exclusion of those aspects of the past that would help people think historically about alternative outcomes or about patterns of power and privilege in society, or about themselves as agents and victims of history."[28] Indeed, Bates never gave the people of the Nova Scotia fishery that chance.

To Bates, the saltfish sector was a "traditional" industry that belonged to the past, not the present. It produced a product that was sold out of wooden barrels in open air markets in poorer southern climates. The companies that operated it were small, undercapitalized, and lacking in technology needed for large-scale harvesting and processing of fish. The people within the industry were missing the spirit of innovation to carry out sweeping changes in the direction of fisheries development. Yet, Bates argued, the saltfish industry was powerful enough to exert its influence and orientation over the entire Atlantic fishery. For that reason, he asserted that the frozen fish industry needed assistance, and that the saltfish industry should take a back seat to the sector he believed could deliver prosperity to the Atlantic fishery.

Apparent throughout the Bates Report is the assumption that the frozen fish sector offered the best hope for a modernized future through the technology employed in the industry, its industrial organization, the potential markets, and the type of consumer product it produced. Indeed, the frozen fish industry, as Bates saw it, was fundamentally North American in orientation. It was beginning to create "newer and more acceptable food products for North American consumers, in the form of fresh, frozen, smoked, and certain salted fish specialities."[29] Throughout the Report, Bates referred to the "revolution" in North American food processing that occurred because of mechanization and consolidation of agriculture and the food distribution industry. "For the housewife," he argued, "this transformation filled her shelves with canned fruits, vegetables, meats, soups, and a great variety of packaged and graded foods that her grandmother never knew."[30] In fact, his frequent references to the "housewife"—the central figure in the iconography of post-war consumer culture—provides a clue to the direction that Bates thought the fishery needed to move.[31] Fish products needed to be more accessible to North Americans, argued Bates, particularly the millions of people who lived inland and were not accustomed to preparing and eating fish. Frozen fish products, found in the frozen food section of the new "supermarkets," with their colourful packaging and cooking instructions on the back, seemed to embody the changes in the way food was bought and consumed in post-war North America.

Technology plays a particularly important place in Bates' schema, mainly because of his emphasis on raising the volume of fish caught and raising fishers' incomes without increasing the overall number of people working in the fishery. The frozen fish industry, which generally employed trawlers, large-scale processing plants filled with air-blast freezers, filleting machines, cellophane wrapping machines, and other equipment, had the potential to consolidate the industry and concentrate it in a few centres. Bates asserted that trawlers could also assist in the integration of processing and harvesting facilities. With a fleet of trawlers regularly landing fresh fish at processing plants, not only would quality improve, he argued, but more regular operations would save on costs of production. As Bates envisioned it, the modern fishery would consist of a number of larger firms in a few main ports, and perhaps a few smaller plants, no more than six, scattered along the Nova Scotia coastline to cater to the shore fishery. Ideas of economies of scale were at work here, with Bates arguing that the larger plants would be more economical, and more able to produce a better quality product.

If fishing operations were going to become larger and more mechanized, the workforce would have to undergo transformations as well. Because Bates found the mental attitudes of the people in the saltfish industry in Atlantic Canada so lacking in the attributes needed for a modern fishery, he placed a special emphasis on education in the future of the Atlantic fishery. Education needed to take place at all levels, from the fishers to the plant workers to the company managers, he argued, if they were to help raise productivity and if they were to have "the imagination and intelligence on which economic progress depends."[32] He suggested that a fisheries "workshop school" might be created, perhaps in Halifax, to provide training in marine engineering and navigation. He warned, however, that in the absence of sufficient primary education, industrial education would have to fill the void: "Industrial education may not be sufficient to create the desired habits of mind in the industry, but the attitude to adopt is not to regard such education as having only a remote chance of creating the desired habits of mind; it is rather the only chance."[33]

For Bates, education involved both technical training and perhaps more importantly, the development of a new mindset that would enable the fishing industry to move away from the world of the saltfish trade towards the modern industrial (and North American) world of frozen fish. Until the people of the Atlantic fishery accepted the social and economic values of the larger North American society, they would be doomed to lives of poverty.

Bates clearly believed that an industrialized, modernized, frozen fish sector offered the best future for the Atlantic Canadian fishery. An awareness of the particular context of the time in which he wrote the report, however, further adds to our understanding of why his vision for the future took the form it did. Bates wrote his report during the latter stages of the largest war in recorded history. The fishery itself underwent massive changes, with increased demands for products such as frozen fish, rising prices, and technological innovations that made mass production of fish for consumer use possible. Yet, just a few years before, the Atlantic fishery had been racked by a world-wide depression that left it impoverished by low prices and scarcity of credit. Bates insisted that fundamental changes needed to be made to the fishery because he worried that the prosperity brought by the war might only be temporary. Unless dramatic steps were taken, he warned, the fishery would slip back into its pre-war conditions.

Bates, however, did more than warn about the possible effects of a post-war recession. Throughout his report, he hinted that the post-war fishery, and the post-war economy and society, would be fundamentally different from what had previously existed. In the matter of fishing methods, he tentatively suggested that "while the physical changes have been slight, it may be that the war has brought changes in men's minds and outlook."[34] These changes included the growth in trade unionism on the east coast, a greater willingness to try new methods, and an acceptance by "some of the sterner individualists" that co-operative principles might be beneficial for the fishing industry.[35] Besides changing attitudes within the industry, Bates argued that those returning fishers or plant workers who had served overseas might well have different ideas about how they wanted to work and live, and this might have an impact on the fishery as a whole. Bates speculated: "More may soon understand that a previous boast, 'we have done it this way for 40 years,' is fast becoming a self-criticism."[36] In the post-war fishery, Bates suggested, people would look away from the past and its way of life, towards a new economy and society. Bates also indicated that the role of government in the post-war economy would be different from the past, and public opinion might favour assistance to the Atlantic fishery.[37] For one thing, Bates argued, developing an offshore fleet would make better use of the large fisheries resource, as well as providing a reserve of experienced seafarers for the post-war navy. For another, he claimed that reducing unemployment, economic inequality, and poverty would likely be one of the main economic goals of democratic countries after the war. He referred to the Rowell-Sirois Commission in Canada, which not only suggested that some areas of the country were indeed in need of economic assistance, but also recommended that it was in the national interest to provide it. Perhaps the public, less tolerant of regional disparity, might accept government intervention in the Atlantic fishery. Bates was looking ahead to the day when welfare state principles would be embraced by politicians and population alike, and he assumed that assisting the fishing industry would merely be one facet of the nation's evolution.

Bates, as Deputy Minister of Fisheries in the crucial post-war years, and as author of the Bates Report, was undeniably one of the most significant players in the shaping of post-World War II fisheries policy. Indeed, his approach to fisheries development, and his dualistic world view, are reflected in the writings of other fisheries planners in the 1940s and after, who were looking at the problems of the Atlantic Canadian fishery.[38] We need to be aware, however, that it was not merely the opinions and actions of a single man that moved fisheries development in this direction. Bates' ideas about what the future society and economy would be like were shared by most development economists and policy advisors in the western world following the war.

Many of Bates' views regarding the direction of fisheries policies for Atlantic Canada came to fruition in the years following the war. Although there were other factors that influenced the scale and pace of fisheries development in the post-war period, such as political pressures from the provinces, the market situation, and international involvement, Bates' model for the fishery generally prevailed. The federal government did favour the development of the frozen fish sector, raising incomes by increasing the productivity of individuals in the fishing industry through improved technology and educating the new fishery workforce. The Department of Fisheries started taking steps in this direction shortly following the war, carrying out experiments in fishing technology, providing funding for provincial fishers' education programs, and offering subsidies for vessels 45-65 feet in length. It was not until the 1960s that the government provided the degree of assistance to the industry that Bates had envisioned, with loans made to the frozen fish companies for trawlers and plant expansions.[39] The saltfish industry, however, received little attention from the federal government and its significance decreased in the years after World War II.[40]

As we look back, we can begin to see why the ideas that the Bates Report represented were attractive to federal fisheries planners. A centralized, industrialized, highly technical frozen fish industry seemed to promise a path towards higher productivity, higher efficiency, and higher returns to labour and owners. An educated, "forward-thinking" workforce, from fishers to plant workers to company owners, who had shed the shackles of "traditionalism," suggested a prosperous future for the entire Atlantic region. Additionally, the frozen fish industry seemed more closely connected to the North American economy and way of life than did the saltfish industry with its ties to the world of transatlantic trade. Catering to the "housewife" and providing a product that was not out of place in the supermarket, the frozen fish industry was integrating into North American consumer culture. Although none of those early fisheries bureaucrats, including Bates, foresaw the crises experienced by the Atlantic fisheries in the 1970s and after, their views provide some insights into the origins of the current problems. Their approach to development favoured improving technological efficiency to produce massive volumes of frozen fish. They created a bureaucratic structure that privileged the advice and opinions of fisheries "experts," including economists, scientists, and administrators, thus ignoring warnings of the declining state of the stocks made by fishers and others in the industry. Perhaps above all, Bates and the other fisheries "experts" believed that they knew how to build a modern, prosperous fishing industry. The weight of these fisheries advisors' opinions, which were connected to the North American industrial capitalist framework, played a role in precluding alternative forms of fisheries development. Experiments with co-operative ownership of fishing enterprises in Newfoundland in the 1950s, for example, received little government support. This is not to say, however, that the recent crises in the east coast fishery were solely the result of the actions and inactions of a federal government department in the first 20 years after World War II. Other factors such as the pace of capitalist development within the industry, demands of organized fishery workers' groups, inter-regional rivalry for federal assistance, and international involvement in the Atlantic fishery also helped shape the course of development. Yet, examining the role of people such as Stewart Bates in moving the Atlantic fishery in a particular direction will bring us to a more complete understanding of the roots of the problems of the late-20th-century Atlantic Canadian fishery.

Notes

1.   I acknowledge the financial assistance of SSHRC and ISER. I would like to acknowledge the assistance of Gregory Kealey and Robert Hong who commented on earlier drafts of this paper. I would also like to thank Ian McKay, who supervised my M.A. thesis, from which this article is derived.

2.   Stewart Bates, "The Report on the Canadian Atlantic Sea-Fishery," in *Nova Scotia Royal Commission on Reconstruction and Rehabilitation* (Halifax: 1944).

3.   In the post-war period, two general approaches were applied to economic problems by western governments: Keynesianism and modernization. Both involved the fostering of capitalism, but Keynesian policies were applied to areas already industrialized, whereas modernization policies tended to be used in underdeveloped regions. Policies inspired by the economist John Maynard Keynes included both social welfare and fiscal policies, to raise the level of aggregate demand by encouraging a higher degree of consumption through higher wages and social assistance, and providing incentives to invest for private enterprise. See Robert Campbell, *Grand Illusions: the Politics of the Keynesian Experience in Canada 1945-1975* (Peterborough: Broadview Press, 1987); David Wolfe, "The Rise and Demise of the Keynesian Era in Canada: Economic Policy, 1930-1982" in *Modern Canada*, Michael Cross and Gregory S. Kealey, eds. (Toronto: McClelland & Stewart,

1984)). The modernization approach, on the other hand, as it was practised in the post-war period, encompassed a dualistic vision of the world, which consisted of "modern" and "traditional" societies. According to this paradigm, "traditional" societies needed to be infused with capital, technology, education, and forward-looking, progressive attitudes before they could attain a higher level of social and economic existence as "modern" industrial capitalist societies. Modernization had its roots in 19th-century evolutionary theory, and was also informed by early sociologists. The paradigm reached its height in the 1960s, when it was popularized (and many would argue, distorted) by M.I.T. economist and U.S. government advisor W.W. Rostow with his book *The Stages of Economic Growth* (Cambridge: Cambridge University Press, 1960). Although modernization, as it was interpreted and applied by western countries in the post-war period, was attacked for its simplistic, linear concept of change which ignored such factors as class, gender, ethnic, and imperial relationships in economic disparity in the world, it remained a popular approach for underdeveloped regions in both western and so-called "third world" countries. For background on the modernization paradigm, see David Harrison, *The Sociology of Modernization and Development* (Boston: Unwyn Hyman, 1988).

4.   For background on the rise of the importance of the role of the "expert" civil servant in policy-making, see J.L. Granatstein, *The Ottawa Men* (Toronto: Oxford University Press, 1982); Doug Owram, *The Government Generation: Canadian Intellectuals and the State 1900-1945* (Toronto: University of Toronto Press, 1986).

5.   Canada, *Department of Fisheries Annual Report, 1950* (Ottawa: 1950).

6.   R.N. Wadden, *The Department of Fisheries of Canada 1867-1967* (Ottawa: 1967); *Trade News* 16: 11 (May 1964), p. 10.

7.   Owram, ch. 8.

8.   Stewart Bates, *The Financial History of Canadian Governments* (Ottawa: 1939).

9.   Kenneth Johnstone, *The Aquatic Explorers: A History of the Fisheries Research Board of Canada*, (Toronto: University of Toronto Press, 1977), p. 189.

10.   David Alexander, *The Decay of Trade: An Economic History of the Newfoundland Saltfish Trade 1935-1965*, (St. John's: ISER, 1977).

11.   *Bates Report*, p. 81.

12.   *Bates Report*, p. 113.

13.   *Bates Report*, p. 26.

14.   See *Bates Report*, pp. 41-43 for more on the Nova Scotia trawler controversy of the 1930s. Bates claims that by 1939, only 3 trawlers were fishing out of Nova Scotia ports.

15.   National Archives of Canada, RG 23, Bates Papers, vol. 505, box 3, file "Wartime Conditions and the Atlantic Fishery."

16.   *Bates Report*, p. 33.

17.   *Bates Report*, p. 44.

18.   Ian McKay, *The Quest of the Folk: Antimodernism and Cultural Selection in Twentieth-Century Nova Scotia* (Montreal and Kingston: McGill-Queen's University Press, 1994).

19.   McKay, p. 39.

20.   *Bates Report*, p. 107.

21.   *Bates Report*, p. 82n.

22.   *Bates Report*, p. 26.

23.   *Bates Report*, p. 107.

24.   *Bates Report*, p. 82n.

25.   *Bates Report*, p. 107.

26.   *Bates Report*, p. 61.

27.   *Bates Report*, p. 106.

28.   McKay, p. 40.

29.   *Bates Report*, p. 108.

30.   *Bates Report*, p. 11

31.   For background on gender and fisheries modernization see Miriam Wright, "Women, Men and the Modern Fishery: Images of Gender in Government Plans for the Canadian Atlantic Fisheries" in

Their Lives and Times, Women in Newfoundland and Labrador, eds. B. Neis, M. Porter, C. McGrath (St. John's: Killick Press, 1995).

32. *Bates Report*, p. 108.

33. *Bates Report*, p. 22.

34. *Bates Report*, p. 90.

35. *Bates Report*, p. 90.

36. *Bates Report*, p. 90.

37. *Bates Report*, p. 111.

38. Miriam Wright, "The Smile of Modernity: The State and the Modernization of the Canadian Atlantic Fishery, 1945-1970" (M.A. Thesis, Queen's University, Kingston, 1990).

39. The Fisheries Development Act of 1966 was the first large-scale assistance program to the fishing companies by the federal government. Direct assistance to frozen fish companies had a longer tradition in Newfoundland. The Commission of Government set up a loan program during the war, a program that the provincial government continued after Confederation.

40. See Alexander, *The Decay of Trade,* for a study of the decline of the saltfish industry in Newfoundland and the federal government's reluctance to assist.

chapter fifteen

THE INTERNATIONAL FISHERY OFF CANADA'S EAST COAST IN THE 20th CENTURY

Raymond B. Blake

Introduction

Fishing nations have been sending their ships and sailors to distant water in search of fish for centuries. It was international interest in the rich fishing grounds off North America that stimulated the first European economic activity in what eventually became Canada. Even after five centuries, international participation remains a prominent feature of the fishery off Canada's east coast. What began in the 16th century as an economic pursuit by Spanish, Portuguese, French, and English merchants and fishers, had, by the late-20th century, become a truly international activity as ships and sailors from nearly two dozen nations fished the famed Grand Banks. By the 1960s, when more than 2,000 ships and some 50,000 fishers worked the banks, it was little wonder that Canadian fishers often complained that at night the Grand Banks were lit up like a Christmas tree. The great fishing grounds had never seen such a fishing effort in all the years that Europeans were harvesting the wealth that the waters off North America had to offer. While the number of ships and fishers engaged in the fishery is one of the dominant features of the international fishery, the increasing deployment and destructive use of technology, particularly after 1950, is another. A third theme is the failure of the participants to implement a regulatory regime for the sustainable development of the marine resources in the northwest Atlantic. By the early 1990s, fishers had virtually destroyed every viable fish stock on the Canadian continental shelf.

Changes in Technology

When the 20th century began, most fishers engaged in the international fishery off Newfoundland sailed to the Grand Banks in schooners, known as bankers, to fish from dories, small flat-bottomed boats between 12 and 13 ft. in length that narrowed at one end. To capture the fish, fishers relied primarily upon trawl lines introduced in the mid-19th century. Trawl lines consisted of a main or ground line to which a number of shorter hooked lines, commonly called ganglings, were attached at regular intervals of 3.5 ft. The trawl, baited and neatly coiled in tubs each morning before the dories left the schooner, was set with each end of the line anchored and marked with a buoy and distinctive flag. The trawl usually ran about a mile and was left for a couple of hours before fishers pulled up one end to remove the fish and rebait the hooks. When bait was in sufficient supply and the weather cooperated, the trawl was set four times each day. At the end of the day's fishing, the fish was cleaned and salted.[1]

The introduction of the steam engine allowed fishers to utilize more sophisticated and effective catching technologies. The bankers were gradually replaced by steam trawlers, which were successively displaced by larger and more efficient crafts such as side trawlers, stern trawlers, and factory freezer trawlers. The typical early trawlers, often called side or beam trawlers, first used on the Grand Banks by the French in 1907, used the otter trawl, a large cone-shaped net or bag pioneered in England in 1894. Dragged slowly through the water, the otter trawl was kept open horizontally by rectangular 'doors' attached to the cables or warps towing the net. A vertical opening was maintained by weights on the bottom, floats on the top, and the pressure generated from towing. The net rolled along close to the bottom with the aid of bobbins.[2] Because the otter trawl had to be put in the water and retrieved over the side, the vessel had to come to a stop and turn broadside to the wind. In heavy weather this was a dangerous operation that frequently forced the dragger to cease fishing. Even so, side trawlers represented the finest in fishing technology on the offshore banks for nearly half a century.

In 1954 the British successfully introduced stern trawling and the first factory-equipped freezer trawler, altering forever the nature of fishing on the banks off Canada and, indeed, throughout the world. Distant fishing, necessitated largely by the destruction of coastal fisheries in Europe and the growing world demand for protein, had led to the development of factory freezer trawlers. The British *Fairtry*, more than 280 ft. in length and 2,600 in gross tonnage, was the first trawler that combined stern-trawling with onboard filleting machinery, freezing capability, and a fish reduction (fishmeal) plant; it became a model for a generation of fishing vessels. And, like so many fishing ships before it and thousands since, the *Fairtry* headed to the Grand Banks which British side trawler owners had eyed for years but found the distance too great for a viable fresh fishery. The *Fairtry* changed all this, of course, as distance was no barrier and the rewards were great; it rarely took the *Fairtry* more than 40 days to fill its fish hold with 600 tons of processed cod in its first year of operation.[3]

With open access to the groundfish stocks outside the Canadian three-mile limit, factory freezer trawlers enabled the Europeans to exploit the Grand Banks as never before. With the development of the highly successful midwater trawling in the late 1960s, these vessels became extremely efficient for capturing pelagic species such as herring, mackerel, and caplin, as well as the traditional groundfish. The midwater trawls resembled otter trawls in that they were conical and constructed of webbing. However, they had fewer weights and could be adjusted for towing at various depths by increasing the vessel's speed and by increasing or decreasing the length of the warp between the vessel and the net. Although factory freezer trawlers proved remarkably successful and allowed the Europeans to process their catch without salt, Canada did not license its first factory freezer trawler until 1986 when the *Cape North* was commissioned to meet the growing consumer demand for top-quality products processed or frozen at sea. However, Canada had adopted the large stern trawlers between 1966 and 1968, about ten years after the arrival of the modern European fleet,[4] but they preserved their catch for up to two weeks in crushed ice and operated in relatively close proximity to fishing ports.[5] International factory trawlers continued to fish the north Atlantic without any reservation until Canada and the United States imposed 200-mile exclusive economic zones in 1977.

Participants in the International Fishery

Throughout the early-20th century the Canadian Department of Fisheries attempted to monitor the level of international fishing, though it was difficult to gather exact information on the number of ships fishing on the Grand Banks. From 1908 to 1914, France sent an average of 236 sailing vessels and 25 steam trawlers across the Atlantic. By 1922, 111 sailing vessels and 26 steam trawlers operated from France.[6] During the 1929 season, fisheries' officials spotted 34 French trawlers, two Portuguese trawlers, and two Spanish trawlers on the Grand Banks. In 1934, a Canadian fishing captain counted 40 American trawlers, 35 French, 6 Spanish, and 3 Canadian.[7] A year before the outbreak of World War II, a departmental memo noted that at least 34 European trawlers were sighted on the Grand Banks, including one from Italy that had entered the Newfoundland fishery when Italian fishing interests purchased the *Gatooma*, Britain's largest steam trawler.[8] By the 1960s, many other nations began to fish the Canadian continental shelf. Japan assembled the largest fishing fleet in the world and regularly sent its fishing vessels to the Grand Banks after 1962. Poland, East Germany, Romania, Spain, Portugal, and France all built larger vessels to harvest the fish on the fishing grounds off Canada. Moreover, Greece, the Netherlands, and Belgium were all capable of fishing in the Atlantic by the late 1960s; even Israel operated a number of freezer trawlers in the central and north Atlantic.[9] Romania began exploratory fishing in the southern part of the continental shelf in 1964-65 and was making regular voyages by 1967. The Norwegians came to the Grand Banks in the 1970s in bulk carriers equipped as floating fishmeal factories in search of caplin and, together with other nations who fished the species, caused a precipitous decline in caplin stocks, a key source of food for cod. Italian factory freezer trawlers came for squid, German ones came in search of herring after the decline of North Sea herring stocks, and two American freezer trawlers turned to the Grand Banks. By the 1980s vessels registered in South Korea, Panama, Honduras, Morocco, and Venezuela had joined them.

Many of the Europeans were encouraged by bounties paid by their home governments.[10] The Spanish fishery, for instance, was revived through state assistance. Although Spain had nearly 100 trawlers fishing off the east coast in 1959, its government realized that the country was still deficient in animal protein. To address the problem in the most cost-effective manner, it chose to encourage the expansion and modernization of its distant deep-sea fleet rather than initiate an intensive exploitation of its own coastal fisheries. Spain then began an aggressive shipbuilding program by offering entrepreneurs loans of up to 80% of the total cost of production, and encouraged the construction of factory freezer trawlers.[11] Spain's production of frozen fish subsequently increased from 4,000 tons in 1961 to 500,000 tons in 1972, as Spain assembled the third largest fishing fleet in the world behind the USSR and Japan. In the northwest Atlantic, the fleet caught more than 200,000 tons of fish each year between 1961 and 1972.[12]

The Spanish fleet became one of the most productive on the Grand Banks because of the *parejas*, or pair trawlers, where a single net is towed between two ships. This technique was first used by Spanish fishers in the Mediterranean in the early 1700s, but it was only after World War II that the *parejas* moved from their traditional grounds off the Spanish coast in search of more productive fishing areas. First, they exploited areas off the Canary Islands and along the west coast of Africa, but in 1948 a lone pair of trawlers headed across the Atlantic to the Grand Banks.[13] Pair trawling, one of the most dangerous and difficult methods of fishing, demands considerable practice and skilled seamanship, especially when used in the rough waters and dense fog of the Grand Banks; it is also one of the most effective methods of fishing. Before the two ships approached each other to start the operation, one of them would have placed its net in the water. Tradition dictated that the ship that would take the second cable of the net (already in the water) had to throw a *tirador*, a light heaving line with an eight-ounce slug of lead, the *pina*, attached at the end, aboard the sister ship. With the *tirador* securely tied to the cable, the crew then pulled it back to their ship and attached the cable to the stern. When this was done, the two ships together pulled the net through the water. During the drag the vessels maintained a distance of a half-mile between them and a speed of one-and-a-half knots. Pieces of chain held the net close to the bottom and a continuous string of floats raised the headline to give the net a tremendous mouth. The towing of the net between the two ships kept the net open. The mouth of the net was wider than the standard trawl, though it was not as deep. Taking back the net also required special skills. When the ships began to 'haul' or take back the cable, they moved together until they were about 25 yards apart and until the net was near the stern of both vessels. Then the cable was transferred back to the other ship. (The ships usually alternated in using their nets and taking the fish on board.) When the net was at the side of the ship, the crew opened the cod-end and removed the fish. When only a small quantity of fish remained, the whole net was hoisted aboard.[14]

Soviet fishers also came to the Grand Banks as a result of state planning and support. The Soviet Union started to develop its fishing industry in the 1930s when it established several state-run institutes to train fishers. The state eventually constructed several hundred trawlers at shipyards in Russia and Germany and brought experienced fishers from England and other countries to provide instruction in fish processing and curing. The initial results were meagre.[15] After World War II, state planners realized that the country's agricultural production, especially animal protein, was insufficient to meet the people's dietary needs. They realized, too, that Russia had failed to develop its marine resource. After a lengthy investigation, Soviet economists concluded that the oceans could supply the necessary protein foods more cheaply than the land. Soviet leaders were subsequently convinced that the most effective way to satisfy the state's growing demand for animal protein was to exploit the marine resources at home and further afield. Poland, East Germany, and other nations reached similar conclusions.[16]

Once the Soviets had decided that they would obtain their necessary protein from the sea, fisheries' expansion became a national policy, meticulously planned and successfully executed.[17] By 1948, the Soviets had built several large trawlers to begin experimental distance fishing. Three years later they purchased several distant-water side trawlers from a British shipyard and used them for exploratory fishing on the Grand Banks. These experiments went largely unnoticed, but they must have been successful as the Soviets became interested in the *Fairtry* and placed a tender in 1953 for 24 factory freezer trawlers. After lengthy discussions between the British shipyard and Soviet officials, the negotiations broke off, but not before the Soviets secured the blueprints for the *Fairtry*. Meanwhile, the Soviets continued with their plans for a distant water fleet and had 24 factory freezer trawlers that coincidentally resembled the *Fairtry* constructed in the Howaldtswerke shipyards in Kiel, West Germany.[18] In 1956, the Soviet Union made its debut on the Grand Banks with two factory freezer trawlers, the *Pushkin* and the *Sverdlovsk*, which were almost exact replicas of the *Fairtry*, and within two years there were 35 other Soviet factory freezer trawlers on the banks. By 1965, the Soviet fishing fleet in the northwest Atlantic had grown to 106 factory freezer trawlers of the *Pushkin* class or larger, 30 mother or factory ships, and 425 side trawlers that harvested 886,000 tons of fish, one-third more than the combined catches of Spain, Portugal, and France; in all, the Soviets had a fishing fleet of 2,370 vessels for deployment around the world. One of the vessels in the Soviet fleet was the *Professor Baranov*, a factory ship 543 ft. in length, that processed the catch of a fleet of 20 SRT-class trawlers, each between 130 and 180 ft. long. In a single day, the vessel could salt 200 tons of herring, reduce 150 tons of fish and offal into fishmeal, fillet and freeze 100 tons of groundfish, and manufacture 5 tons of fish oil. It also had the capacity to produce 20 tons of ice and 100 tons of distilled water daily. By 1974, the Soviet fleet had grown to more than 4,000 vessels.[19]

The Soviets were not the only ones constructing large factory freezer trawlers and sending them to the Grand Banks. In 1957, one year after the Soviet arrival, West Germany's first experimental stern trawler, the *Heinrich Meins*, made its debut on the banks, and it was soon followed by 30 more factory freezer trawlers. The West Germans quickly earned a reputation for finding the fish with the aid of sophisticated electronic equipment. In 1969, they perfected the mid-water trawl which allowed the vessel to tow at any depth by "a combination of airfoil-shaped doors, changing ship speeds, and weights on the towing cables."[20] Some of the mid-water trawls were 1000-ft. long, the length of three football fields. Moreover, the Germans attached underwater sonars to the headlines which gave them a constant reading of the depth of the net, the topography of the bottom, and the fish entering or leaving the trawl. These sonars, which were able to scan up to two miles ahead, made the mid-water trawl extremely effective.[21] In 1974 some 1,076 fishing vessels crossed the Atlantic from western Europe and the Communist bloc to fish in the waters off Canada and the United States.[22]

Unlike many other countries, Portugal retained its traditional line fishery, prosecuted from one-person 16 feet dories, *a pesc à linha*, long after other nations had turned to trawlers. By the 1950s Portugal also had a fleet of about 25 otter trawlers. The trawlers made two voyages a year, one from February through June and the other from July or August to November. The line-fishing fleet was limited to a single voyage. The fishers in this fleet usually rose at four in the morning and after breakfast rowed away, or sailed (if winds were favourable) from the mother ship for a day's fishing with handlines and baited trawl. Between two and four in the afternoon, the dories returned to the vessel and unloaded their catch. Some of the newer vessels also carried larger three-person boats that were equipped with hauling devices and usually carried three fishers. After a hearty dinner, the fishers had to clean and salt the day's catch. Most of this work was done on the deck of the ship and they commonly worked past midnight to gut, behead, split, wash, and salt the cod. In the 1950s, many of the ships were equipped with loudspeakers that piped music and songs to the fishers as they worked. This was obviously a morale boaster, but many of the captains believed that it increased the productivity of the fishers as well. After a day of solitary fishing in a small tossing dory, the crew were eager to gossip, but the captains thought that the more the men talked the more slowly they worked. Once the music was turned on, the men ceased talking and subconsciously adjusted their actions to the rhythm of the music: the faster the tempo the faster they worked. A Portuguese captain told an official from the Canadian Department of Fisheries that "we get 30% more work out of the men with music." The fishers received an advance against their wages when they joined the fleet as well as payments for individual jobs such as splitter or salter, and additional wages based on their personal catch. The captain estimated each fisher's production when the dories returned to the ship. All the fish were taken back to Portugal wet-salted, to be dried there either in the sun or in modern drying plants.[23]

This adherence to traditional fishing in the age of modernization can be explained largely by the philosophy and corporate structure of Portugal under Antonio de Oliveira

Salazar, who held power from 1926 until the Revolution of 25 April 1974. It was an "authoritarian, nationalistic, Christian, anti-capitalist and anti-communist state.... [It feared] liberalism, democracy, populism, mass society, and industrialization."[24] Industrialization and modernization of the fishery were inconsistent with the government's philosophy. The state consequently offered the cod fishers numerous incentives to maintain the labour supply for the traditional fishery. The government created an organization, known as the Gremio, to control and regulate the fishing industry and develop a modern fleet.[25] The Gremio provided a series of social welfare measures for fishers and their families. It assisted with the establishment of mutual insurance groups, regulated labour contracts, and provided medical assistance. The Gremio also established fishing schools along the Portuguese coast and undertook an extensive housing program for fishers, constructing what amounted to planned settlements for them. These settlements usually included the Fishermen's Institutes, schools, dispensaries, medical centres, recreation facilities, maternity centres for fishers' wives, and public nurseries. The Gremio also controlled the fishing activity, drying, and the sale of the final product.[26] Many fishers joined the Newfoundland fleet to escape military service and earn enough cash to build a home and perhaps purchase a small fishing boat to participate in the coastal fisheries in Portugal. Shortly after Salazar died in 1968, the state was unable to maintain its control over the economy and the fishery was modernized along with many other sectors of the economy.[27]

Beginning in the mid 1950s, and especially in the 1960s, more European nations moved into the northwest Atlantic as the resource was depleted in the northeastern sections closer to Europe. Despite initial rapid growth, the yields for cod and haddock from the main fishing areas of the Grand Banks failed to keep pace with the increase in fishing activity. This trend had been evident from the mid 1950s in the major cod fishing areas of the northeast Arctic, Iceland, and West Greenland, and by the mid 1960s was being reflected in the Newfoundland-Labrador area. Between 1960 and 1968, the catches of all groundfish off Canada's coast increased by 55%, and by 1968—when world fish production peaked—international fishing fleets were catching more than 80% of the groundfish on Canada's east coast. It was becoming clear that the sector was being heavily overfished.[28] Yet, despite Canadian and American concerns about overfishing, the effort continued to increase.[29]

The new participants in Canada's east coast fishery not only competed with the traditional fishing nations for the conventional catches of cod, redfish, haddock, and flatfish but also developed new fisheries, usually with disastrous consequences. Large international factory freezer fleets frequently targeted a particular species and then moved on to another as soon as the catch per unit fell to an unprofitable level. The USSR, for instance, was among the first to develop small-mesh gear fisheries on the Grand Banks. The International Commission on the Northwest Atlantic Fisheries (ICNAF) was powerless to halt such fishing practices. A series of ICNAF regulations introduced in 1953 governing the net size of bottom-trawls were never ratified by member-states. When the Soviets began to develop a fishery for small-bodied silver hake, they introduced nets with meshes of 40 mm. (ICNAF regulations called for large mesh nets of at least 114 mm. for the groundfish sector). Young cod and other traditional species were taken, incidentally, in the small mesh-gear fisheries, which no doubt contributed to the decline of traditional fish stocks. The Soviets also developed an offshore herring fishery in the 1960s to support a fishmeal industry that, combined with the vigorous development of a Canadian herring fishery, devastated the stock. Once that happened and the herring fishery was no longer commercially viable, the international fleets turned to other pelagic species, first mackerel, then caplin and squid; all of the species subsequently went into steep decline.[30]

Regulation Without Cooperation

As the fishing nations shifted to more versatile and sophisticated fishing technologies, often at considerable capital expense, economics and profit, not conservation, drove the industry worldwide. There was increasing pressure from owners and governments—which often granted huge subsidies to the fishing industry—for year-round fishing operations, prompting seasonal shifts from one species to another.[31] Although some scientists, such as J.J. Cowie of the Canadian fisheries department, believed in the 1920s that the resources in the oceans were infinite, other scientists and officials from many of the fishing nations were warning about the consequences of overfishing. Even after the fishing nations agreed to protect the stocks in the 1940s, too many fishers, often with the support of their governments, continued to pursue the fish with little regard for the rules. Canadian Fisheries Minister Jack Davis warned ICNAF members in 1971 that "the catching capacity had increased and...nature, it seems, cannot keep up with us. The regeneration of fish stocks is not sufficient to withstand this attack from outside. We have now, or will soon have, too many fishing vessels chasing too few fish. The Northwest Atlantic is over-manned today." However, many of the warnings about resource depletion were largely ignored until the stocks virtually collapsed in the late 1980s.

Increased European activity in the waters off Newfoundland and Canada during the interwar period had alerted some Canadians and Newfoundlanders to the dangers of overfishing. However, it had become clear in Newfoundland as early as the 18th century that there were limits to the number of fish that could be caught in a given area. The traditional response to poor catches, declining yields, and overcrowding was to move to unsettled and unexploited areas in search of new sources of fish. With the introduction of steam trawlers and year-round fishing in the early-20th century, the prospects of overfishing took on new meaning. In 1911, Canada and Newfoundland called, unsuccessfully, for a ban on the use of steam trawlers, arguing that the new technology threatened the fish stocks by destroying the young fish, polluting the bottom by discarding unwanted species, and destroying the gear of inshore fishers. London reminded both countries that the whole issue of trawling was "very complex" and that any measures to deal with trawling outside territorial waters would be ineffective without international agreement. Nevertheless, Nova Scotia limited the number of trawlers in the province, prompting historian Ruth Grant to argue in 1934 that such a policy only discouraged large-scale capital investment in the fishery and prevented the expansion of fish products into the lucrative Canadian and American markets. Moreover, she suggested that conserving the fish stocks was an international problem that had to be addressed by all fishing nations; it was naive, she maintained, for Canadians to limit the number of trawlers to four to protect the fish stocks while other countries sent their fleets to fish off the Canadian and Newfoundland coasts without any restrictions.[32]

In 1920 Canada, Newfoundland, and the United States created the North American Council on Fishery Investigations to undertake a series of scientific investigations of fish stocks on the banks adjacent to their shores. Yet, even when it became clear that the stocks were threatened, the three countries failed to ratify an agreement that would have limited the mesh size for otter trawls. There were simply too many nations with too many different interests to adopt a common strategy for conservation. In 1946, the fishing nations bordering the North Atlantic did agree, however, to separate the North Atlantic into eastern and western sections for the purposes of conservation.[33]

A widespread belief in the efficacy of internationalism, perhaps best demonstrated by the creation of the United Nations, emerged after World War II. In a spirit of cooperation rare among the fishing nations, Canada and the United States played leading roles when 11 countries who fished the Grand Banks came together in January 1949 and created the International Convention for the Northwest Atlantic Fisheries (ICNAF) to usher in a new era of management and conservation. ICNAF came into force on 3 July 1950 with responsibility "for the investigation, protection, and conservation of the fisheries of the northwest Atlantic in order to make possible the maintenance of a maximum sustained catch from those fisheries." By promoting and coordinating the results of scientific investigations, the commission was to ensure the wise use of the commercial fish stocks.[34] Each member state had one vote, and ICNAF made decisions based on a two-thirds majority. The northwest Atlantic was divided into sub areas, each defined by an alphanumeric code, and panels were established to monitor the groundfishery in each area and to make recommendations to the commission. However, ICNAF had only the authority to recommend; it had no power to enforce its regulatory measures. It soon became clear that while ICNAF was a step in the right direction, it was ill-equipped to conserve fish stocks.[35]

By the mid 1960s, conservation and overfishing had become perennial subjects for ICNAF annual meetings. Members agreed that the problem of managing the northwest Atlantic fishery in the face of a continually increasing fishing effort was, indeed, a pressing one, but little was accomplished. Even an ICNAF regulation passed in 1953 calling for minimum mesh size for otter trawls continues to be ignored by some member states. By the late 1960s, ICNAF warned that fishing activity was approaching or had already reached the level giving maximum sustainable yields for the main cod and haddock stocks in the region.[36] In 1967, Stanley A. Fish, the Assistant Secretary for Fish, Wildlife, and Parks in the US Department of the Interior, said fishing nations had recognized as early as the 1940s that some of the stocks in the northwest Atlantic showed signs of depletion, but even with the creation of ICNAF they had done little to deal with the problem. Overfishing was becoming more serious, he warned, as fishing nations continued to increase their catching capacity. He suggested that additional regulatory action and greater international cooperation were necessary, adding that an acceptance of the much-discussed 200-mile economic zone around coastal states might be necessary to protect stocks. Modern technological developments no longer allowed a leisurely approach to fisheries conservation because a stock could be depleted quickly. He called upon ICNAF members to cooperate to solve the technical problems of regulating the fishery and to demand that their governments accept ICNAF's recommendations more quickly.[37]

While ICNAF members clearly recognized the imminent threat to the Atlantic fishery if the level of fishing were maintained, they saw no easy way to protect the stocks. All of the fishing nations had invested heavily in the industry and none of them were eager to reduce their fishing efforts. Not surprisingly, one of ICNAF's first steps in addressing the problem of overfishing was to establish, in 1966, a working group to examine the economic effects of any conservation measures. The group of economists and biologists found that the cost of harvesting the annual catches was substantially greater than necessary because too many nations had constructed too many vessels for too few fish. It even suggested that a moderate reduction in fishing activity might result in an increased catch per vessel. Unless restrictions were imposed on fishing activity, it warned, fish stocks would come under additional pressure. A report from the Organization for Economic Cooperation and Development had indicated that fishing fleets engaged in the northwest Atlantic were expected to grow rather than shrink.[38] Because of government support for most national fishing fleets, the ICNAF group saw little likelihood of any self-regulating mechanism to maintain capacity at the current levels; in other words, it was unlikely that economic forces alone would prevent further increases in fishing effort in the groundfishery. The group recommended a quota or catch limit to reduce the fishing effort in the northwest Atlantic. This would set the amount of fish to be harvested by member nations from a given stock over a specified period of time. A national quota, moreover, would allow each country to regulate its quota in a manner best-suited to its internal economic conditions. ICNAF adopted the recommendations of the group in 1967 and established a Standing Committee on Regulatory Measures to examine the administrative, economic, and practical aspects of fixing annual catch quotas.[39]

It was 1970 before ICNAF established its first quota or what it termed the Total Allowable Catch (TAC). The quota was based on the concept of Maximum Sustainable Yield (MSY) or the maximum amount of fish that could be taken from a stock without depleting it. The first quotas were issued for haddock stocks on Georges and Browns Banks, but Canada advocated the implementation of TACs for all stocks in the northwest Atlantic in order to control the excessive fishing activity of some member states. Jack Davis warned ICNAF in 1971 that "our catching ability is now beginning to outstrip our resources. There are no longer enough fish to go around [and] there is a very real danger of overfishing in the northwest Atlantic in the 1970s." Later that year, ICNAF agreed to allocate shares between member countries, and by 1974 TACs were established for most species of fish. ICNAF also instituted closed areas and seasons for threatened stocks, but took into consideration the historic participation of individual states and the special interests of coastal states like Canada. In allocating national quotas, ICNAF allowed each country to plan its fishing activities and invoke some order in the fishery by avoiding the intense competition created by a first-come first-served fishery. However, because ICNAF considered economic factors as well as biological ones in establishing TACs, the quotas were often too high.[40]

Although ICNAF adopted a more interventionist and aggressive stance in the 1970s, its policies could, at best, only slow the decline in the stocks and could do little to ensure that fish populations could be rebuilt to levels that might provide higher sustained yields. Despite ICNAF's attempt to impose a ceiling on catches, member states regularly ignored the quotas which, under the existing rules, were strictly voluntary. Each member state was expected to police itself, and not surprisingly, the overfishing that took place in the 1960s before TACs were established continued.[41] By the 1970s, fish stocks were in serious trouble; the Canadian groundfish industry virtually collapsed as the total catch of cod in the northwest Atlantic fell from 2,000,000 metric tonnes in 1965 to less than 500,000 metric tonnes in 1977. At the 1975 ICNAF meetings there was general acceptance of Canada's call for stricter conservation measures and reduced fishing effort. However, member states were unwilling to take fewer fish to allow the stocks to recover. Portugal, for instance, claimed that it was not responsible for the decline, and Spain said that any reduction in its quota would mean great economic hardship for its people. Spain can certainly be chastised for its position, but the Spanish representatives were the only ones honest enough to admit the truth about their situation. Profit, which had come to mean jobs in the fishing industry, was a key impediment to meaningful conservation, and had taken precedence over conservation in Spain, Canada, and elsewhere.[42] Hence, ICNAF's efforts to establish order in the fishery and to protect the dwindling stocks were defeated by its own member states who refused to accept quotas that might have allowed stocks to recover. Member states often submitted incomplete and inaccurate catch reports, particularly with respect to incidental catches and discards. The scientific and biological data were often inadequate and imprecise, and the TACs were often allocated based on outdated statistics. Moreover, the majority of the ICNAF members were not really interested in conserving the resource—a Protocol to the 1964 ICNAF Convention allowed countries that lodged a formal objection to ignore the conservation measures adopted by the commission.

Even while working within ICNAF, Canada and other coastal nations adopted a more aggressive role to claim and manage the fish stocks in the waters off their shores, although conservation was not their only guiding principle. Coastal nations had also become interested in the mineral wealth that lay under the ocean's floor. By 1970, Ottawa had proclaimed a 12-mile territorial sea and, in 1971, it declared the Gulf of St. Lawrence and the Bay of Fundy exclusive fishing zones, bringing them under complete Canadian management.[43] Even as Canada extended its jurisdiction it only slowly phased out the fishing operations of those nations with historical fishing rights on the east coast.

Although the Third Law of the Sea Conference had failed to reach an agreement on the right of coastal nations to control the resources off their shores, a consensus had clearly emerged that coastal nations had the right to take such action. In 1977, Canada and several other coastal nations declared 200-mile economic zones extending from their coastlines.[44] Canada claimed jurisdiction over most of the fish stocks on the continental shelf, but the southern and eastern tips—known as the nose and tail of the Grand Bank—as well as the Flemish Cap lay outside its jurisdiction. The Canadian delegation had argued unsuccessfully at the Law of the Sea Conference that coastal states should be permitted to extend their jurisdiction to the edge of the continental shelf to protect the fish stocks that migrated to the breeding areas just outside the limits of a country's jurisdiction. Ottawa wanted the Canadian continental shelf treated as a single ecosystem, but failed to win international support for its position. Consequently, Canada could not effectively manage all the fish stocks on the Grand Bank, and an international regulatory body continued to regulate international fishing in the northwest Atlantic in areas outside the 200-mile limit.[45]

Nations fishing in the northwest Atlantic created a new body, the Northwest Atlantic Fisheries Organization (NAFO), to replace ICNAF after Canada extended its jurisdiction in 1977. It was responsible for studying, managing, and conserving the northwest Atlantic fishery resource outside Canada's 200-mile fishing zone, and its members included Bulgaria, Canada, Cuba, Denmark, the European Community, Iceland, Japan, Latvia, Lithuania, Norway, Poland, Romania, and the Soviet Union. Following Canada's adoption of a new regulatory regime for the stocks within the 200-mile territorial limit, NAFO also abandoned the concept of Maximum Sustainable Yield. It was superseded by a new regulatory regime based on a concept known as $F_{0.1}$, a measure of fishing mortality that set quotas at about 40% below maximum sustainable yields, permitting a "healthy and steadily growing stock and a TAC that increased proportionally in successive years."[46]

Despite the overfishing, Canada remained committed to the concept of internationalism and tried to avoid inflicting undue hardship on countries that had traditionally fished in Canadian waters. Moreover, Article 62 of the Law of the Sea Convention had stipulated that coastal states had to determine their capacity to harvest within the established TAC limits in their 200-mile economic zones, and if the TAC were greater than their ability to harvest, the coastal states had to give other countries access to the surplus. Canada subsequently signed 13 bilateral fishery agreements, including one with the European Economic Community on behalf of four countries, which permitted their vessels to continue to fish, at reduced levels, within Canadian waters for surplus species.[47]

Canada's licensing program for international vessels was a key element of its management of the 200-mile zone, dictating exactly where, when, and how each could take its quota or allotment of fish. Canada determined the quotas annually and allocated to foreign fishing fleets only the fish stocks surplus to its own needs; those nations had to sign bilateral treaties with Canada and adhere to Canadian fisheries' laws. As well, they had to comply with conservation measures and other conditions established by Canada and NAFO. The Department of Fisheries and Oceans together with the Departments of Transport and Defence monitored fishing activities within the 200-mile limit through air and sea patrols, but the area was simply too large for effective surveillance. The bilateral treaties also demanded the countries' adherence to NAFO quotas beyond the 200-mile limit and were often tied to the maintenance of a satisfactory trade relationship with respect to the purchase of Canadian fish and other products.[48] International vessels were to report to Canadian authorities prior to entering and departing the zone and had to report their catches on a weekly basis. Moreover, they were required to submit to detailed inspection of their operations, catches, and records by Canadian surveillance and enforcement personnel. Canadian observers were placed on board international as well as domestic vessels to observe fishing and handling methods and to monitor quota management requirements.[49] Under this regime, the East Bloc countries were clear winners, as they were given 52% of all foreign allocations in Canadian waters. If there was a loser, it was Spain, still among the top three deep-sea fishing nations and with a large domestic market for fish products, but which lost much of its traditional fishing grounds when Canada declared its 200-mile limit.[50] While Spain and other fishing nations turned first to other fishing grounds, many of those nations that did not have bilateral treaties with Canada or were dissatisfied with their allotments turned to the nose and tail of the Grand Bank just outside the 200-mile limit.

The 200-mile limit and the new regulatory regime were intended, in part, to conserve and restore fish stocks, and at first they appeared to be working. There was a dramatic increase in Canadian groundfish catches after 1977, and Canada's share of the catch reached 73% in 1979, up by more than 21% over 1977 levels. Much of this, of course, was attributable to increased fishing. Even as the yields reached an historic high on the east coast in 1986, the World Commission on Environment and Development warned that the world's oceans were in trouble: "Today [1987], the living marine resources of the sea are under threat from over-exploitation, pollution, and land-based development. Most major familiar fish stocks throughout the waters over the continental shelves, which provide 95% of the world's fish catch, are now threatened by overfishing." By the early 1990s, the cod fishery on Canada's east coast had collapsed.

The Collapse of the East Coast Fishery

The emotional response of most Atlantic Canadians is to blame the international fishery for the decimation of the resource, though inshore fishers are as quick to blame Canadian trawlers as international ones. What role did the international fishing fleets play in the destruction of the cod and other groundfish stocks? The evidence suggests that they played a prominent one, though all those involved in the fishery must shoulder a share of the blame. Even within the 200-mile limit Canada failed to maintain a sustainable fishery, but it was virtually powerless to control fishing outside the 200-mile limit. As early as May 1979, Gordon Slade, the Deputy Minister of Fisheries for Newfoundland, warned Ottawa at a Federal-Provincial Atlantic Fisheries Committee meeting that unless some remedial action were taken, international fishing on the breeding grounds on the nose and tail of the Grand Bank and the Flemish Cap would destroy the east coast fishery. In fact, overfishing in international waters had always been a serious threat to any attempt at conservation.[51]

This problem had also bedeviled ICNAF. Indeed, many nations, such as the Soviet Union, Romania, and Cuba, had conducted experimental fishing off the east coast before they applied for membership in ICNAF in the 1960s. In 1963, non-members caught 2,000 metric tonnes of fish; in 1966, it was 98,000 metric tonnes. ICNAF had noted in 1968 that fishing by non-member states was still on the rise, but it could do little to stop the illegal practices.[52]

After Canada's declaration of the 200-mile limit, non-NAFO members continued to fish illegally and some NAFO members, dissatisfied with their quotas, flew flags of convenience (registering their fishing vessels in countries not members of NAFO) to avoid NAFO obligations as a condition for obtaining Canadian licenses. During April and May 1979, for instance, Canadian aircraft and patrol vessels spotted at least six Spanish vessels that were registered to either Mexico, Venezuela, or Panama. NAFO eventually passed a resolution calling on its members to prevent their vessels from flying flags of non-member states. Moreover, NAFO informed Mexico, Panama, and Venezuela of the dangers their actions posed for conservation measures. By the 1980s overfishing on the nose and tail of the Grand Bank had become a major problem and it marked the beginning of serious stock management difficulties in the NAFO area. These areas made up a small area of the vast continental shelf, but they were important spawning and nursery grounds for a variety of species including cod and caplin. Between 1986 and 1991, non-NAFO vessels allegedly caught more than 200,000 metric tonnes of fish from the straddling stocks.[53]

Canada continued to have a particularly difficult time with Spain, as it refused to adhere to signed agreements on the importation of Canadian fish products and follow Canadian and NAFO fishing regulations. Canadian officials maintained that the Spanish were not serious about adjusting their fleet to the new fishing realities, a fact borne out by Spain's insistence that its share of the cod allocations should be returned to their levels before Canada declared the 200-mile limit. Such an arrangement would have given Spain 27% of the total international catch rather than the 13% it was usually allocated after 1976. After the new management regime was adopted in 1976, Spain attempted to have ICNAF introduce a new quota system that would allocate cod on the basis of historical performance, and in 1977 it attempted to force Canada to increase its allocation by threatening to withdraw from ICNAF and to fish indiscriminately in the area immediately beyond the 200-mile limit.[54]

Despite the problems with the fishing activities of non-member states and Spain, NAFO worked reasonably well until 1985 when Spain and Portugal joined the European Union (EU). The EU then argued, for the first time, that TACs should be set well above the previous levels. Canadian diplomats have alleged that when the EU expanded in 1986, it placed tremendous pressure on existing fisheries' arrangements within Europe, for which the easy solution was to find fishing grounds outside its own waters for Spain and Portugal. Because those two nations had traditionally fished in Canadian waters, it was perhaps easy to accuse Canada of being unnecessarily conservative and ignoring the socioeconomic problems that its cautious approach created for Spain and Portugal. Hence, the EU proposed that TACs be based on the maximum sustainable yield, which had failed to protect the stocks during ICNAF's tenure, instead of on Canada's and NAFO's more conservation-oriented $F_{0.1}$ strategy. The EU received little support for its position, but it formally objected to many NAFO decisions and opted out of the conservation process. The objection rule, a relic from ICNAF days, allowed any member to file an objection to a NAFO decision within 60 days and not be bound by the commission's decision.[55]

From 1986 to 1992, the EU set unilateral quotas for itself, much higher than NAFO TACs and often without any historical or traditional basis. In 1986, NAFO set a quota of

25,665 metric tonnes for EU vessels on the nose and tail of the Grand Bank, but it reported catches of 110,000 metric tonnes; the EU exceeded its quota by a large margin in each of the next four years. In fact, from 1986 to 1991, fishing fleets from the EU reported catches of cod, flounder, and redfish totalling 590,000 metric tonnes, five times the NAFO quota. In some instances, the EU catches even exceeded the quota for the entire NAFO membership.[56] This did not include unreported catches nor those taken by non-NAFO states that were frequently fronts for EU vessels. Even though scientific evidence showed a serious decline in fish stocks in areas outside the 200-mile limit, Spanish vessels turned in even larger numbers to some of these areas after Namibia expelled them from its territorial waters in 1990 for overfishing.[57] Between 1988 and 1991 the EU catch of northern cod increased from 26,500 metric tonnes to 42,000 metric tonnes; Canada's catch fell from 120,000 metric tonnes to 60,000 during the same period.

The EU was not solely responsible for the devastation of the cod stocks. It was late 1991 before Canada and NAFO finally acknowledged that the cod stocks on the continental shelf were in grave danger of extinction. Despite tough Canadian rhetoric about international overfishing, the Atlantic Groundfish Management Plan failed to reduce significantly the TAC for the 1991 winter fishery in the Gulf of St. Lawrence. Even after a scientific report was presented to the NAFO annual meeting in 1991 warning of declining stocks and calling for a partial moratorium on some fishing methods and a drastic reduction in cod and redfish quotas, the 11-country organization voted to leave nine of 10 quotas unchanged. Both Canada and NAFO member states maintained high quotas largely because they did not know how to deal with the massive unemployment that would have resulted from shutting down the fishing industry.

In February 1992, federal Fisheries Minister John Crosbie began to reduce the quota for catches of northern cod inside the 200-mile limit that would shut down the whole east coast fishery within two years. Even after Crosbie announced in July 1992 a two-year moratorium on the catching of northern cod, some foreign fleets continued to exploit the resources just outside Canada's 200-mile economic zone. Most of these vessels were registered in Panama, Korea, Honduras, Venezuela, Morocco, and Sierra Leone. In December 1992, there were 15 Panamanian vessels and three Korean vessels fishing on the nose and tail of the Grand Bank outside Canada's 200-mile limit.[58] Despite constant urging from conservationists and some fishing interests, Ottawa refused to extend its custodial management to the disputed waters on the nose and tail of the Grand Bank. Some progress was made in 1992 when the 180 countries attending the Earth Summit in Rio de Janeiro passed a resolution committing themselves to the principle of sustainable fishing on the high seas. They also approved a follow-up conference to work out details for a new fishing regime and to focus on fish stocks straddling 200-mile limits. After considerable posturing throughout most of 1992, the EU finally agreed to curtail its catch. Moreover, the EU agreed to allow Canada to set quotas for the scarce northern cod, about 5% of which was found in international waters. In return, Canada allowed EU vessels back into east coast ports that had been closed to deter overfishing. EU fishers also gained access to secondary species such as silver hake and grenadier in Canadian waters. Even after the EU announced that it would suspend fishing off the Grand Bank, Spain removed only slightly more than half of its pair trawlers, leaving 34 others in the disputed zone. Moreover, some EU trawlers continued to fly flags of convenience to avoid monitoring by NAFO.

In early 1995 Canada brought world attention to the depleted fish stocks by arresting a Spanish trawler outside the 200-mile limit. This action marks a new course for Canada, away from a dependence on internationalism that goes back more than a generation, to a course where Canada will play the role of environmental police even if it means stretching the rule of international law. Unfortunately, Canada's tough stand on conservation may have come too late for a resource that sustained an international fishery since the 16th century.

Notes

1. Ruth Fulton Grant, *The Canadian Atlantic Fishery* (Toronto: Ryerson Press, 1934), p. 71-72.
2. Ibid., p. 97; and Joseph R. Smallwood, ed., *Encyclopedia of Newfoundland and Labrador*. Volume 2 (St. John's: Newfoundland Book Publishers (1967) Ltd., 1984), pp. 165-66.
3. Ibid., pp. 44-47.
4. W.B. Scott and M.G. Scott, *Atlantic Fishes of Canada* (Toronto: University of Toronto Press, 1988), p. xv.
5. Department of Fisheries and Ocean, *Discussion Paper: Factory Freezer Trawlers* (Ottawa: Department of Fisheries and Oceans, 1985), p. 3, p.16; and Scott and Scott, *Atlantic Fishes of Canada,* p. xv.
6. National Archives of Canada, Records of the Department of Fisheries and Oceans (RG23), volume 1310, file 728-5-2 [1].
7. Ibid., volume 1292, files 728-3-8 [1], 728-3-8 [2].
8. Ibid., volume 1292, files 728-3-8 [1], 728-3-8 [2].
9. ICNAF, *Annual Proceedings* 17 (1966-67), pp. 81-82; Peter R. Sinclair, *State Intervention and the Newfoundland Fisheries* (Aldershot, England: Avesbury,1987), pp. 84-85; William Warner, *Distant Water: The Fate of the North Atlantic Fisherman* (Boston: Little, Brown and Company, 1983), pp. 52-53; and Bern Keating, *The Grand Banks* (Chicago: Rand McNally, 1968), pp. 81-82.
10. Memorial University of Newfoundland, Maritime History Archives, John Cheeseman Collection, "French Fishing Industry" (no date).
11. Warner, *Distant Water*, pp. 137-38; and Joseba Zulaika, *Terranova: The Ethos and Luck of Deep Sea Fishers* (St. John's: Institute of Social and Economic Research, 1981), p. 4.
12. Ibid.
13. *Trade News*, 10 (1) (July 1957), pp. 3-4; and Warner, *Distant Water*, p. 119.
14. *Trade News*, 10 (1) (July 1957), p. 4; and *Trade News*, 12 (6) (December 1959): pp. 3-6.
15. Grant, *The Canadian Atlantic Fishery*, pp. 59, 68.
16. C.P. Idyll, *The Sea Against Hunger* (New York: Thomas Y. Crowell Company, 1970), pp. 114-16; and James R.Coul, *The Fisheries of Europe: An Economic Geography* (London: G. Bell & Sons, Ltd., 1972), p. 2.
17. Quoted in Warner, *Distant Water*, p. 49; and Idyll, *The Sea Against Hunger*, pp. 115-17.
18. DFO, volume 1127, file 721-45-3[1]; and Warner, *Distant Water*, pp. 50-52.
19. ICNAF, *Annual Proceedings*, 17 (1966-67), p. 81; Idyll, *The Sea Against Hunger*, pp. 115-19; and Warner, *Distant Water*, pp. 48-50, 53, 58.
20. Warner, *Distant Water*, p. 14; See *Trade News*, 12 (7) (January 1960), p. 17.
21. Ibid.
22. ICNAF, *Annual Proceedings*, 17 (1966-67), p. 81; Idyll, *The Sea Against Hunger*, pp. 115-19; and Warner, *Distant Water*, p. 48ff.
23. *Trade News*, 7 (1) (July 1954), p. 4; and *Trade News*, 12 (4) (October 1959), pp. 6-7.
24. Alan Villiers, *The Quest of the Schooner Argus: A Voyage to the Banks and Greenland* (London: Hodder and Stroughton, 1951), p. 43.
25. Ibid.
26. G.A. Frecker, "Portugal, Land of Promise," in *Journal Do Pescador* (May 1955), pp. 34-35; and Villiers, *The Quest of the Schooner Argus*, pp. 48-49.
27. Sally Cole, "Cod, God, Country, and Family: The Portuguese Newfoundland Cod Fishery." *MAST: Maritime Anthropological Studies* 3 (1) (1990): p. 11ff.
28. DFO, Acc 83-84/120, Box 369, file 728-5-2 (sub. 1); ICNAF, *Annual Proceedings*, 17 (1966-67), pp. 14-15; and *Trade News*, 17 (5), pp. 11-12.
29. DFO, volume 1276, file 728-3-1 (38).
30. Scott and Scott, *Atlantic Fishes of Canada*, pp. xv-xvi.

31. See, Department of Fisheries and Oceans, *Discussion Paper: Factory Freezer Trawlers.*

32. Grant, *The Canadian Atlantic Fisheries*, pp. 91-96.

33. Cabot Martin, "The 200-Mile Limit," (Unpublished paper, Centre for Newfoundland Studies), pp. 3-5; and Court of Arbitration, Delimitation of the Maritime Areas Between Canada and France, *Memorial Submitted by Canada*, 1 June 1990, pp. 55-56.

34. ICNAF, *Annual Proceedings*, 17 (1966-67), p. 13.

35. Claude Emery, *Overfishing Outside the 200-Mile Limit: Atlantic Coast* (Ottawa: Library of Parliament, Research Branch, 1992), p. 3.

36. ICNAF, *Annual Proceedings*, 16 (1965-66), pp. 24-25.

37. ICNAF, *Annual Proceedings* 17 (1966-67), pp. 14-15.

38. Ibid., pp. 21, 51-56.

39. Ibid.

40. DFO, volume 1860, file 1165-36/558, pt. 1; and M.P. Sheppard, "International Competition for Fisheries Resources," *Proceedings. Government-Industry Meeting on the Utilization of Atlantic Marine Resource* (Montreal: Department of Fisheries and Oceans, 1974), pp. 217-18.

41. Department of Fisheries and Oceans, *Development in Foreign Allocation on the Atlantic Coast Since the Extension of Jurisdiction* (Ottawa: Department of Fisheries and Oceans, 1985), pp. 1-2; and Court of Arbitration, *Memorial Submitted by Canada*, pp. 62-63.

42. ICNAF, Seventh Special Commission Meeting, September 1975.

43. DFO, Acc. 83-84/120, Box 369, file 728-5-2 (pt. 7); *Daily News*, 15 April 1970; and see Cabot Martin, "200-Mile Limit." Incidentally, Iceland had taken similar action 11 years earlier following the collapse of the 1958 Law of the Sea Conference.

44. Emery, *Overfishing Outside the 200-Mile Limit*, p. 4.

45. Department of Fisheries and Oceans, News Release, 21 December 1992; and Scott and Scott, *Atlantic Fishes of Canada*, p. xvii.

46. Leslie Harris, *Independent Review of the State of the Northern Cod Stock* (Ottawa, Department of Fisheries and Oceans, 1990), p. 9. See, Alan Christopher Finlayson, *Fishing for Truth: A Socio-logical Analysis of North Cod Stock Assessment from 1977-1990* (St. John's: Institute of Social and Economic Research, 1994).

47. Department of Fisheries and Oceans, *Developments in Foreign Allocation on the Atlantic Coast Since the Extension of Jurisdiction*, preface.

48. DFO, Acc. 86-87/149, Box 15, file 1165-N109 (pt. 2); and DFO, volume 1968, European Economic Community (May files).

49. Department of Fisheries and Oceans, Information Leaflet, May 1981.

50. DFO, volume 1968, European Economic Community (May files).

51. DFO, Acc. 86-87/149 Box 15, file 1165-N109 (pt. 2).

52. ICNAF, *Annual Proceedings* 17 (1966-67), p. 8.

53. DFO, Acc. 86-87/149 Box 15 file 1165-N109 (pt.2) and (pt. 3); and see Emery, *Overfishing Outside the 200-Mile Limit*, p. 7.

54. DFO, volume 1929, Canada-Spain (Tansley files).

55. Emery, *Overfishing Outside the 200-Mile Limit*, pp. 5-6.

56. Ibid., pp. 6-7.

57. Department of Fisheries and Oceans, *Developments in Foreign Allocation on the Atlantic Coast Since the Extension of Jurisdiction*, pp. 4-6.

58. Department of Fisheries and Oceans, News Release, 21 December 1992.

chapter
sixteen

THE FISH KILLERS*

William W. Warner

The middle of May is a time of hope in Newfoundland. The hope is that winter will at last disappear. Although snow may yet fall, alder and birch are beginning to bud in inland valleys. Out on the barrens the larch trees are still bare, but horned larks and buntings fly bravely, low to the ground, over wind-rippled clumps of tussock grass. Along the coast the first warblers—blackpolls, myrtles, and yellowthroats—begin to invade sheltered pockets of dwarf spruce on the lee slopes of towering headlands. In each outport, as the more isolated fishing communities are generally known, laughing troops of boys and girls eagerly clutch long bamboo poles as they strike out over the barrenlands in search of their favorite pond. If the wind drops and the sun comes on strong, insects will rise, birds will sing, and hungry little brook trout will fill their creels. But, of course, the gray clouds in the west can at any moment race across the moors, spitting wet snow or cold rain, and force the children to run home. It does not much matter, really. Tomorrow or the next day will do as well. Spring in Newfoundland is not a recognizable climatic season. Rather, it is a state of mind. Hope identifies and sustains it throughout, a sure hope that winter will inevitably relax, that children will soon be free to play long hours out of doors, and that men can return to the sea.

Once, some years ago, I stayed with a fishing family in a small outport on Newfoundland's north coast at just this time of year. For the men it was a busy period. Their rickety stages, or mooring platforms, each year mauled by winter ice, needed much repair. Inside the nearby twine lofts were miles of moldering long lines and gill nets to check for serviceability. Some of the fishermen talked of setting out the first cod traps, but others thought it too early. There was, after all, a big iceberg off on the horizon, not to mention some few pans of residual local ice. Might not the berg, before it grounded, drift down on a trap and carry it away? That was a serious question. A Newfoundland cod trap is in reality a large underwater pound net suspended from the bottom by an intricate combination of floats, steel chain, and heavy grapnel anchors. A fully rigged trap costs over $6,000; to set it in place takes two boats "with a good crowd of hands" working for four or five hours in reasonably calm weather. It is not, therefore, something that you can move around on a moment's notice. Taking all this very much into account, most of the trap crews thought it prudent to wait. In the meanwhile they could always put out a few lobster pots. Thus, every morning I either watched or accompanied the fishermen as they went off lobstering in their handsomely fashioned lapstrake trap skiffs. Often their work took them uncomfortably close to the bases of dizzying cliffs, where heavy ground swells exploded as booming surf on wet-black rock. In many places, just outside the churn of the surf, were rocky half-tide ledges— "sunkers," they call them in Newfoundland's lugubrious maritime vocabulary—

on which the seas broke heavily. Those who set their pots as close as they dared to the sunkers, it seemed to me, got the most lobsters.

One Sunday morning, remarkably clear and warm, my hostess suggested we "ride out on the high road" to go visiting here and there by car. A vigorous woman in her late 50s, she had been active in local politics and was thus welcome in many places normally reserved only to men. Secure in this knowledge, she took me down to the village's main wharf. Here, as in every outport, older or retired fishermen gathered on any fine Sunday to watch the weather and retail the latest news.

The talk that morning covered many subjects. There was a new boat in the harbor, a big longliner just come up from St. John's. Some of the men were very worried about her. They wondered if she would stay in local waters throughout the summer voyage, as they called their brief inshore fishing season, and thus be competing with them. Then there was the man that came ashore at Green Island Light, hadn't we heard? Just two days ago, it was. A handsome enough young fellow, most likely off one of the Norwegian seiners. It took me some moments to understand what everyone else took for granted; namely, that "to come ashore" in these icy north-coast waters meant as a corpse, washed in by the tide. Newfoundland speech can be disturbingly direct.

But the oldest men in the group much preferred to talk of the past, particularly of the halcyon days of the annual schooner trips to the Labrador coast. Their eyes brightened under the visors of their Irish-style cloth caps as they competed to tell me of this once great vernal event.

"Yes, my dear, we fared well in those days. Down to the Labrador we went. Just this time of year it was, and you don't see us home until October month! A thousand cantles [quintals or hundredweights] we fished, and we were all in the family for crew."

"That's right, Captain Aaron. All in the family except one and, you know, I'm thinking that was the cook. Yes, it was a good share we got those times. We fared well."

Never, in fact, did the conversation stray very far from this single and sorrowful theme, or an almost rhythmically recited threnody for the better days of the men's youth, which few of them thought would ever return. Now, everyone agreed, the problem was mainly the squid and the caplin. Squid were nowhere as plentiful as before, and none of the men could ever forget the first dense shoals that moved inshore every summer. So thick they were that the whole town went out to fish them. You used little lead weights—squid jigs, you called them—painted red and crowned at their bottoms with 20 or more tines, bent upward and needle-sharp. You jerked a string of these weights up and down, attracting the curious squid, and pretty soon, sure enough, you hooked one on the tines. Sometimes, in fact, you even got two or three of the creatures on a single jig. And wasn't that great sport and a bright time for all!

Yes, the men kept saying, you filled your dory easily in the old days and always had more than enough bait for the season. (Squid has long been an essential cod bait in Newfoundland; today it is used mainly for longlining, fishing with a long central line to which hundreds or even thousands of hooked snoods, or drop lines, are attached.) It was a happy time, no doubt about it. There were many more scoffs and soirees, as the oldest men called the attendant parties and celebrations.

Then there was the caplin. With every mention of the subject the men's heads nodded in unison. The caplin, they insisted, were simply not coming ashore the way they used to. Their numbers were way down, and they had been that way for the last several years.

The Atlantic caplin is a small pelagic fish seldom exceeding eight or nine inches in length that swims in large schools in the coldest waters of the north Atlantic. For reasons not fully understood by scientists, one part of the northwest Atlantic caplin population spawns far out at sea near the eastern edge of the Grand Banks. Another part comes close inshore—right up to any suitable sand or pebble beach, in fact—to cast their milt and roe in the roll of the surf. The "scull," or arrival, of the inshore spawners is an eagerly awaited event in nearly all Newfoundland outports. It can come very quickly at any time from the first half of June to the end of July, depending on slowly rising water temperatures. Suddenly, one day, the green wall of every wave will show itself dark with swarming fish. It is a dramatic sight and the word that "the caplin are rolling" travels fast. Men, women, and children run down to the beaches, since fresh-cooked caplin are a highly prized dish and the little fish also "pickle" well when salted in barrels. The men wade out into the waves, expertly throwing circular castnets, to be sure they get their share of the first runs. Behind them women and children soon do just as well with baskets or pails. Nearly always there will be a wave-sculptured hollow on the beach shingle; inevitably, many caplin strand in these hollows, unable to wriggle back in the draw of the surf. These fish need only be picked up, as fast as small hands can work. So it happened, the men told me, that the big runs of former years were another happy occasion, another time for the scoffing and the soirees.

There was, however, a more important consideration. Biologists, fisheries officials, and generations of fishermen could all agree about it. Soon after the caplin come the cod. Caplin and cod; the two always go together. Practically speaking, for Newfoundland's inshore fisheries, the one does not exist without the other. So important are caplin as a prey fish that draw cod inshore that they largely determine the success or failure of the smaller outports' fishing year. If the caplin run thick and strong, the trap crews, the gill netters, the longliners, and even the few dory fishermen who still row out to jig the cod will all have a good summer voyage. If not, they will sink deeper in debt to the buyers. The problem, as every man on the wharf that Sunday morning repeated again and again, was that "caplin is nowadays everywhere scarce." I had heard this opinion many times in outports extending from the Labrador coast to Newfoundland's south shore. The men were not sure why it was so, but many had heard the big foreign ships were beginning to take caplin and thought that probably had something to do with it.

When at last we left the wharf to continue our Sunday visits, my hostess turned to me and asked: "Now, you seen the gentleman with the gall by his neck? Well, he was a great fish killer! Yes, by deed, a great fish killer, he was!"

Unlike many Newfoundland localisms, fish killer is easily translated. It means what the rest of the English-speaking fishing world calls a high liner, a top-earning fisherman. I was to hear the term often in the next ten minutes.

As we followed a crooked street between the little cracker-box homes that are the trademark of Newfoundland outport architecture, my hostess identified each house's owner and told me in lengthy detail of his or her relative standing in the community. "You see the house all pretty yellow, with the blue trim?" she might ask. "Belonged my grandfather, on my dear mother's side. And wasn't he the great fish killer! Had two schooners going down to the Labrador. *Two* schooners, by deed, he did!"

Only when we passed the last house and the high road arched straight over the rolling barrens did she become momentarily quiet.

"Truth is they was all great fish killers, those days," she said after a pause. "A whole generation of them, you could say."

A week later I was in St. John's, the capital city of Newfoundland and the most frequented port of call for the factory trawler fleets of many nations. Fishing is the principal business of St. John's, and St. John's takes its business very much to its heart. Unlike other ports, which shunt fishing vessels to the most undesirable locations, hidden behind waste-lands of rotting wharves, warehouses, and steel-link fences, the city offers foreign fisher-men its best waterfront. Their ships berth comfortably alongside a long quay with park benches and a scenic parallel avenue, known as Harbour Drive. Walk a short block up from Harbour Drive on any one of a number of steep and narrow side streets and you are in the centre of St. John's business and shopping district. Thus it may happen that a visitor dining in one of St. John's downtown restaurants can look out a harborside window and find a giant ship nudging up to the quay, directly at eye level and close enough to hear shouted orders from the bridge.

It was under exactly such circumstances that I first saw a modern factory trawler close at hand. First to glide into view in the frame of the restaurant windows was a huge flared bow. Painted on its side were Cyrillic characters that seemed to spell out, rather incongru-ously, the word *Rembrandt*. Next came towering kingposts, cargo masts of the type one more commonly associates with freighters, and a multi-windowed bridge painted fresh white and extending across the full width of the ship. Aft of the bridge was a boat deck with two large lifeboats hung on strong davits and a lengthy promenade that gave promise of cruise ship comforts. A large placard between the lifeboats repeated the vessel's name in the Roman alphabet. There was no mistake. The ship was named the *Rembrandt*. Uni-formed officers, replete with gold braid and white hats, peered from the bridge wings. A large crew, including some women, lined the rails. All appeared stolid and impassive, yet it was clear they watched everything on shore with the utmost attention, like cautious visitors from another world. In a very real sense, they were. Towering above them, dominating the *Rembrandt*'s superstructure, was a large funnel painted with a wide red band. In the middle of the red band were a golden hammer and sickle.

Down on the quay, standing in the shadow of the *Rembrandt*'s high walls, I judged her length at over 300 feet and her capacity at something beyond 4,000 gross tons. Such di-mensions are, of course, quite small by contemporary merchant and tanker fleet standards. But viewed against the backdrop of St. John's small fjordlike harbor and compared with any-thing most of us picture as fishing vessels, the *Rembrandt* looked every bit the *Queen Mary*.

Curiously, nothing in the ship's forward half suggested anything of her function. The foredeck resembled a freighter; the superstructure, a passenger liner. Only back aft did the *Rembrandt* reveal a more workmanlike character. Here large and heavily rusted steel plates—extra doors for mid-water trawling, I later learned—hung on the stern rails, just above where the ship's name was repeated, along with the name of her home port, Murmansk. Near the stern quarters rose a high two-legged gantry mast for heavy lifting. From it and a smaller mast at the forward end of the main deck hung many strong steel cables, each eventually leading to one or another winch. Most noticeable amid the clutter of gear, stand-ing out by their very size alone, were the nets. Great mounds of them rested on the trawl deck or were jammed into odd corners in the lee of the superstructure. Each of the net mounds looked as large as the average Newfoundland outport home. In fact, it was easy to imagine them engulfing a dozen such homes or an entire small outport—as well they could in the case of the mid-water trawls, at least, when fully opened in a proper set. Almost as surprising was the view from astern of the *Rembrandt*. Here the dominant feature was the stern ramp. At a distance it seemed only a gash, an ugly wound that parted and disfigured the normally pleasing contours of a ship's stern. Up close it was a monstrous chute, plung-ing steeply from the main deck some 20 feet down to the waterline. One might understand the need for such a device to haul great whales up from the sea. But, like everything else on

226

the *Rembrandt*, it seemed way out of scale or too gross a conduit for anything the size of ordinary fish.

Further up the quay were smaller vessels, some tied two or three abreast. Although not more than 150 feet in length, they were in every way as unusual in appearance and as baffling to the novice as the giant *Rembrandt*. They bore such names as *Morebas, Uksnøy*, and *Garpeskjaer*; their home ports were invariably Ålesund, Tromsö, or Hammerfest. Sturdily built with high bows and elegant sheer, there was about them an extraordinary compactness. Odd assemblages of equipment—tripod lookout masts, large rubber-sheaved blocks, metal chutes, and bulbous suction pumps—crammed every square inch of their deck and superstructure space. Along their sides back aft were horizontal steel rods holding scores of steel rings, like jousters' lances heavy with trophies. Only a few of the boats carried their nets on deck. They were made of fine black twine, and of very small mesh.

A Canadian Fisheries and Marine Service inspector doing his rounds of the quay told me these boats were purse seiners, designed to hunt and encircle schools of fast-swimming surface fish. "They come over from Norway with a big factory ship for the spring caplin spawn," he added. "The factory ship's anchored way out. She converts the caplin to fishmeal, some 300 tons a day, I'm told."

Were the reader to have a similar experience—that is, to pass rapidly from the rural outports to the harborside scene at St. John's—he might imagine that he had witnessed two worlds of fishing, perhaps centuries apart in methods, existing side by side in Newfoundland. If so, no well-informed scientist or fishing industry spokesman would ever dismiss such a thought as in any way exaggerated. Nothing less than a latter-day industrial revolution, the experts might claim, separates the outport fishermen with their longlines, traps, and other methods that are indeed centuries old from the multinational citizens of the electronically guided factory trawler fleets that pursue many of the same fish in much the same or at least contiguous waters. A revolution, some might add, that has radically altered the fortunes of nearly all fishermen, the seas where they fish, and, ultimately, the laws by which man governs the world oceans.

A ship like the *Rembrandt,* in the vocabulary of the trade, is correctly termed a factory-equipped freezer stern trawler. "Factory trawler" is the more commonly used short form; quite appropriately it emphasizes the fact that vessels of this type both catch and process their fish. Factory trawlers are thus distinguished from factory, or "mother," ships, which perform only the latter function, after obtaining their raw material from smaller catcher boats.

Common to all factory trawlers are four essential elements that set them apart from the generations of fishing vessels that preceded them. These are a stern ramp or slipway for the rapid recovery of nets from astern (rather than over the side), a sheltered belowdecks factory section with assembly-line machines to gut and fillet fish (as opposed to cleaning by hand on an exposed main deck), an ammonia or freon refrigerating plant for the quick freezing and frozen storage of fish (in place of heavy and space-consuming chopped ice), and equipment to make fishmeal (to utilize both the factory leavings and trash or nonmarketable fish).

The idea for a ship that might combine such advantages was born in the United Kingdom in the late 1940s. Not surprisingly, at least to those acquainted with the traditional resistance to change among fishermen and fleet owners alike, it did not originate within the fishing industry. Rather, it first took shape from whaling or, more specifically, from the concerns of a Scottish general shipping and whaling firm known as Christian Salvesen Limited. Although this company was at the time a world leader in modern factoryship whaling, Captain Harald Salvesen, its chairman, was much worried about diminishing whale stocks. Captain Salvesen himself, in fact, had tried to do something about them as early as the 1930s, through industry or self-imposed quotas on certain species. But his industry colleagues were as yet in no mood to police themselves, and nothing significant in the way of whale conservation occurred for more than a decade, until the establishment of the International Whaling Commission in 1946. By that time, however, Captain Salvesen was convinced the initiative was already too late. The international control systems would be ineffective, company officials remember him predicting, and there were too many new entrants, most notably the Soviets and the Japanese, competing for stocks that continued to decline very rapidly. Rather than wait for the death of the whales and the industry, Salvesen's would use its whaling experience in new directions. The most obvious was fish.

Among the concepts borrowed from whaling was a ship with the size and the range to go for lengthy voyages, to be equipped with a large stern ramp to haul up the great quantities of fish expected to be found in new and more distant waters. The application of a stern ramp, so commonplace in whale factory ships, might not at first seem a significant technological breakthrough. But in the tradition-directed world of commercial fishing, it was no less. To understand why, one must first have some knowledge of how the side trawlers that then dominated world fishing went about their business. These vessels, as their name clearly implies, carry out all their working operations over the side. To recover their nets, for example, they must stop dead in the water, broadside to the seas. The forces of wind and current will then cause the side trawler to drift rapidly downwind and stretch out the gradually rising net abeam, perpendicular to the hull, where it will not foul the rudder or the propeller. The net is then brought in over the rail, an action that can become both wet and dangerous when high seas crash aboard during heavy rolls and flood the main deck. A stern trawler is spared this ordeal, since it can maintain some forward progress and a safer bow-to-the-wind heading throughout most of the haulback. As a consequence stern-ramp trawlers can continue fishing in heavier weather, when side trawlers must stop, although fishermen argue endlessly about the exact point in the Beaufort scale where this occurs.[1] Another consequence, about which there is no argument, is that stern trawlers can both set and haul back their nets faster than conventional side trawlers. Fishing time, in other words, is significantly increased. For those who first thought of applying stern trawling to large vessels, therefore, it seemed to have all the advantages.

All except one, that is, which critics were quick to point out. The collecting bags, or cod ends, of ground trawls would surely burst, the critics insisted, with every good catch. This was because any trawl hauled up a stern ramp would have to come aboard all of a piece or by continuous heaving; there was no way of dividing a large catch with the time-tested methods used by side trawlers, which we will examine presently. The dissenting critics' most serious objection—it later proved to have some foundation—was that no known netting material could stand the strain of such total-catch hauling. (Synthetic twine had not yet been introduced; at the time all nets were still made of manila or other natural fibers.) If nets did not burst, the detractors further claimed, crews would at the very least have to "cut in" or rip out panels of netting to spill fish and ease each load, thereby losing any time advantage. True, stern trawling was not new. It had long been practiced successfully by small wooden draggers and at least one experimental steel vessel in New England. But these boats simply hauled in their relatively small catches over a roller on the stern transom or up a short ramp, and they operated for the most part in calmer inshore waters. To bring in a cod end with 30 tons of fish over the waterline edge of a bucking and heaving stern ramp in the giant seas of open oceans was quite another matter. In such event one had to consider the powerful inertial forces generated by the extreme pitching of a large vessel.

On ships over 250 feet in length stern pitching was known to reach 30 degrees. The vertical acceleration forces so produced might mean that the 30 tons of a flat calm would become 60 in heavy weather. Even if a trawl were somehow successfully recovered under such conditions, the critics finally insisted, the fish would be badly bruised and crushed in the process. In England and other European countries, where fish quality is important in the marketplace, this seemed another very real objection.

By contrast the largest side trawlers of the day had none of these problems. If their crews were lucky enough to see a swollen cod end with 30 or more tons broach the surface in a boiling, heaving mass—always a thrilling spectacle—there was no need to worry about the strain of taking it aboard all at once. Side trawler crews first winched the mouth and foreparts of their trawl onto the deck and then used a rig known as a splitting strop or doubling becket to "split" or divide the contents of the sacklike cod end. The strop was simply a noose of cable laced into the girth of the cod end; by heaving in on another cable attached to it which ran along the top of the net, variously known as a bull rope or bag becket, the noose immediately tightened and cut off or cinched the fish in the rear of the cod end into a compact ball of four or five tons. After this ball was winched aboard, crew members quickly untied a slip or purse knot at its bottom, let the fish fall out on deck, retied the knot, and then heaved the empty cod end back over the side. (During all this time, we must remember, the mouth of the net remains on deck, thus effectively closing off to the remaining fish any escape route.) It was then only necessary for the skipper to send his ship "slow astern" for a brief moment to collect more fish in the butt of the cod end, after which the stropping process could be repeated as many times as necessary. In this manner side trawlers were able—and are still very much able today—to bring aboard heavy catches in discrete amounts. This option, of course, would be impossible with a long net trailing well astern in the wake of a big ship with forward progress. Side trawlers might require a little more time to haul back, to be sure, and there was always a possibility a man might get knocked off his feet or even swept overboard by a boarding sea. But at least there was no danger of losing a whole bag of fish. For the owners of the conventional sidewinders, as side trawlers are called in Great Britain, this was reason enough not to try anything different.

"Yes, there was a great debate in those days as to the practicability of stern trawling, especially for a ship of the size we had in mind," recalls L. M. Harper Gow, a Cambridge graduate and army commando officer, known to his associates as Max, who began his career with Salvesen's on an Antarctic whaler and later served 17 years as the company's chairman. "The majority thought it impossible, but we had at least learned enough from our experiments with the *Fairfree* not to be overly discouraged. So Salvesen's took the big leap forward. We decided to build a factory-equipped stern trawler from the keel up."

The *Fairfree* was the former H.M.S. *Felicity*, a minesweeper of the Algeria class purchased and rechristened by Salvesen's in 1947 to provide a floating laboratory for Sir Charles Dennistoun Burney, R.N., an enthusiastic yachtsman, aircraft engineer, and inventor of the paravanes used in World War II minesweeping. Sir Dennis, as he was better known, was mainly interested in a radically new design for trawl doors, or otter boards— "parotters," he called them, in obvious acknowledgement of their evolution from his experience with paravanes—which he hoped might lift ground trawls free of the bottom and permit fishing at various depths. The parotters, however, did not work. Try as Sir Dennis might, the *Fairfree*'s trawls either refused to budge from the bottom or at best took off for only brief bursts of mid-water flight. But even in failure Sir Dennis must be given due credit for his early vision of the great potential in variable-depth fishing. When mid-water trawls were finally perfected a decade later, they immediately took their place alongside electronic fish finders as one of the most significant technological advances of 20th century fishing.

Meanwhile, although mid-water fishing would have to wait, Salvesen's gained considerable experience with stern trawling, which was the method used throughout the *Fairfree* experiments. As a result company officials came to believe it could be tried on a much grander scale. The plans for the stern trawler Salvesen's had in mind, in fact, ultimately called for a ship of 2,600 gross tons, with an overall length of 280 feet, and accommodation for a crew of 80. Thus, by dimension alone she would indeed be a big leap forward. The largest British side trawlers then in use averaged 185 feet in length and 700 gross tons. Their normal crew complement was 20.

Salvesen's placed the order for their new leviathan with J. Lewis and Son, a small shipyard in Aberdeen, early in 1953. Although the Lewis firm had never attempted so large a vessel, work proceeded on schedule and the ship was launched with appropriate fanfare in March of 1954. In a felicitous choice that honored both her immediate ancestry and her pioneering role, she was christened the *Fairtry.* Or as maritime historians later came to know her, the *Fairtry I,* since she soon spurred the construction of two sister ships, the *Fairtry II* and *III.*

Size alone, however, was not all that surprised the crowds that watched the *Fairtry*'s launching. There was also her unseemly appearance. As nearly everyone commented, she looked more like an ocean liner than anything ever built for fishing. To the layman's glance the *Fairtry*'s superstructure seemed almost entirely given over to living accommodation; there were no fewer than 22 cabin portholes on each side at the fishing-deck level and an equal number just below. All were well above the waterline, which meant the luxury, unknown on low-freeboard side trawlers, of being able to keep them open in all but nasty weather. Inside were comforts never dreamed of by previous generations of fishermen. A photograph in a special *Fairtry* issue of London's *Fishing News International* showed a typical four-man cabin in which contented crew members reclined in sturdy double-decker bunks or wrote letters at a common central table. The caption glowed:

> Individual lockers are given each man and there is an air of attractive roominess about each berth. Selection of mates is left to the men themselves. Officers have single berths, petty officers are in double berths, and other hands in four-berth cabins. Ample showers (hot and cold) are provided throughout the ship, and cinema is regularly shown in the crew's mess.

"The floating Ritz, we called her," a member of the original crew has told me. "Showers we never had before, and some of the old sidewinders had no toilets. Over the side you went, summer or winter. A shocking thing it was, but little the gaffers [owners] cared!"

"We planned the better accommodations, the movies, and such from the beginning," says Ted Sealey, a veteran Salvesen's official. "We were accustomed to handling problems of keeping crews out of mischief. After all, the whaling voyages lasted eight months, and we knew that an idle mind is an evil mind, as we used to say."

But the true test of the *Fairtry*'s future would lie more in the equipment concealed deep within her hull than in external design or creature comforts, although the latter quickly proved their value in attracting and keeping good crews. The most important single unit of equipment—central to all Salvesen's planning and the key to potential profits—was her quick-freezing plant. The reason for installing one was quite clear: at the time of the *Fairtry*'s commissioning British fishing had already reached a point of no return in searching out new and more distant grounds. This search, nearly always successful, had long been the particular pride and mainstay of the British high-seas fishing fleet. It had started as early as 1891, when a few steam-powered trawlers first fished the rich offshore grounds of Iceland. Not long thereafter the British fleet extended the quest for better fishing north of the Arctic Circle to the North Cape of Norway, Bear Island, Spitsbergen, and the Barents Sea. By the 1930s, as these Arctic grounds (especially Bear Island) began to supplant Iceland in importance, the term "distant water fishing" had become firmly established in British vocabulary and British fishermen had grown accustomed to the endless night of their winter trips to the Norwegian Arctic. After World War II occasional journeys to West Greenland—the round-trip distance was 4,500 miles—were not uncommon. British skippers and industry chiefs might be justifiably proud of these achievements, which other nations sought to follow. Nevertheless, two nagging problems remained. First, as distances increased, more and more ice had to be carried, until ice and not fish came to be the heaviest item in the fishholds of returning trawlers. The farther you went, in other words, the less was the absolute limit of fish you might expect to bring home. Second, the time expended in getting to and from the more distant grounds—"steaming time," as it is still called today—was seriously prejudicing and sometimes totally destroying the quality of fish. A typical trip to the Barents Sea will readily show why. Given a little luck, a trawler fishing the Barents could fill her hold in 10 or 11 days. But five and a half to six days of steaming were first required to get there, as well as to run home, making a round trip of 22 to 23 days. This meant that the fish caught the first day on the grounds were already 16 days old upon reaching port. Sixteen days, as it happens, is the generally accepted maximum-preservation period for iced fish, beyond which it is judged putrid or inedible. Even 6-12 day-old fish are considered by food technicians as "spoilage affected" in both firmness of flesh and flavor. Thus the "freshers," as ice-filled side trawlers are known in Great Britain, rarely brought home a truly fresh product. Their owners suffered accordingly on the marketplace, since British auctioneers, unlike their American counterparts, grade the catch of every vessel and sell it in lots, according to freshness. But far worse was in store for any trawler forced to heave to and lose precious fishing days riding out gales and storms, which on winter trips was more the rule than the exception. The added days meant that one-third to one-half of a catch might be condemned, to be sold, if at all, for fishmeal. Sometimes, in fact, total-catch condemnations might occur, especially on the West Greenland run. As C. L. Cutting, one of a small band of distinguished British fishing historians, said of the times, the ever-lengthening trips were sealing the doom of fish quality in Great Britain because "icing at sea has been carried up to and beyond its natural limits."

New methods of preservation, quite obviously, would be needed to go beyond these limits and to continue without risk the search for more profitable grounds. Salvesen's thought quick-freezing the only workable alternative. By fortunate coincidence frozen-food technology had advanced to the point where high-capacity freezing units small enough for shipboard use were being perfected at about the same time the plans for the *Fairtry* were taking shape. This was so largely because of the long and untiring efforts of the American inventor Clarence Birdseye. More than a quarter century earlier Birdseye had begun experimenting with frozen foods—with frozen fish, in fact, using New England redfish and not the vegetables for which he was later to become more famous—in a small laboratory in Gloucester, Massachusetts. After 10 years of effort, Birdseye and his associates had invented in 1933 what is now known throughout the industry as the multiplate freezer, a device in which stacks of metal plates, through which ammonia refrigerant ran in small tubes or passages, could be opened to receive prepackaged foodstuffs and then closed to the desired pressure by hydraulic action. Originally developed to be mobile or at least capable of being trundled around without too much difficulty to follow vegetable harvests for "field-fresh" pickings, the multiplate freezer had a number of advantages over the quick-freezing methods that preceded it. The most common of the earlier methods was cold air or blast freezing, which required large "rooms," or freezing chambers, and was also much the slowest process of the time. Another was brine immersion, which many experts, Birdseye included, disliked because of the flavor changes it inevitably produced. The multiplate freezer effectively solved these difficulties. But it was not until after World War II (industry historians usually cite 1946 as the turning point, or critical year) that frozen foods gained widespread public acceptance in the United States and multiplate freezers became a standard of the industry. Public acceptance was much slower abroad, but word of Birdseye's imaginative advertising campaigns—he is as well remembered in the industry for his perseverance in marketing as for his inventive genius—eventually trickled across the Atlantic. Cautiously, a number of European countries began to inquire about licensing arrangements for the manufacture of Birdseye-patented freezers.

One of the first was England. Salvesen's had itself begun experimenting with a freezer designed by Sir "Dennis" Burney—he seems to have enjoyed working in many different directions at once—aboard the *Fairfree* in 1947. Dubbed by him the "Fairfreezer," it combined both blast and contact-plate freezing. Once again, however, his invention had only theoretical and not practical merit. After sharp scrutiny Salvesen's engineers found it too big to carry its weight, so to speak, or to allow of economical operation within shipboard confines. Much more to their liking were the weight and space saving features Clarence Birdseye had worked out for on-site vegetable and fruit freezing. From orchards and truck farms, therefore, the multiplate freezer went to sea.

The third essential element in the *Fairtry*'s planning, the idea of using automated filleting machines, had not been previously tried on the *Fairfree*. As usual, many experts thought it impossible. Such fish processing machinery as then existed in shore plants depended on delicate and accurate pre-positioning of fish in guides or holding brackets. Thus the conventional wisdom was that at sea fish would be constantly slipping or slithering out of the guides with any kind of ship motion. There were troubles enough in high-seas fishing, the experts maintained. To add to them with a roomful of complicated machines that might spill fish with every pitch or roll did not make sense.

"Yes, here again there was much debate," Max Harper Gow remembers. "The big question we faced was whether to fillet or simply to freeze whole fish. But since we were going to make long voyages, we thought it most economical to bring home a maximum tonnage of fillets and also to use the offal for fishmeal. We wanted no waste."

"No, you didn't want to go to all that trouble just to be storing bones," Ted Sealey agrees. His remark is well put. The yield from machine-cut fillets is approximately one-third the "round," or whole weight, of a fish. But the remaining two-thirds of the carcass, bones and all, may be ground into fishmeal of good quality. The decision to do this—to use all parts of the fish, as well as to fillet at sea—seemed doubly advantageous.

To make it a reality Salvesen's turned to Nordischer Maschinenbau Baader, a small firm in Lubeck, West Germany. During the 35 years of the company's existence its founder, Rudolf Baader, had invented about a dozen machines for gutting and cutting fillets from cod, herring, and a few other fish of comparable size and shape. Ingenious as they were, the machines had not, however, had a great success with land-based processors. The weight of tradition—in this case a strong preference for hand labor in the cleaning of fish—was as heavy in Germany as anywhere else. Nevertheless, among Baader's early wares was the Model *99 Schwanzlaufer*—the "tail grabber," most fishermen now call it—a complex filleting mechanism that was finding some clients. The machine was so named because in its first action it grasped cod or similar fish of about two to four feet in length by the tail in a viselike clamp. In the next step the holding brackets or guides measured the fish for size and transmitted the information so obtained to the internal elements of the machine. Then, in a split-second sequence too fast for the eye to follow, the fish was pulled into the machine's central section, where revolving knives sliced out fillets along both sides of its backbone. Water shot in jet streams to cool all cutting edges, and loud shrieking noises might occasionally emanate from the machine's innards when knives sawed into the vertebrae of carelessly positioned fish. But in the end, most of the time, out came neatly cut fillets, 48 to the minute. Simultaneously, fish with only their heads, vertebral columns, and tails remaining—indeed, they looked like they had been eaten by discriminating cats—dropped from the machine's nether parts onto conveyer belts to be ground into fishmeal. In short, the Model 99 *Schwanzlaufer* was precisely the kind of mechanical wonder that would have transported the late Rube Goldberg into ecstasies of delight. More to the point, however, it caught the eye—with cautious approval, if not transports of joy—of Salvesen's far-searching engineers.

Still, there were problems when the *Schwanzlaufer* went to sea. It proved rugged enough, but salt water caused many of the moving parts to seize up. For this reason, in fact, the machine had to be almost totally redesigned.

"But I was satisfied the Model 99 was reliable," Max Harper Gow recalls. "Although there were many difficulties in the beginning, the rolling of the ship, curiously enough, was not one of them. We found the freshly caught fish had much firmer meat than fish brought back after long trips to be processed on shore. The firmer meat prevented slipping. In time we worked out all the problems and added new machines, including skinners and a special line for redfish. The key to success was that Baader sent their mechanics to sea to work with us. We had a real hand-to-hand partnership all through the early years. In fact you could say we gave Baader their future."

The statement is no exaggeration. The Baader firm went on to become the IBM of automated shipboard processing, training operators afloat or ashore, and selling, renting, or licensing its machines to all the world's fishing fleets. Today the complete Baader line includes over 60 machines, both general and specialized, and their work speed has increased dramatically. Not 48 per minute, but two fillets per second now shoot out from their fastest units.

The objective of no waste through using the remains of filleted fish for meal (and also incidental or by-catch fish species normally thrown overboard) was much more easily realized. Ever since the 1920s, when the fishing nations of northern Europe began to discover the great efficiency of fishmeal as a growth agent in chicken and livestock raising, "industrial fishing," the capture and conversion of herring or other abundant small fish to meal and oil, developed rapidly, especially in Norway. The machinery to do this—it consists essentially of dehydrating presses, grinding mills, and drying ovens—is not complex and may readily be designed for a wide variety of capabilities. Salvesen's, therefore, simply went to the best of the Norwegian manufacturers and ordered a scaled-down unit to fit a convenient space on the *Fairtry*. Other than exercising strict precautions against fire—fishmeal, like flour, is extremely combustible—there were no special problems. Thus equipped, the *Fairtry* and all the factory trawlers that were to follow her became "the vacuum cleaners of the sea," as envious small-boat fishermen say of them, using all the fish their trawls swept up from the sea bottom.

Stern ramps, multiplate freezers, Baader machines, and a compact fishmeal plant. The *Fairtry* incorporated all of them and in so doing proved to be the complete factory trawler. Of all her remarkable attributes none stands out more clearly than the fact that practically speaking she allowed of no further evolution. The *Fairtry* inspired generations of factory trawlers; all followed her basic design and all included the four key elements she first took to sea. Naval architects and engineers might later experiment with many different systems of rigging, superstructure configuration, and deck layout. But belowdecks all the succeeding generations had the same essential features (often with equipment of identical manufacture) placed in approximately the same relative position as in the *Fairtry*'s original design. This is as true today as it was 30 years ago.

But advanced design and the *Fairfree's* trials notwithstanding, the *Fairtry*'s first months at sea were not easy. Unanticipated problems broke out everywhere—on deck, in the factory, or back aft in the curious little second bridge or pilothouse, perched above the stern ramp, that was first used to steer the ship during the setting or hauling of nets.[2] Meeting these problems with great patience and perseverance was the late Captain Leopold Dixon Romyn, the *Fairtry*'s first captain. In no way typical of British fishing skippers, Romyn was a public-school-educated man widely regarded in the fleet for his knowledge of engineering and general erudition. ("He keeps up-to-date by reading technical magazines of many countries in the original," *Fishing News International* said of him, "but he never lets up in the search for fish.") In addition, Captain Leo, as he was best known, had a gift for command and was very popular with his men, although some are said to have objected to the Sunday morning church parades he occasionally held at sea, in best public-school fashion. He was also a good writer. His graphic logbook entries and special reports on the *Fairtry*'s initial difficulties are required reading for anyone who would understand the early trials of stern trawling from a large ship.

Ironically, the first and most serious problem was that the *Fairtry* caught too many fish. Bags of 20 or more tons regularly inched up her stern like swollen sausages too big for their skins, as photographs in Salvesen's archives painfully attest, only to get stuck in the rampway. If, that is, they had not already burst, just as the critics had predicted, or caused damaging strains on winches or other gear.

"No method of splitting the bag has been found; consequently all the fish has to come in with one heave," Captain Romyn wrote rather wearily in one of his first reports. "This is possible with new gear and fine weather up to a weight of about twelve tons, but anything more than that is a source of great anxiety and often damages the gear severely, even if it does not tear it clean off...and slither cod ends, fish, and all down the ramp into the sea to disappear forever. Nothing is more provoking than to see this happen or the cod end burst and the sea covered with dying fish. Especially is it vexing to have an accident of this kind while a side trawler is heaving in a similar amount of fish safely, bag after bag, alongside one's ship."

But in time remedies were found. The *Fairtry* was fortunate in having as her first mate Jim Cheater, a native of Twillingate on the rugged north coast of Newfoundland, who had fished in England after the war and had married a Grimsby girl. Cheater (pronounced *chay*-tor) had what is called in fishing "a pure nose," the unfailing, seat-of-the-pants instinct to know where the fish are, which, although impossible to define, is enviously acknowledged by all skippers and fleet owners. He was also a seasoned fisherman, skilled in the art of jury-rigging, making on-the-spot solutions to gear problems. Between Cheater's practical knowledge and Captain Leo's technical expertise, the *Fairtry* was blessed with a formidable command. One by one, the vexations gradually vanished.

"We done a thousand things, learning from experience," Cheater says of the process. "First we learned to cut down on tow time. You let your net down one hour and it was more fish than you could handle. That's how it came about we made tows of 20 or 30 minutes, no more. Even so, the trawls were loaded! These times it's hard to believe the fishing we had. Then we fitted out the winches with springs and hauled back very slow, with slack wire. That way there was no surge [sharp rise and fall] of the cod end in heavy weather. And then we found ways to stop the fish from squatting [being too much crushed]. We made a double-wing cod end. The net was double-tailed at the end, same as your trousers."

On her maiden voyage and nearly all subsequent trips, the *Fairtry* went to the Grand Bank of Newfoundland. The choice was obvious. In the 1950s, in spite of having been intensively fished for the salt cod trade by the sailing fleets of many nations for over four centuries, the Banks were still one of the worlds most productive fishing areas. Its best grounds, however, required round trips of over 5,000 miles from England. For ice-provisioned side trawlers this was too far, the barrier beyond which they could not profitably fish. To be sure, a few British "freshers" periodically attempted the trip after World War II, but optimum fishing and steaming time frequently combined to more than 30 days, and too many of the Newfoundland catches were condemned. For the *Fairtry*, of course, with her freezing units and 70-day cruising range, the Newfoundland barrier did not exist (nor any others, for that matter, in the world oceans), and she thus enjoyed some remarkable fishing. Some of her first-year trips produced such astounding results as 650 tons of cod fillets (from 2,000 tons of round or whole weight fish) in 37 days of fishing, a catch rate that makes present-day factory trawler captains shake their heads with envy. Occasional side trips to West Greenland for the highly prized Atlantic halibut, which were then abundant there, did almost as well. Seldom, in fact, did the *Fairtry* need to use her full 70-day range. As often as not she crossed the Atlantic, filled her fishhold, and was bound home in half that time, without any provisioning calls at North American ports.

Among other factors contributing to these early successes was the relative absence of competition, not only from rival British trawlers, but from all other sources as well. For a limited time, in fact, the *Fairtry* had the field pretty much to herself. The few other fishing vessels then on the Grand Bank—the side trawlers that Captain Romyn enviously watched bringing in their catches safely while he was sometimes losing his—mainly carried the flags of France, Spain, and Portugal, nations that had fished there more or less continuously since the discovery of the Banks in the 16th century. The ships they employed were large "salters," designed especially for Newfoundland service, with cavernous holds that entombed mountains of salt to make the split and salt-dried cod so much esteemed in those countries. These ships, well designed and sea-kindly, represented the ultimate in the evolution of conventional side trawlers; a number of them measured over 250 feet and were thus only 15 feet shorter than the *Fairtry*, although only half her tonnage. The Spanish, for reasons no one can remember, called them *bous*. To the French, much more logically, they were *chalutiers terre-neuvas* or "Newfoundlander trawlers"; to the Portuguese, *arrastoes* or simply "draggers." In addition Portugal had her "great white fleet," of about 50 dory schooners, which arrived in April and left in October, and Spain was beginning to have great success with her lethal pair trawlers, small twin vessels that towed a widely stretched common net. Like the Portuguese, many went home for the winter months. The offshore reaches of the Grand Bank was, in fact, a rather lonely sea, with far fewer fishing vessels than in the great age of sail.

All this changed, and very rapidly, beginning in 1956. As the *Fairtry* fished in solitary splendour on the ground known as the tail of the Bank on a fine July morning of her third summer, Acting Captain Jim Cheater got the surprise of his life. A large vessel suddenly appeared on the horizon. She was totally different from any of the company the *Fairtry* occasionally kept, yet there was something strangely familiar about her general outlines. As she came closer, the mystery cleared very quickly.

"I couldn't believe my eyes," Captain Jim recalls. "The ship was the *Fairtry* exactly! Only the name was different. She was called the *Pushkin!*"

Captain Cheater's startled reaction was all too accurate. The Soviet Union was making its debut in distant water fishing. It was doing so, moreover, with a ship that naval architects would later describe as a "slavish copy" of the *Fairtry.*

As with all things in the Soviet Union, the debut was the result of years of careful planning, both to obtain the *Fairtry*'s blueprints and to secure fishing rights to the offshore Canadian waters, which were already internationally controlled (loosely speaking, some would insist) by ICNAF. For the latter purpose, unknown to the *Fairtry,* a sister ship named the *Sverdlovsk* came over later that summer and put in at St. John's. Aboard her were Alexander Ishkov, the USSR's minister of fishing industries, and a host of lesser functionaries and fishery biologists. The *Sverdlovsk*'s visit was fully authorized. More than that, it was in fact a second and reciprocatory event in a government-to-government exchange aimed at possible fishing industry cooperation. A year earlier Canada's Minister of Fisheries, James Sinclair, had been invited by Ishkov for a grand transcontinental tour of the Soviet Union. Now Sinclair was returning the favor. "I'm looking forward to showing the Russian visitors how fishermen in a democratic country such as ours live," he is quoted as telling the *Canadian Fisherman.* "I think it will be an eye-opener for them." The tour would take Mr. Ishkov from St. John's to Vancouver, the *Fisherman* added, and include "everything from champagne to Cadillacs."

The visit of the *Sverdlovsk* became front-page news in St. John's, especially when after some hesitation the Russians allowed both public and press to come aboard. Reporters marvelled at her modern deck gear, the ingenious Baader machines, and, in one case, at least, "the plentiful washroom facilities....including the luxury of Turkish baths."

Throughout the visit Minister Ishkov and his aides stressed two points, one in private and the other very much in public. Privately, the Soviets expressed great interest in becoming a member in good standing of ICNAF (in which Canada, as much as the United States, played a dominant role) and then carrying out some experimental fishing from which "both countries can learn much from each other....[through] an exchange of scientists, fishermen, and periodicals," as minister Ishkov put it. In its public utterances, however, the visiting Soviet delegation repeatedly held out the lure of a new export market, thus touching on the raw nerve of the Canadian fishing industry's long-standing concern over heavy dependence on sales to the United States. "Unlike Canada, nearly all our catch is used on the home market," Ishkov's deputy, Igor Semenov, is quoted as saying. Although the Soviet Union was second only to Japan, with an annual catch of two and three-quarter million tons, he further explained, it still had to look to imports to supply the fish needs of its two hundred million people.

As the initial excitement of the visit subdued, one reporter thought to ask the *Sverdlovsk*'s captain if the USSR was building more such vessels and, if so, where they intended to fish. The captain, described as speaking excellent English, replied without hesitation that the Soviet Union was indeed building more factory trawlers, but that they were designed primarily for Russia's rich home waters, north of the Arctic Circle, although some few might be sent to the south Atlantic for sardines. Similarly, when asked if the Soviet Union might not want to intensify its fishing efforts on the Grand Bank, Minister Ishkov was equally reassuring. "No, we can catch all the cod and haddock we need in many other fishing grounds," he answered. "We are, however, interested in the ocean perch [redfish] that can be caught here."

Even Minister Sinclair joined the chorus. There was no need for concern; he felt sure the Soviets' intentions were mainly experimental and research-oriented, as they had repeatedly stated. "It is more economical for Russia to fish in waters nearer home," he told the *St. John's Evening Telegram* at the time of the visit. "The Grand Banks is a long way from Russia."

That autumn the *Pushkin* and the *Sverdlovsk* went home with a catch of 3,000 tons of cod and 12,900 of redfish, about one-thirtieth the combined catch of the Portuguese, Spanish, and French salter fleets. Perhaps, after all, there was no need to worry.

But the very next summer the *Pushkin* and the *Sverdlovsk* returned with several sister ships. Two years later there were 35 Soviet factory trawlers on the Banks, or all but one of 24 *Pushkin*-class ships and 12 of a newer and larger class. Six years later, by 1965, the Soviet fleet in the northwest Atlantic—now fishing north to Greenland and Labrador and south to Georges Bank—numbered 106 factory trawlers, 30 factory or mother ships, and 425 side trawlers of assorted size and vintage.[3] Its total catch for the year (the fleet now fished winter and summer alike) was 886,000 tons, one third more than the combined catches of Spain, Portugal, and France. Within this total, moreover, were 278,000 tons of the cod and the haddock Minister Ishkov had declared to be of no interest. Still, this was only a beginning.

How the Soviets managed to replicate the *Fairtry* is another story and one almost as deceptive as Minister Ishkov's assurances had proven to be. As early as 1951 a resident Soviet trade delegation incorporated as the Anglo-Soviet Shipping Company of London had begun buying British distant water side-trawlers from the well-regarded Brooke Marine yards in Lowestoft. A small scouting force of these vessels and some Polish-built copies had in fact been used to conduct tests on the Grand Bank, scarcely noticed, two years before the appearance of the *Pushkin* and the *Sverdlovsk*. But by 1953 the events taking place at J. Lewis and Son in Aberdeen were of much more interest to the Anglo-Soviet Shipping Company. Late that year, but still well before the completion of the *Fairtry,* Lewis officials were surprised to receive a tender from the Soviet government for the construction of 24 factory trawlers of the *Fairtry* design. As an aid to speeding negotiations, the Soviets inquired, might not a set of the *Fairtry's* plans be sent in advance for preliminary study?

"I remember strongly advising them [the Lewis yard] to make the Russians pay a substantial fee in advance," states Max Harper Gow. "But they kept telling Lewis their business regulations prevented them from providing any advance. "We merely want to make a prior study of the plans and will settle such other matters later' was in effect what the Russians said."

Gordon Milne, a thoughtful and hardworking Scot who was then managing director of Lewis, well recalls the consternation the Soviet inquiry caused. "We were a small company, the *Fairtry* was the biggest ship we had ever built, and I wasn't prepared to book up the yard to the exclusion of all our other clients," Milne remembers thinking. But there was the satisfactory precedent of the Brooke Marine side trawler sales, and Britain's Conservative government was much interested in expanded trade with the USSR. "Yes, you could say I was urged," Milne says of this. "The government said: 'At least talk to them.'"

The plans of the *Fairtry I* were sent to Moscow.

Milne himself followed within a year. He was pleasantly surprised with the technical professionalism of his Russian counterparts, but not with their requisites for doing business in the Soviet Union. Contract terms, it appeared, would have to be written according to Soviet law. All Milne's suggestions for outside or third-party arbitration in case of dispute—the Hague Court, for example—were rejected. These and other preconditions, along with the burden the large order would place on the Lewis yard, ground the negotiations to an inflexible halt.

Milne returned home, thoroughly discouraged. The plans of the *Fairtry* did not. The Soviet authorities were too busy shopping for their soon-to-be-born distant water fleet to bother with so trifling a detail, much less such troublesome capitalist niceties as patents or the determination of naval architects' fees.[4] Very soon thereafter, in fact, Minister Ishkov and colleagues succeeded in placing their order for their 24 *Pushkin*-class ships with the Howaldtswerke yards in Kiel, West Germany.

Thus occurred what was probably one of the most important and certainly the fastest transfers of technology in the history of commercial fishing. With virtually no prior tradition of high-seas fishing, the Soviet distant water fleet was off and steaming.

"The Lewis people weren't cute enough," Captain Jim Cheater has commented on the incident in retrospect. "It was all right they talked with the Russians. But they should have left the plans at home."

Well might he say so. The rush was now on. Not surprisingly, West Germany was next in line. In the summer of 1957 or one year after the debut of the *Pushkin* and the *Sverdlovsk*, the *Heinrich Meins*—the Federal Republic's first experimental stern trawler—appeared on the Grand Banks. Although small and equipped for whole-fish freezing rather than complete factory processing, the *Meins* was the forerunner or test ship of what soon became a highly efficient fleet of about 30 factory trawlers much envied for their electronic fishfinding wizardry and over 50 large fresh-fish stern trawlers which made a specialty of fast trips to Iceland and East Greenland. A year earlier, far away in the Pacific, Japan commissioned her first experimental stern trawler, the *Umitaka Maru*, "following closely the design of a European stern trawler and a study of European stern trawling techniques," as one Japanese authority has diplomatically acknowledged the *Fairtry* influence. Thereafter, relying heavily on ever-larger factory trawlers, Japan rapidly built up the world's leading distant water fleet.[5] Soon other major European fishing nations followed the British and West German initiative. Although some might hesitate or adopt a wait-and-see attitude before investing the heavy sums necessary to factory trawler construction—the *Fairtry* had required a capital expenditure of £1,000,000—the strong demand and steadily rising fish prices in European markets along with the decline of fish stocks in home waters made the factory trawlers, costly as they might be, an economically viable alternative that was hard to ignore. As the 1960s drew to a close, Poland, East Germany, Rumania, Spain, Portugal, and France all had sizable fleets of the new leviathans, fishing mainly on the Banks or other northwest Atlantic grounds. Waiting a little longer or only building smaller freezer stern trawlers primarily for use in other or nearer waters were Norway, Italy, the Netherlands, Belgium, Iceland, and Denmark. And, paradoxically, the United Kingdom. In England the sad truth was that the brilliant successes of the *Fairtrys I*, *II*, and *III* at sea found no counterpart at home on the marketplace. So strong was consumer resistance to frozen fish—stemming largely from the introduction of poorly made supplies to the armed forces during World War II—that 10 years elapsed before other companies followed Salvesen's lead and cautiously began to build their first "filleters," as the British commonly call factory trawlers. Nevertheless, overall, the factory trawler—the big leap forward—was fast coming of age.

Nowhere, however, was the growth process so spectacular as in the Soviet Union. By 1974, in what some experts have called the most rapid and successful development of a specialized fleet in the history of merchant shipping, the USSR had by far the world's largest fishing fleet. Within its ranks were 2,800 side trawlers and other smaller fishing vessels of all kinds, 103 factory or mother ships, and 710 factory trawlers. The latter, as always, were the heart of the fleet, accounting for 70% of its tonnage and approximately two-thirds of its catch. Sustaining this armada, moreover, was a large train of fleet support vessels. Like the trawlers, many had been purchased abroad and even greater numbers had been built in the Soviet Union. They included not only the giant factory ships, but also refrigerated fish carriers for offloading at sea, oceangoing tugs with well-equipped repair shops, research and scouting ships, food and fresh-water supply ships, tankers, and, ultimately, combined tanker-fish carriers. All were considered necessary because the Soviet government wanted its fishing fleet to be completely self-sufficient at sea, since it was unwilling to spend hard-earned foreign currencies for provisioning or repairs abroad or, for that matter, to have its trawler crews with too much time on their hands in foreign ports. By 1974, in fact, the worldwide fleet support train had grown to 510 ships of 3,000 tons or more.[6] By this time, too, the efforts of Soviet fishermen completely encompassed the world oceans. Although the North Atlantic remained the area of most intense fishing, large deployments of the distant water fleet also ranged the incredibly rich grounds of the northern Pacific and the almost equally productive coasts of South Africa. Smaller detachments went even farther—to the southwestern Atlantic, the southwest Pacific, and Antarctica. Only in tropical latitudes did the big factory trawlers suffer failures. As early as 1962 the Soviets had attempted expeditionary fishing in the Gulf of Mexico and the Caribbean, only to find too many different kinds of fish and not enough of any one for efficient factory processing. The problem remains, not only for the Soviets in one particular area, that is, but for all factory trawlers in almost all tropical waters.

In two decades after the *Fairtry I's* launching, in these ways and by such shipbuilding programs, factory trawlers had moved to a dominant position in world fishing. What this single fact meant was that for the first time in history fishing vessels of one nation might steam to the shores of any other in the world within two or three weeks, remain there indefinitely with either crew or vessel replacements at sea, fish in all but the worst weather, and be rewarded with an hourly catch rate that surpassed the best efforts of conventional side trawlers by 25 to 50%.

Equally important, the new fishing leviathans soon found themselves freed from the restrictions of fishing only along the bottom. For a number of years their owners had watched with mounting envy as purse seiners rapidly increased their catch rate tenfold, making record hauls of herring, mackerel, and other surface-swimming fish, all as a result of a single invention known throughout the world as the Puretic power block.[7] But factory trawlers suddenly acquired the ability to catch these fish, too, when in 1969 the West Germans perfected the mid-water trawls

Once armed with the new free-flying nets, moreover, the factory trawlers immediately held one significant advantage over purse-seining vessels. Since the latter make their catches solely on or near the surface, their skippers are often frustrated by tantalizing views on their sonars of great schools of fish swimming placidly along at depths just below the "hang," the vertical reach of their nets. (Or, worse, by frightened fish that dive for the bottom before encirclement is complete or the purseline at the bottom of their seines can be drawn tight to prevent downward escape.) Fishing with mid-water trawls has no such problems. Properly trimmed and handled, the nets can be made to tow anywhere from a few meters off the sea floor, where they may take bottom fish hovering over grounds too rough for weighted trawls, up to the light-filled waters of the surface, where they will easily overtake pelagic species in direct or straight-line pursuit.

Factory trawlers, in other words, became all-purpose. As their mid-water nets grew ever larger, they found they could take the coveted herring in bigger hauls than the purse seiners. Or, for that matter, any other fish in the oceans, with the single exception of the fastest-swimming members of the tuna family.

The cumulative effect of so many technological advances coming within so short a historical span created problems, of course, not all the problems that now plague world fishing, that is, but many crucial and long-lasting ones, the exact nature of which it is important to understand. Contrary to the complaints of many American and Canadian fishermen, factory trawlers were not the sole cause of the depletion of fish stocks in the north Atlantic. Rather, what the factory trawlers did do was to place an additional or extra burden on stocks that were in many cases already being fished close to the maximum sustainable yields by vessels far smaller. (In the case of the Atlantic halibut in New England waters, the species had long before been commercially fished out.) That this additional burden soon became intolerable is mainly due to the factory trawler fleets' remarkable ability to mass their force very quickly, to "target," in the language of fishery biologists, on specific fish in specific areas. Factory trawlers could focus their efforts very efficiently, in other words, on the often precise localities where fish gather seasonally in greatest number. To give the reader some idea of what this meant or how elements of a large fleet might be deployed in the actual practice of massed or highly targeted fishing power, we must go to the eyewitness accounts of those who first monitored the factory trawlers in the northwest Atlantic. To the time, that is to say, when free fishing, meaning either the nonexistence or the nonobservance of international controls, was the order of the day.

"I remember flying surveillance out of North Carolina in the winter of '68, flat-hatting in a Grumman *Goat* 300 feet above the water," says Charles Philbrook, a veteran National Marine Fisheries Service enforcement official. "Often you could count as many as 200 Communist-bloc trawlers within a 20-mile area off Hatteras Island. Every one of them would be wallowing—filled to the gunwales, you might say—with herring."

What Philbrook saw was a good example of what came to be known as pulse fishing. Simply defined, pulse fishing means the calling together of all ships in a given fleet—even those at some distance, that already might be enjoying fine fishing—to combine their efforts on a particularly good find. In the above case the find was late-winter or pre-spawning schools of Atlantic sea herring and two species of anadromous herring that would soon enter the Chesapeake Bay or the North Carolina Sounds (where they are the object of important local fisheries) on their way to fresh-water spawning creeks. In this as in all examples of pulse fishing, the strategy was to keep fishing until the schools were wiped out, after which the trawlers might disperse in more random patterns. Special scouting ships then took the pulse, so to speak, of other waters until new concentrations were found and all ships again called in.

The sad result in the example reported by Philbrook would be that a month or two later the Chesapeake Bay watermen who fish for herring with pound nets might find their annual spring runs greatly reduced. "Herring is everywhere scarce," the watermen would then say. And wonder why.

Not 200, but a much smaller number of factory trawlers might have the same effect on a particular year class, as scientists call the crop and the year in which fish are born. Dr. Wilfred Templeman, dean of Canada's fishery biologists, remembers one such case from a cruise in July of 1960 of the Canadian government's veteran research ship, *A. T. Cameron*, to a Grand Bank ground known as Southeast Shoal.

"The *Cameron* was fishing 30,000 lb. of haddock in 30 minutes, at the height of the caplin spawning," Dr. Templeman recalls. "They were mainly small fish from the 1955 and 1956 classes, the last good year classes of Grand Banks haddock. The *Cameron*'s skipper, the late Baxter Blackwood, was much given to radio chatter, like most Newfoundland captains. Excited, he started shouting the news to all his friends. The next morning the *Cameron* was surrounded by eight Soviet factory trawlers and four Spanish pairs."

"The new Soviet effort that followed, added to the already existing Spanish and Canadian haddock fisheries, was," Dr. Templeman ruefully concludes, "the start of our most serious haddock decline."

For many years after the first factory trawler invasions, Canadian and American fishery management officials clung to the hope that the fishing fleets of both countries—large in number of vessels and fishermen employed but small in tonnage—could somehow keep pace with the big visitors from abroad. At times, in fact, there was reason to believe that the patient work of the U.S. and Canadian ICNAF delegations in strengthening that body's conservation measures might reverse the tide of overfishing. Or that the United Nations' Law of the Sea Conference, then trying to settle all problems of extended national jurisdictions over coastal-shelf waters for offshore oil drilling, seabed mining, and fishing, might eventually succeed in its visionary effort to construct a rational worldwide fishery conservation scheme. Time, however, seemed to be running out. The haddock was down to an all-time low throughout its northwest Atlantic range, so low, in fact, that fishery biologists feared the end of a commercially viable fishery, if not species extinction. Even the bountiful cod—the sacred cod of America's first industry and the chief sustenance of Newfoundland for over four centuries—was showing serious declines. So, too, were the redfish, yellowtail flounder, and many other prime market fish.

The year 1974 marked the turning point. In that year alone 1,076 Western European and Communist-bloc fishing vessels swarmed across the Atlantic to fish North American waters. Their catch of 2,176,000 tons was 10 times the New England and triple the Canadian Atlantic catch.[8] For anyone who cared to analyze this catch figure or compare it with previous years, moreover, a picture far more disturbing than sheer volume would emerge. Huge as the total catch might seem, the catch per vessel was down and the fish were running generally smaller than before, even though the foreign vessels fished longer hours with improved methods over a larger range for a greater part of the year. In fact, the foreign catch had been better for five of the preceding six years—slowly declining from a peak of 2,400,000 tons in 1968—with fleets of equal or lesser size. When this happens or when the yield per unit of effort starts to fall off, as management experts prefer to say, it is nearly always an early and sure warning of the general decay of major fishing areas. First observed in the Old World with the herring and sole of the North Sea and the cod of Norway's Arctic grounds, the phenomenon now seemed to be repeating itself in the northwest Atlantic. Everywhere the factory trawlers went, in other words, more were fishing for less.

Back on the Grand Bank a new fishery was developing very rapidly to satisfy a worldwide boom market for fishmeal. Because herring was scarce on both sides of the Atlantic, power-block purse seiners accompanied by mother ships and factory trawlers equipped with the new mid-water nets were now zeroing in on the little caplin that came each spring to spawn far out near the edge of the Banks at Southeast Shoals. In 1974 their catch was 300,000 tons and rising fast. As yet Canada's fishery biologists could not detect any effect from this offshore catch on the inshore or beach spawning caplin. The two populations, in fact, were considered separate stocks.

But to the Newfoundland fishermen the signs were already there. By 1974 you could hear the refrain in every outport. The caplin sculls were not as before. Nor the cod, of course, that followed them in. This was a serious matter. In over 270 Newfoundland outports with no other employment than fishing, the fishermen's average annual income was down to $4,500. If things got any worse, the men told you, a man just couldn't make it on the summer voyage.

A new generation of fish killers had come across the Atlantic. In 20 years in North American waters, they had by their own account taken over 72,333,000,000 pounds of fish. Very few among them could see that this was too much or that they had fished too well for their own future.

* From William W. Warner, *Distant Water: The Fate of the North Atlantic Fisherman* (Boston: Little, Brown, 1983).

Notes

1. A standardized scale of wind speeds and sea conditions.
2. This structure was originally thought necessary because the *Fairtry's* long superstructure blocked any view of fishing operations astern, but in practice the officers of the watch found it easier to steer from the wings of the forward bridge, with shouted guidance relayed from the crew aft.
3. Although relatively few in number, the 106 factory trawlers accounted for over 70 percent of the fleet's gross tonnage. Soviet trawlers first extended their operations to U.S. offshore waters in 1961; four years later 60 percent of the Soviet northwest Atlantic catch came from Georges Bank alone.
4. Patent infringement suits were never seriously considered. Expert opinion usually holds that only parts or features (for example, the Baader machines) and not the total concept of a new ship are patentable. Besides, the Soviet Union at the time still had an ambiguous view of patent law in general and did not establish its first permanent patent office until a year later.
5. In volume of catch, that is, averaging ten million tons a year by the 1970s. The USSR developed a considerably larger fleet, but remained a constant second to Japan in world catch rankings, by about one to two million tons.
6. Fishery statisticians, lacking any better yardstick, measure fishing efficiency by comparing total fleet tonnages with catches. The Soviet Union's numerous support ships have customarily been added to her fishing-fleet tonnage; this is one of the reasons why the Japanese, who use a very small support train, have always been considered as "catching more with less" or "most efficient."
7. Named after its inventor, Mario Puretic, a Yugoslavian-born fisherman of San Pedro, California. Before it was patented, the large seines used in the southern California tuna fishery were pulled in hand over hand through one or more blocks (pulleys), a process that normally took five or six hours. Puretic applied hydraulic power to the rotating internal element, or sheave, of a much enlarged block; the block was thus transformed from a static load point to a dynamic mechanism, which itself transported the net back into the boat, in minutes instead of hours.
8. These and all other catch figures are for fish only and do not include crustaceans, mollusks, and other invertebrates, which by 1974 formed 16 percent of the New England catch and 10 percent of the Canadian in the northwest Atlantic. The latter term, as defined by ICNAF and here used, means all waters north of 39°, the latitude of Cape May, New Jersey, and west of 42°, the approximate longitude of Cape Farewell, Greenland.

chapter
seventeen

FISHERS' ECOLOGICAL KNOWLEDGE AND STOCK ASSESSMENT IN NEWFOUNDLAND*

Barbara Neis

In his classic study of indigenous fishers in Palau, *Words of the Lagoon*, Johannes shows that gathering the ecological knowledge of such fishers can add dramatically to our often impoverished understanding of marine ecosystems. So called traditional fishing peoples differ in terms of the extent and character of their ecological knowledge, as well as their adherence to a conservation ethic. In the case of Micronesia, for example, Johannes found examples of waste as well as restraint. He concluded, however, that "the existence of the former does not diminish the significance of the latter."[1] Although not necessarily developed to meet explicit conservation goals, he noted that local management systems often work more effectively than those introduced through colonization and government initiatives.

Until recently fishers' ecological knowledge has been largely neglected by social and natural scientists. While accounts of both might contain fascinating references to fragments of information, disciplinary boundaries that accorded to science the study of nature and to the social sciences the study of human societies marginalized such knowledge.

This paper contributes to the growing literature on the nature and importance of fishers' ecological knowledge and its relationship to western fisheries science. The analysis is based on a reassessment of existing anthropological and folklorist research on the Newfoundland inshore fishery, as well as information on fishers' ecological knowledge contained in archival sources at Memorial University. Most of the archival and anthropological sources date from the 1960s and 1970s, a period when the handline and trap fisheries were being eroded by the expansion of offshore fishing and the introduction of gill nets and longliners. None of the above research had, as its primary goal, the collection and analysis of fishers' ecological knowledge. It does contain, however, some important insights and contradictions. When these are put together with the results from eight recent interviews with trap and handline fishers in Petty Harbour, four with fishers in St. Shotts, recent scientific research on cod, and some identified weaknesses within stock assessment science, the results are illuminating. Looking at stock assessment science from the perspective of these fishers suggests, I will argue, that the problems this knowledge poses for science are not dissimilar to those posed by the ecosystem itself. Furthermore, an examination of current stock assessment science highlights important biases in favour of the offshore trawler fishery contained in its models and methods. In short, on the basis of a preliminary assessment of fishers' ecological knowledge in Newfoundland and Labrador, it seems that efforts to integrate this knowledge into our understanding of fisheries ecology and stock assessment should be both a rewarding and an unsettling exercise.

Fishers' Ecological Knowledge and the Anthropologists

The scope and nature of the ecological knowledge of Newfoundland inshore fishers have never been the direct focus of either scientific or social scientific research.[2] Not even folklore, which has made local understandings its focus, has systematically addressed this.

Handline and trap fisheries dominated much of the northeast coast cod fishery from the late-19th century, and, where they still exist in combination, have undergone much more limited technological changes than others. In addition, they are fisheries that have generally been carried out on the same grounds, by successive generations of fishers. They are also fixed gear fisheries and hence rely more on "harvesting" rather than "hunting" the resource. For all of these reasons, these fishers are probably the most likely group to have developed a relatively rich understanding of ecological relationships such as those between cod movements and abundance, the presence of other species, as well as climatic and oceanographic changes within the micro area of local fishing grounds.[3]

Handline and trap fishers in Petty Harbour were also one of the first groups to insist that scientific assessments regarding the health of the northern cod stocks were incorrect. They played a leading role in the formation of the Newfoundland Inshore Fisheries Association (NIFA) in 1986, an organization that has challenged, from the beginning, the accuracy of stock assessments. Cabot Martin, a leading member of this organization, developed his concern about the state of the stocks through interactions he had with trap fishers in Bay Bulls during the early 1980s.[4] As early as 1983, fishers reported feeling skeptical about the accuracy of scientific claims that the northern cod stocks were growing.

Ethnographies of Newfoundland fishing communities have documented the existence of local management systems in the inshore fishery that limit access to the resource, minimize gear conflicts, and distribute effort. They involve such mechanisms as patrilineal transfer of knowledge about the grounds, regulations regarding the distance between trap berths, patrilineal inheritance of trap berths or the allocation of trap berths by draw, and the creation of gear "sanctuaries" for handlining and trawling.[5]

Anthropological accounts from the 1970s generally maintained that these management regimes reflected efforts to manage space (i.e., reduce gear conflicts) rather than conservation goals. Thus Andersen and Stiles argue, in perhaps the most widely cited article on this subject, that "any attempt to understand the fisheries in Newfoundland, inshore *and* offshore, in terms of their approach to *resource management* must confront the very elementary fact that Newfoundland fishers do *not*, as a rule, manage their *resources*, but rather manage *space*—that is, points of privileged *access* to the resource" (emphasis added). In the 1970s, according to Andersen and Stiles, fishers tended to see resources as "more or less infinite."[6] One consequence of this treatment of resources as infinite was a failure on the part of fishers to perceive overfishing in the 1960s and 1970s as an ecological crisis that might require restraint for them as well as others.

Kent Martin's detailed research on the Fermeuse fishery supports the view that management of space was a central concern of trap and handline fishers during the 1970s. Quotes such as the following also call into question the conservationist orientation of these fishers: "The idea is to get a fish where ya can and how ya can and let tomorrow take care of itself."[7] However, management of space and not the validity and nature of fishers' ecological knowledge was Martin's preoccupation. As has been the case with anthropology elsewhere, Martin and others appear to have had limited knowledge of fisheries biology and their primary interest was in documenting the impact of the environment on culture and not identifying what we might learn about fisheries ecology from fishers.

Significantly, in this anthropological research, there are also references to comments and concerns on the part of fishers that appear to reflect a commitment to husbanding local fish resources. The reasons used by some fishers to justify their opposition to gill nets noted in Andersen and Stiles and other ethnographic research of the period illustrate this commitment. Bonnie McCay, for example, in her ethnography of longlining on Fogo Island, documents inshore trap and handline fishers' and longliner owners' concern about the ecological and not simply the spatial effects of gill net technology.[8] She notes that "the relationship between the use of gill nets and fishery decline is not clear, although fishers have a variety of theories which try to make sense of the correlation."[9] These theories include the view that lost or abandoned gill nets (so-called "ghost nets") continue to fish on the bottom of the ocean, a theory that was supported by a Department of Fisheries and Oceans (DFO) study of ghost nets in the Trinity and Bonavista Bay areas in the 1970s.[10]

Another criticism of gill nets common among handline and trap fishers then and now has to do with the fact that they catch large numbers of so-called "mother" or "bottom" fish that are relatively untouched by other gear. These fish should, the fishers feel, be left in the water to spawn and breed.[11] The following excerpt from a recent interview with Dave Hearn, a highliner trap and handline fisher in Petty Harbour, illustrates the inductive reasoning, taxonomy of codfish, and assumptions about the prerequisites for a healthy fishery underlying his opposition to gill nets.[12]

> Gill nets should be banned completely. There's no fisherman, maybe two or three fishermen in all Petty Harbour out of probably 130 fishermen, believe in using gill nets, the rest of us are dead against it.
>
> Q. Have you always felt this way?
>
> Oh yes, as long as I've been fishing and my father and my grandfather before that....You know, they catch that bottom fish we call it and those were the breeding fish, the mother fish we call it and definitely that's what's after happening.
>
> Q. Seems to me you could have caught some of those fish with a handline?
>
> Well obviously, I'm sure we caught lots of big fish that were breeding fish, mother fish...those years we used to gut all our fish ourselves....You could tell by the roe in the fish whether it was a he fish or a she fish. Days we caught some of those big fish, cut them open, big cod roe in them, almost feel ashamed for catching them. We'd almost feel like letting them go again, but not too many do that either. Obviously we know we caught some of the breeding fish, we had to catch some of them but not near as much as the gill net caught.
>
> I remember one time I was out there, they had a gill net in our waters and the handline fishery was over. I was just out hunting ducks and I see the gill net, a balloon there so I hauled it up by hand and when I hauled it up it was mad alive with the biggest kind of fish, twenty-thirty pound fish and to put a cod jigger down there or even a bait at the same time you wouldn't catch one, there was no hope, the fishery was over. I believe that's what we call bottom

fish, fish that don't really eat. They're there but they don't really eat bait. They're just all bottom fish moping around the bottom, you know, probably eating crab or whatever and the only way probably to catch those is with a dragger or a gill net, and I think a lot of those are the breeding fish.[13]

For David Hearn, then, it was important not to catch all of the breeding fish. While they might catch some of these on the handline, the fact that these fish did not tend to "eat bait" (as attested by the observation that by this time in the fall, the handline fishery was pretty well over) meant that handliners would catch some, but most would remain. Gill nets were a different matter.[14] Located on the bottom, where these fish congregated, gill nets would and, he believes, did catch them all. This fits with McCay's observation in the 1970s that fishers believed gill nets could "fish out" grounds.[15]

Whether or not it is possible to indicate clear conservation objectives in traditional systems of management, it is still important to assess the extent to which they embody practices that are conservationist in their impact. Some such practices include closed areas, closed seasons, or bans on fishing during spawning, allowing a portion of the catch to escape, holding excess catch until needed, bans on taking small fish, restrictions on numbers of traps in an area, protection for species when they are particularly vulnerable (as when in agglomerations), gear restrictions, and efforts to prevent waste.[16]

The trap and handline fisheries are characterized by closed seasons in that they are largely passive fisheries that wait for and are dependent on a migrating fishery resource rather than actively hunting it throughout the marine ecosystem. Fish migrations are seasonal and uncertain. To the extent that cod spawn offshore, these fisheries also do not disrupt the complex spawning process.[17] This is not the case with the Atlantic Canadian trawler fisheries that fish year-round and on spawning concentrations.

The trap and handline fisheries are spatially segregated. Traps are placed in berths and minimum distances between berths are specified. Handline and trawl berths are located in the most productive areas where ledges and shoals tend to coincide with concentrations of cod. Fishing is concentrated in these berths rather than encompassing the entire local marine resource. As noted above, specific gears such as jiggers and gill nets have sometimes been excluded either by legislation or local practice from these grounds.[18] Specialized handline grounds have generally been identified as a mechanism for regulating access to space. However, by limiting the effort applied to larger fish, they may also have been an important conservation practice.

In the past, there were also other mechanisms that acted to limit effort. Until the 1950s and 1960s, the east and northeast coast Newfoundland cod fishery was based on the household production of saltfish. In this household form of production, the volume of saltfish produced was limited not only by the amount harvested but also by the labour and space available for processing catches. Landings from cod traps could fluctuate dramatically during the relatively short six-to-eight-week cod fishery. In some communities trawl and handline fishers would abandon their fishery and help cod trap crews process their catch in return for a portion of the trap landings. This informal system of bartering labour for fish was efficient in that it helped match labour to supplies of fish, minimizing both effort and waste.[19]

Members of inshore fishing communities did not always fish as much as they could, nor, it seems, did they always take all that they could get. There are several communities on the Avalon Peninsula, for example, where historically trap fishers did not participate in the fall fishery. In Bay Bulls, the local merchant chafed at this limitation on landings during the period of saltfish production prior to the 1950s.[20] In St. Shotts, the short fishery season seems to have been linked to a heavier reliance on farming for subsistence than was found in many other communities.

Interviews also suggest that in the past, when handline fishers encountered small fish on the grounds, they would move elsewhere rather than catch and discard these fish. In more recent years, declining catches and technological changes in fish plants seem to have discouraged fishers from leaving any fish because they can now sell smaller fish and because, much more than in the past, they cannot be assured of future catches. With the displacement of saltfish production by the sale of fresh fish to plants for processing, the fishery was also more likely to extend into the fall because it was no longer necessary to take time out to process saltfish.

The gear combination of trap and handline tended to distribute effort over a fairly wide range of year-classes of fish. Younger, smaller fish were more abundant in trap landings and larger, older fish predominated in the fall handline fishery. Prior to the introduction of offshore trawling and gill nets, this helped maintain a stock consisting of many different year-classes and a relatively large population of older fish. In general, more fish were extracted from the younger year-classes in traps, and less from the older year classes because of the technological constraints of the handline fisheries and limitations imposed by migration and weather. The older year-classes were also protected, somewhat, by periodic annual bait shortages that reduced the fall handline fishery in many communities in the past.

Berkes argues that a Cree whitefish fishery in northern Canada that acted to maintain a stock consisting of a broad range of year classes was ecologically sensible. The preservation of older fish in which natural mortality tends to be quite low can, Berkes says, "be considered insurance against the variability of the physical environment which, in some years, results in complete reproductive failure."[21] Cod stocks, particularly those in more southern areas, may be somewhat more resilient than subarctic whitefish and hence less reliant on the preservation of a "standing stock" of older fish. However, a harvesting strategy that preserves year-class diversity would seem to be ecologically wise in any fishery, particularly one such as the northern cod fishery that includes more vulnerable northern stocks and small bay stocks (to the extent that any still exist). The known correlation between fecundity and length, together with the slow growth rate of more northern stocks, supports this view.[22]

Other practices associated with the handline and trap fisheries, although not broadly conservationist in their orientation, reflect trap and handline fishers' application of their understanding about the dynamics of fish behaviour in an effort to manipulate that behaviour and enhance the local fishery. An example of "husbanding" in this sense includes handliners' practice of "baiting the grounds." Handline fishers in Petty Harbour and Fermeuse dumped their leftover bait overboard at the end of the day to "build up the grounds" and encourage the fish to stick around. In so doing, they "complemented the habits of the codfish."[23] Given the dependence of inshore fishers on migratory fish stocks and the limited daily catching capacity of the longline fishery, mechanisms that might hold codfish to the grounds longer into the fall were important. They were another reason for opposing the use of technologies such as gill nets and bottom trawls that did not use bait.

Transcripts from the 1960s also describe the practice of "trouncing," i.e., attaching iron rings to a rope and dragging this along the bottom. The ringing noise would "make the fish move...like a crowd of sheep in the garden. You get em all goin' the once, and drive 'em."[24] Extrapolating from this practice, one might hypothesize that the pattern of fish always swimming towards the net in the otter trawl fishery could be the result of the noise created when the so called "doors," heavy weights attached to the trawl cables, drag along the ocean bottom. Like these fishers' concerns about preserving the "breeding" fish, and well-known concerns about the impact of trawling on the spawning grounds, these practices reflect the focus on ecological relationships in their knowledge about the cod resource.

Stock Assessment Science and Fishers' Ecological Knowledge

Until recently, stock assessment in Newfoundland has primarily relied on data drawn from the offshore commercial trawler fishery and from research vessel surveys, also carried out using trawler technology. Scientists used several reasons to justify the neglect of inshore data in the stock assessment process. These included: the large number of fishers in the inshore; the complexity of the inshore fishery in terms of gear, local oceanographic variations, and climate; and the absence of any measurement of catch per unit of effort for the inshore.[25] In contrast to the data from the offshore commercial fleet, which were relatively easy to quantify and perceived as "rational," inshore information was "largely opaque to statistical analysis" because "results are not evaluated objectively but as an irreducible part of an individual's social and cultural reality."[26]

Finlayson argues that the basis for fishers' dissatisfaction with the claims of scientists' was their experience with declining inshore landings and smaller fish. A closer look at the concerns of Petty Harbour fishers suggests, however, that the observed changes were somewhat more complex. I will use testimony by David Hearn, a trap and handline fisher, and Bernard Martin, a handline fisher, to explore the observed changes in their fisheries that made them skeptical about scientists' claims that the northern cod stocks were recovering. Hearn began fishing handline in about 1960, Martin in the early 1970s.

Hearn fished handline for about 15 years. In 1975, he shifted to traps. After doing well with their trap the first year, he and his brother bought a second trap. Two or three years later, costs for gear were increasing and fish "seemed like it was getting a little scarcer so it seemed like there was nothing else to do only get another trap." They had no established berth for the third trap, so they would just throw it out where they could find a spot. They fished three traps for four-to-five years:

> and then it seemed like that wasn't enough, we were catching no more fish and we were increasing our gear and it seemed the only way to improve was to get another trap, just keep on getting gear, more gear. Started off one, now we got four. Even though we were increasing our gear the fish were getting much smaller. Fish wasn't half the size the last few years as they had been years before that. We had to catch twice as much fish for the same amount of weight.

> We even had to change the twine size in our traps. One time we used to use eight inch twine in the side of the trap and the drawing twine would be three-and-a-half, now we got to put five inch or four inch in the side to stop a lot of those smaller fish from going out through the side of the trap. But, since the early 1980s, I'd say, we noticed the fish getting much smaller.

> You've got to punch in a lot more time and effort at this than in the past, something like 15-18 hours a day and that's where we're at now.

In the minds of the general public and scientists, trap fish is generally assumed to be small fish. This was not, however, always the case. Hearn describes trap fish in 1965:

> I'll tell you I was looking at a picture last night, it was taken I think in 1965 and it showed these fishermen with a cod trap, trapfish. Jeezus, what a difference in the run of fish. There was twenty and twenty-five pound fish in it and the small ones were as big as the big ones we catch now. I can remember my uncle bringing in fish from the cod trap. I used to be cutting out the tongues, twelve, thirteen year old, I couldn't lift them out on the table. They were bigger than me. They came in out of a cod trap. There was no such a thing as a tom cod. You wouldn't get one small enough to eat out of it. If they are there you can catch big fish in the cod trap.[27]

In the past decade, trap fishing effort and competition for berths in Petty Harbour have intensified to the point where the community and DFO decided recently that because of space shortages there would be no new trap fishers in the community. Like Hearn, other Petty Harbour trap crews have bought more traps and many crews have shifted from regular to Japanese traps. Japanese traps can be placed with less risk in berths with rougher bottoms and where, owing to tides, it was difficult to hold fish in the past. They have a smaller mesh, are easier to haul, and have a roof; it is more difficult for the fish to find their way out. Japanese traps opened up new areas to the trap fishery but by the late 1980s, even these areas were filled.[28]

Effort, competition, and handline fishers' conviction that the stocks were in trouble also increased in response to smaller fish, longer fishing days, and a significant reduction in the length of the fishing season. Smaller fish add dramatically to the effort required of handline fishers. They have to catch more fish to make poundage and, in addition, smaller fish do not bite as readily or as firmly. Instead they "pick" at the bait. Bernard Martin estimates it takes twice as much bait to catch the same number of small fish.

Handline fishers also started to fish longer hours in the 1980s. This was facilitated by the introduction of sounders, which allowed them to locate handline berths before sunrise. Sounders have also reduced the effort involved in testing for fish where none actually exist. Because they can see whether or not there are any fish on the grounds, handline fishers tend to move more quickly between grounds, so long as there is space. Competition between these fishers seems to have been exacerbated by reduced landings on some grounds.

Hearn handlines in the fall. He commented on the declining fishery on some grounds:

> There is a certain piece of ground out here, I mean guaranteed, if you anchored there you would catch all big fish, back in the late 1970s, early 1980s. But the last 5-6 years, it's a waste of time to even anchor there. There's no such thing as catching a big fish. If you caught one 40-50 lb, you'd be telling about it, but I mean 10 years ago if you came in with a couple thousand pound of 40-50 fish there'd be nobody, you know they'd be telling about it but, nothing unusual. You can count on your fingers the number of big fish that you catch today.

A handline season that used to run from June to November declined, in the 1980s, to a season that started in July and ended in mid-September. Bernard Martin describes this change:

Back when I started fishing first, people started handlining in June, and they expected a couple of good weeks in June, but nowadays it seems that period when you would expect to catch a few fish for a couple of weeks doesn't happen any more. For the past two years there has been very little fish caught on bait in June month. And so the trend seems to be that the handline season is getting shorter and shorter. On the other end, when you get up in September months, well usually the fish starts eating pretty good right around the end of July or early August, and then you have got about four or five weeks of like peak handlining and after that, around the middle of September, it's just downhill from there. It dwindles away very quickly.

The shorter fishing season poses particular problems for inshore fishers because they are only eligible for unemployment insurance (UI) from November until the beginning of May. A shorter season means months without income from either fishing or UI.

In summary, the changes observed in the handline and trap fisheries in Petty Harbour in the 1980s that contributed to fishers' perception that stocks were not recovering included smaller fish, intensification of effort, shorter seasons, greater competition for space, and related initiatives to limit access to the resource. It is perhaps not insignificant that precisely the same changes were documented by Martin in his study of the Fermeuse fisheries in the early 1970s.[29] It is well known that the cod stocks were in trouble at this time. Recent scientific research has confirmed that there is a relationship between the density of a cod population in a particular inshore area and the range covered by this population. In other words, reduced inshore migrations could be expected to coincide with increased competition for berths as the range of area occupied is also reduced.[30]

Both David Hearn and Bernard Martin attribute the problems in their local fishery to gill nets and offshore draggers. They put together local trends with reports of Newfoundland trawler workers about the discarding of small fish, reports from the media and fishers from the west coast about the dramatic decline in stocks there that was not prevented by either scientists' assessments or stock management, and long-standing concerns about the negative effects of trawling on spawning fish and the spawning grounds.[31] Like others in their harbour, they also compare the abundance and composition of the cod and caplin migrations of the 1980s with those of the past. There was a consensus among older fishers in Petty Harbour and in St. Shotts that the fishery would never again be like it was in the past. The days when cod would leap out of the water in search of caplin (so-called breaching) were acknowledged as over. As suggested by Billy Hewitt in St. Shotts: "Never again, you never see another fish over the water now, not around here. Couldn't last anyhow. Couldn't last."

"Anecdotal" Ecological Knowledge

It should be the experience that leads to a modification of knowledge, rather than abstract knowledge forcing people to perceive their experience as being unreal or wrong.[32]

According to Johannes, the neglect of fishers' ecological knowledge among scientists has been encouraged by ethnocentrism and the class backgrounds of scientists.[33] In addition, the organizational structure of western fisheries science and an approach to doing science that is experimental, laboratory-based, and wedded to mathematical modelling has encouraged the relegation of fishing households to the role of passive audience—suppliers of the fish needed for experimentation and samples of rare species found in their nets. As suggested by Johannes, "scientists tend to have an attitude problem," in that they generally dismiss fishers' knowledge as "anecdotal" or "unsubstantiated," despite the historical depth and richness of experience on which it is often based.[34] Scientists who disparage fishers' ecological knowledge identify it as "mumbo jumbo"; the more sympathetic tend to see it as not amenable to quantification.[35] As argued by one assessment scientist:

> to separate out testable elements of this view of the fishery system from the fishers' point of view is really difficult. I imagine there is probably integration of all kinds of variables going on simultaneously in any particular fisheries situation on any given day, and also over the years, as people modify the traditional lore. You can't really do a controlled experiment under these situations to say, "we falsified the null hypothesis so now we can move on to the next step in the method." That reductionist approach would seem to me to be different from what you would consider to be traditional lore that integrates a lot of different observations and people's intuitions and gut feelings and is kind of tested but you don't know what kind of testing it's undergone from generation to generation. Have the conditions remained constant over time, or have they been changing? If they have, then how do you know what you are seeing is really the result of the causal mechanism that is attributed to it? So it's basically at odds with scientific method because traditional knowledge has so much more information in it that is unspoken or already subsumed and the scientific method says reduce it and test it at each point and control for all of the other co-occurrent variables. It is hard to integrate those two views of the system.[36]

We have only begun the task of teasing out the "truths" in the ecological knowledge of fishers in Newfoundland and Labrador, and the context out of which they have been generated. Fishers' ecological knowledge seems to reflect the influence of the gear that they use,[37] the local topography,[38] local differences in natural and fishing-induced fisheries ecology, competitive relations between fishers, possibly years of experience in the fishery, and the degree to which they apply themselves to the fishery. Fishers derive part of their understanding of fisheries ecology from social networks, the media, and other general sources. However, as competitive fishers, they are more likely to trust the knowledge derived from direct observation, from other members of their crew, and from friends and family, than other sources of knowledge.[39]

The ecological knowledge of handline and trap fishers has a historical dimension that has been lacking in the data used in stock assessment science. Today's actions reflect yesterday's choices (successes). They have an intimate knowledge of the names and locations of the landmarks (or "marks") that are used to find berths on the fishing grounds.[40] This obviously reduces effort and permits fishers to share knowledge across generations and with each other, thereby permitting them "to accumulate a body of information larger than [they] alone would possess, reducing [their] dependency, real and imagined on good fortune."[41] Butler notes that in the community he studied on the west coast of Newfoundland the marks have been constant for generations. He argues that information that goes into tradition like that is "not haphazard."[42]

The ecological knowledge of these fishers also includes awareness of the relationships between season, winds, tides, water temperature, the presence of other species, and the location and "catchability" of fish. In addition, their belief that caplin tend to spawn on the "spring tides" suggests that lunar cycles may be an element in their understanding of the marine ecosystem.[43] This latter statement contrasts with stock assessment for groundfish, where the timing of research vessel cruises for the same months each year points to reliance on the Gregorian calendar rather than lunar months and associated rhythms.

The relationship between fish size, value, and effort, particularly in the handline fishery, means fishers take note of the size distribution of fish. Fishers maintain, for example, that large and small fish are not found in the same places. This observation has been confirmed in recent acoustic surveys.[44]

The historical dimension of fishers' ecological knowledge is important and one that needs more research, particularly in light of the problems that poor historical data for the offshore fishery create for stock assessment. It is also an important ingredient in shaping these fishers' notions of the meaning of "stock recovery" for the inshore. To some extent at least, stock recovery is assessed in relation to a period of almost unimaginable abundance. While handline and trap fishers may no longer expect a return to the abundance of the past, their assessment of stock recovery will reflect their understanding of the abundance and stock structure necessary to ensure the survival of their fisheries. As suggested below, this is not an integral component of the current stock assessment process.

Fishers of all kinds are located at points of convergence in the biological/ecological and fishery-related social/cultural elements in the marine ecosystem. Both elements play an important role in their understanding of marine ecology. The understanding they have, like that of indigenous peoples, is "essentially of an 'ecological' nature, that is to say, it seeks to understand and explain the workings of ecosystems, or at the very least biological communities."[45] Fishers' ecological knowledge is a form of "vernacular knowledge" or knowledge gained from experience.[46] Franklin argues that the scientific method is, in many ways, antagonistic to vernacular knowledge. This antagonism is rooted in the fact that it is:

> a way of separating knowledge from experience....The scientific method works best in circumstances in which the system studied can be truly isolated from its general context....[Unfortunately] scientific constructs have become *the* model of describing reality rather than *one* of the ways of describing life around us. As a consequence there has been a very marked decrease in the reliance of people on their own experience and their own sense.[47]

This discussion of the tensions between scientific and vernacular knowledge offers insight into the relationship between stock assessment science and fishers' ecological knowledge in Newfoundland. Inshore fishers' knowledge is local and influenced by variable individual experience and practice. The relationship between the local observations of fishers in Petty Harbour and the health of the whole complex of northern cod stocks is unclear. Looked at in this light, it is "anecdotal." On the other hand, the knowledge is informed by an understanding of the relationship between fishery success and social and ecological relationships. The "scientific" data that was an integral part of the stock assessment process in Newfoundland and elsewhere in the 1980s could also be defined as "anecdotal" in that it was separated from the social and ecological relationships out of which it was generated. This decontextualization contributed to errors of interpretation.

As noted earlier, stock assessment science for the northern cod stocks has been based almost exclusively on data from offshore commercial trawlers and from the research vessel survey (also trawling data) carried out annually in October. Neither input from inshore fishers nor data based on the inshore fishery have been used to any significant extent in the assessment process. In the 1980s, DFO scientists overlooked the context of changing technology and skippers' knowledge of the resource in the commercial trawler fishery. Neglect of these changes meant scientists did not take into account their impact on catches. Because it is difficult to reconstruct these impacts "after the fact," these data are now considered to be virtually useless for the purpose of stock assessment.[48] In addition, models assumed that the catchability in this fishery was unaffected by the practice of fishing on concentrations and gear mobility. There is a growing scientific literature that challenges this assumption.[49] Also, research vessel cruises were scheduled according to the Gregorian calendar rather than by lunar months or other factors likely to influence the rhythms of one area. As suggested by Leslie Harris, author of a report critical of stock assessment science:

> to say we are going to do our survey in the first three weeks of October every year and that creates a constant for us is not correct because it's not constant. It's constant in terms of time, our calendar, but that's meaningless to fish who don't use our calendar. They use another calendar entirely, which is based on temperature, food availability, and salinity and a number of other environmental circumstances.[50]

Although science has long recognized that oceanographic and hydrographic changes, as well as the movements and health of other species, influence the catchability of cod in handline and trap fishing communities, the role of these factors in general stock health and their influence on catchability in the offshore trawler fisheries and research vessel cruises seem to have been ignored until recently.[51] During the 1980s, for example, scientists appear to have seen no contradiction in using cold water to explain changes in catchability in inshore fisheries while ignoring the possibility that the same cold water could be affecting the health of the northern cod stocks, thus having another, indirect impact on catchability. They overlooked as well the possibility that such oceanographic factors as water temperature might affect catchability for trawlers as well as for the handline and trap fisheries. In simple terms, the wall of cold water that was reported to be blocking inshore migrations in the 1980s might have resulted in increased catchability offshore by limiting the dispersal of the stocks and concentrating aggregations in a narrow range of warmer water areas.[52] Recent scientific research suggests that oceanographic factors may affect general stock health and catchability in trawler fisheries.[53]

It has taken a succession of stock crises to make fisheries scientists confront their neglect of ecological relationships such as those between species, between stocks and oceanography, and between catches and human and marine ecology in stock assessment models.[54] Analysts maintain that it is precisely these relationships that provide the framework for traditional ecological knowledge.[55]

In her review of over 20 years of interviews with fishers contained in archival records, Alison Carter found that their:

> comments reflected a strong sense of the complexity of the conditions which affect the presence or scarcity of the fish. The winds, climate, currents and tides which affect the temperature and composition of the water, and all of which affect the availability of food sources were recognized as interwoven

variables influencing the movements of the fish. Often these things would be indicated by the presence or absence of birds and other animals. The fishers were quite aware of the predator/prey connection between cod and caplin, between seals and herring, and mackerel and plankton. Some fishers even seem to have included the impact of differences in salinity in their analysis of fish behaviour.[56]

The relationships fishers identify between various elements of the marine ecosystem should be the subject of systematic research and debate. The knowledge on which they are based may not be correct, either in an absolute sense, or from the perspective of fisheries science. However, it could be argued that the problem such knowledge poses for a linear, reductionist science is, ironically, much like the problem the marine ecosystem itself poses for such science. In Daan's view,

> the effects of environmental conditions climb upwards through the system to meet somewhere the effects of fishing cascading from the top end. Obviously, they must ultimately mingle into almost an inextricable ball. At present, it is impossible to separate natural and man induced effects in the changes observed at various levels of the food chain. This situation can only become worse, now that even climate has become affected by man's activities and sea temperature and level are expected to rise as a consequence of the green house effect.[57]

Stock Assessment for Whom?

When stock assessment science is looked at from the perspective of trap and handline fishers, a bias in favour of the offshore commercial trawler fishery is revealed. Fishers' ecological knowledge is a social, technical, and cultural product. The same is true of the models of fisheries scientists.[58]

The shortcomings in stock assessment science noted above were partly the result of the social/cultural and technological factors that structured the relationship between those doing this science, participants in the fishery, and the marine ecosystem. In contrast to fishers, those who design stock assessment models and carry out the analyses often spend little time at sea. They work with "paper fish," computerized data. A stock assessment scientist commented:

> most of the time, by the time I see the data, it is already in a computerized form and I deal a lot of the time with paper fish. So I spend a lot of time looking at computer files and graphs trying to figure out is this a trend going up or a trend going down? My principal tool I would say would be a computer and most of my work is in the office.[59]

The data from the commercial trawlers that were used in stock assessment in the 1980s were not collected by scientists or technicians actually located on the vessels, interacting with skippers and crew. Located onshore and away from the fishing enterprise, these stock assessment scientists did not collect data on changes in technology and skill over time and between vessels. As a result, they overlooked the impact of these variations and the impact of changes in management structures, such as the introduction of "enterprise allocations" (company quotas) in the offshore fishery, on logbook data, and, more generally, on catches. This neglect of changes in technology, skills, knowledge, and regulations made it seem as though catches in the commercial trawler fishery were a reflection of stock health and not a result of these other changes. The negative impacts of this neglect were probably felt most and soonest by inshore fishers. It is thus not surprising that they were among the first to protest.

No fisher can ignore the actions of his/her competitors for the fishery resource. Particularly in a context of resource decline, more competitors and innovations in technology, skill, and knowledge mean more competition for space as the range of fish concentrations declines. However, in the northern cod fisheries the competitors are not equal in either their capacity to respond to these changes or in their options. In local fisheries like the trap/handline fisheries of Petty Harbour, space conflicts are confined to community fishing grounds. They may turn neighbour against neighbour and are a threat to the survival of fishing households. Under ideal conditions, trap and handline fisheries will not make these households rich but, history suggests, they can provide a living. Under conditions of increased competition and resource decline, members of trap and handline households are pressured to increase their investment in technology and hence the economic risks of the fishery. They may have to choose between education and healthy food for their children and more gear. Because they fish locally rather than following the stocks, these fishers can catch only those fish that are not taken beyond the confines of their community.[60]

Competition and innovation in the context of resource decline have a different impact on the offshore and nearshore mobile trawler and gill net fisheries. These fisheries have produced enormous wealth and enormous debt for individuals and corporations.[61] Participants in these year-round fisheries can harvest the stocks before and after their inshore migration and over an enormous range of ocean. As a result, they did not feel the effects of stock decline as quickly or as intensely as those in the inshore fishery. In short, the social impact of the neglect of changes in technology and skill in the stock assessment process was, until quota cuts were introduced in the late 1980s, greater for inshore fishers than for the fishers and corporations fishing offshore. Because stock assessment science derived its data from the offshore fishery and not the inshore, it was handicapped in terms of its capacity to make sense of declining inshore landings in the 1980s. At the same time, sustained landings in the offshore fishery meant that participants in this sector were not, until recently, pressuring science to re-examine its commercial trawler data or its analysis of these data.

DFO has been criticized for concentrating on weight (the total biomass) at the expense of fish size in the stock assessment process. It also has been criticized for managing the northern cod stocks as though they were one unit rather than several more or less discrete stocks.[62] Trap and handline fishers appear to agree with scientific claims that inshore migrations and catchability for particular gears are influenced by oceanographic factors. However, the success of their fisheries also depends upon migrations by large and smaller fish to *their* grounds and a relatively long fishing season. Related to this, trap and handline fishers have an interest in maintaining local stocks, should they exist,[63] and a biomass composed of a broad range of year-classes that is sufficiently large to ensure inshore migrations. The absence of these concerns in stock assessment science meant that it served the interests of the corporations and trawler owner-operators better than those of trap and handline fishers.

The broad range of conservationist practices that characterized the trap and handline fisheries needs to be contrasted with the conservationist practices that have dominated government management of the northern cod stocks since the extension of the 200-mile limit. This system of management has relied almost exclusively on the setting of a total allowable catch (TAC). Expanding reliance on trawler and gill net technology, which select for larger fish, coincided with the virtual disappearance of cod of spawning age (six years and older) in 1991. Prior to the northern cod moratorium imposed in 1992, there was no attempt by managers to protect the larger, older "breeding" fish that inshore fishers like David Hearn think should be protected.[64] In addition, technological change and rapid improvements in the knowledge of offshore and nearshore skippers seem to have produced a situation in which, it was recently argued, areas previously protected from fishing owing to ignorance or technological limits have gradually disappeared, thereby eliminating closed areas and seasons that existed by default rather than being maintained through fisheries policy.[65] In short, the only conservation techniques left in the offshore Newfoundland fishery by the end of the 1980s were the TAC, offshore observers to limit discarding, and mesh size. Recent downward revisions in TACs, culminating in the closure of the northern cod fishery, growing evidence regarding the problems associated with stock assessment, and acknowledgement of the varying requirements of different fisheries alluded to in this paper, highlight the need to consider a more complex approach to management based on a broader range of conservation practices.[66]

The stock assessment models used to manage the northern cod stocks after the extension of the 200-mile limit de-emphasized the role of oceanographic factors and species interaction in stock health. At the same time, scientists used the impact of these factors on the inshore fishery and the complex gear compositions and practices developed by inshore fishers to respond to ecosystem uncertainty as a justification for marginalizing the knowledge of these fishers and data from inshore fisheries in general within the stock assessment process. There is now growing recognition within stock assessment science that the fluctuations and uncertainty observed in inshore fisheries are characteristic of marine ecosystems as a whole.[67] There is also increasing acceptance that overfishing was the primary factor responsible for the collapse of the northern cod stocks.[68]

Trap and handline fishers' recent criticisms of stock assessment science need to be placed in the context of the stereotyped and negative visions of their own competence within that science. From the perspective of yesterday's stock assessment science, fishers' ecological understanding is deficient, local rather than general, stories rather than numbers. The inshore fishers interviewed for this paper did not share with assessment science the confidence that fish stocks could be controlled and manipulated. They did not appear to share the assumption of fisheries models that "changes in variables have predictable outcomes."[69] This uncertainty derived in part from their belief that fish have minds of their own. They saw fish as:

> variable and uncontrollable....This is reflected in how large a role luck and superstition play in fishing. Even if you had a good berth and a good trap, whether or not the fish would "strike" was another thing and quite beyond the control of the fishermen. It was as though the fish had a mind of their own and a rather fickle personality. The fishermen tended to personify the fish in the way they spoke about them [to treat them as subjects rather than objects].[70]

Fishers' ecological knowledge could provide important insights for the development of knowledge about marine ecosystems. At a time when the future of the northern cod stocks is very much in question and the limited nature of current knowledge of cold ocean ecosystems is apparent to all, an approach to stock assessment and fisheries management that starts from local areas and builds on the experience of these fishers must be considered. As argued by Cabot Martin, an outspoken advocate of a new science that includes fishers' ecological knowledge, "it's a savage comment on our system that you would have to try to make the case that such knowledge is important information."[71]

* A slightly different version of this paper was published in *Newfoundland Studies*. I would like to acknowledge the help of Alison Carter and Bev Brown. I would also like to acknowledge the cooperation and active help of some members of DFO, Science Branch St. John's, particularly Larry Coady and George Rose. Chris Finlayson, Ben Davis, and three anonymous reviewers provided useful comments on earlier drafts. The fishers of Petty Harbour and St. Shotts, Newfoundland, endured both my invasive questioning and my seasickness. Bernard Martin accompanied me on some of my interviews to help deal with the inevitable barriers between a shore-based sociologist and fishers. Any errors are my own. This research was helped by a grant from CBC for the production of an "Ideas" programme called "The Science of the Sea" and a small grant from the Canada Employment and Immigration Commission.

Notes

1.   R. E. Johannes, "Traditional Marine Conservation Methods in Oceania," *Annual Review of Ecological Systems* 9, 1978, p. 355.

2.   An exception is an examination of the ecological knowledge of Newfoundland and Labrador salmon fishers contained in Larry Felt, "Two Tales of a Fish: The Social Construction of Indigenous Knowledge Among Atlantic Canadian Salmon Fishers" in *Folk Management of World Fisheries,* edited by C. Dyer and J. McGoodwin (Tucson: Univ. of Arizona Press, 1994), pp. 251-86.

3.   In *Words of the Lagoon*, Johannes argues that the nature and quality of fishers' ecological knowledge is influenced by the fishing technology and techniques used. In the case of Palauans, such knowledge is particularly intimate because they often fish underwater with spear guns. He contrasts the kind of knowledge required in, and garnered from, fishing in this way to that acquired when fishing in a trawler. This should not be taken to mean that trawler skippers would not acquire valuable ecological knowledge. Their success as skippers is dependent on their ability to find fish. However, their knowledge, mediated as it is by vessel mobility, fish finders, sonar, and other technologies can be expected to be quite different from that of fixed gear fishers. See R.E. Johannes, *Words of the Lagoon: Fishing and Marine Lore in the Palan District of Micronesia* (Berkeley: Univ. of California Press, 1981).

4.   Personal Communication.

5.   Raoul Andersen and Geoffrey Stiles, "Resource Management and Spatial Competition in Newfoundland Fishing: An Exploratory Essay," in *Seafarer and Community: Towards a Social Understanding of Seafaring,* edited by Peter Fricke (London: Croom Helm, 1973), pp. 44-66; Melvin Firestone, *Brothers and Rivals: Patrilocality in Savage Cove* (St. John's: Institute of Social and Economic Research,1967); Bonnie J. McCay,"Appropriate Technology and Coastal Fishermen of Newfoundland" Ph.D. diss.(New York: Columbia Univ., 1976); Kent Martin, "The Law in St. John's Says..." M.A. thes. (St. John's: Memorial Univ. of Nfld., 1973); Ralph Matthews, *Controlling Common Poperty: Regulating Canada's East Coast Fishery* (Toronto: University of Toronto Press, 1993).

6.   Andersen and Stiles "Resource Management," p. 45.

7.   Kent Martin "The Law in St. John's," p. 74

8.   Anderson and Stiles ("Resource Management")argue that the effort to exclude longliners from local communities "did not entail a 'conservationist' [strategy] but rather a personal *protectionist* one ... yet some recognition of resource limitation was clearly in evidence." pp. 56-57.

9.   Bonnie McCay, "Appropriate Technology," p. 187.

10.   Eric Way, *Lost Gill-net Retrieval Experiment* (St. John's: DFO, Fisheries Development Branch, 1976).

11.   Martin, "The Law in St. John's," p. 47.

12. All excerpts from interviews have been edited for style.

13. See also interviews with Joe Randell and George Groves (MUNFLA C804 70-04, 1969; C1969 73-46, 1971); and Drew.

14. A recent examination of the ecological knowledge of Saami fishers in a Norwegian community found that they maintained that gill nets would not catch the biggest cod with the highest reproductive capacity "simply because it's too big to get stuck" Einar Eythorsson, "Saami Fjord Fishermen and the State: Traditional Knowledge and Resource Management in Northern Norway (Winnipeg: University of Manitoba Occasional Paper, forthcoming). The difference between the two communities is interesting. However, so is the similarity in that both expressed concern about preserving large fish. As noted below, this has not been a preoccupation of stock assessment science in Atlantic Canada.

15. McCay, "Appropriate Technology."

16. R.E. Johannes, "Integrating Traditional Ecological Knowledge and Management with Environmental Impact Assessment," in *Traditional Ecological Knowledge: Concepts and Cases*, edited by Julian T. Inglis (Ottawa: International Development Research Centre, 1993).

17. Oceans Ltd. "Reproductive Success in Atlantic Cod (Gadus Morhua L.): The Potential Impact of Trawling" (St. John's: Report submitted to NIFA, 1990).

18. Martin, "The Law in St. John's."

19. Cornelius O'Brien Interview, 91461 C14591-4, Folklore Archives, Memorial University, (MUNFLA) St. John's.

20. Cornelius O'Brien Interview, MUNFLA.

21. Fikret Berkes, "Common-Property Resource Management and Cree Indian Fisheries in Subarctic Canada" in *The Question of the Commons: The Culture and Ecology of Communal Resources,* edited by Bonnie J. McCay and James Acheson (Tucson: Univ. of Arizona Press, 1987), pp. 66-91.

22. Art May, "Fecundity of Atlantic Cod," *Journal of the Fishery Research Board of Canada* 24, 7, (1967), pp. 1531-51.

23. Martin, "The Law in St. John's," p. 47; Petty Harbour Interviews.

24. MUNFLA, Ab Tulk Interview, C809 70-4, 1969; see also C83, 1964.

25. Department of Fisheries and Oceans, "The Science of Cod: Considerations in the Scientific Study and Assessment of Cod Stocks in the Newfoundland Region," *Fo'c'sle* 8, 2 (1988), pp. 1-29.

26. A.C. Finlayson, "The Social Construction and Reconstruction of a Scientific Crisis: A Sociological Analysis of Northern Cod Stock Assessments from 1977 to 1990." M.A. thesis (St. John's: Memorial Univ. of Nfld., 1991), p. 180.

27. Cornelius O'Brien, a former fish plant owner, reported that in 1954 he counted the average number of fish in 10 boxes of dried fish. The majority of this fish would have been from cod traps. He found that there were about 100 salted cod per 100 pounds. In "round" (i.e., fresh) form, these fish would have averaged five pounds each and, he maintains, would have been over 24 inches in length. In 1991, he claimed, 30% of the trap fish were under 16 inches (MUNFLA, 91461 C14593). The percentage of trap fish landed in the Bonavista fishery that was under 41 cm increased from less than 5% in 1950-1953 to 20% by the period between 1972-1974. See The Templeman Records of the Bonavista Longliner Study, DFO St. John's.

28. Sam Lee, Personal Communication. Interviews with inshore and nearshore fishers in Bonavista and Trinity Bays carried out in 1995 revealed similar trends in increased effort in these areas in the 1980s. See also Jeffrey Hutchings and Ransom Myers, "What Can be Learned from the Collapse of a Renewable Resource? Atlantic Cod, *Gadus morhua*, of Newfoundland and Labrador," *Canadian Journal of Fisheries and Aquatic Sciences* 51(9)(1994), pp. 2126-46.

29. George Groves, a Bonavista fisherman, commented in 1971 that it seemed as if the fish didn't come any more in the fall of the year (MUNFLA, C1969 73-46). Fermeuse fishers noted that the spring fishery had declined.

30. G.A. Rose and W.C. Leggett, "Effects of Biomass-Range Interactions and Catchability of Migratory Demersal Fish by Mobile Fisheries: An Example of Atlantic Cod (*Gadus Morhua*)," *Canadian Journal of Fisheries and Aquatic Sciences* 48 (1991), pp. 843-48.

31. Henry Baels, "The Stock of Sea Fishes" in *Study of the Sea: The Development of Marine Research Under the Auspices of the International Council for the Exploration of the Sea,* edited by

E.M. Thomasson (Farnham, Surrey: Fishing News Books, 1981), pp. 132-38; Oceans Ltd. "Reproductive Success in Atlantic Cod (*Gadus Morhua L.*): The Potential Impact of Trawling," Report submitted to NIFA. St. John's, 1990.

32. Ursula Franklin, *The Real World of Technology* (Toronto: CBC Enterprises, 1990), p. 40.

33. Johannes, *Words of the Lagoon*, p. ix.

34. Johannes,"Integrating Traditional Ecological Knowledge," p. 37.

35. As a result of a recommendation in the *Independent Review of the State of the Northern* Cod *Stock* (Ottawa: Department of Fisheries and Oceans, 1990) some attempt has been made in recent years to identify ways to use inshore data and the insights of inshore fishers in stock assessment. One product of this initiative is the Sentinel fisheries that now exist in several parts of Atlantic Canada.

36. George Rose, a DFO scientist whose specialty is oceanography, has developed and tested a number of theories about fish movements on the basis of ideas he acquired from inshore trap fishers. His research findings have generally concurred with the views of fishers (Interview, 30 July 1991). A study designed to systematically collect the ecological knowledge of fishery workers and integrate this knowledge with science is currently under way in Newfoundland.

37. Martin, "The Law in St. John's."

38. Gary Butler, "Culture, Cognition, and Communication: Fishermen's Location-Finding in L'anse-a-Canards, Newfoundland," *Canadian Folklore Canadien* 5, 1-2 (1983), pp. 7-21.

39. Gary Butler, Personal Communication.

40. Butler, "Culture"; Martin, "The Law in St. John's."

41. Butler, "Culture," p. 18.

42. Butler (Personal Communication). Butler ("Culture") uses the concept "cognitive map" to analyse inshore fishers' knowledge. Eythorsson uses a similar concept, "mental map," in his discussion of the ecological knowledge of Saami fishers. He emphasizes the importance of intimate knowledge of tidal currents as critical to the success of fishing enterprises.

43. Johannes (*Words of the Lagoon*) provides a fascinating discussion of the lunar rhythms of fish and fishing in Palau and their ecological bases. There has been some scientific investigation of the relationship between caplin spawning and lunar (tidal) cycles in Newfoundland.This research links the timing of spawning to tides, water temperature, winds, and light levels. Templeman's quotes from fishers imply that some link the timing not only to tides but also to the structure of local beaches. See Wilfred Templeman, *The Life History of the Caplin (Mallotus villosus* O.F. Muller) *in Newfoundland Waters, Bulletin* (St. John's: Newfoundland Government Laboratory, 1948), p. 41; Kenneth Frank and William Leggett, "Wind Regulation of Emergence Times and Early Larval Survival in Capelin (*Mallotus villosus*)," *Canadian Journal of Fisheries and Aquatic Sciences* 38 (1981), pp. 215-23.

44. George Rose, Personal Communication.

45. Milton Freeman, "Nature and Utility of Traditional Ecological Knowledge," *Northern Perspectives* 20 (1) 1992, p. 9.

46. Franklin, *The Real World,* p. 36. There are also elements of "extended reality," i.e., "that body of knowledge....we acquire that is based on the experience of others"; perhaps to a lesser extent "constructed reality," for example, the media; and "projected reality," their sense of the future in their thinking (36-8).

47. Franklin, *The Real World*, pp. 37-38.

48. Jean-Jacques Maguire, Personal Communication.

49. Rose and Leggett, "Interactions and Catchability."

50. Personal Communication.

51. Wilfred Templeman, *Marine Resources of Newfoundland* (Ottawa: Fisheries Research Board of Canada, 1966), p. 47 ff.

52. Department of Fisheries and Oceans, "The Science of Cod."

53. Philippe Cury and Claude Roy, eds. *Pecheries Ouest Africaines: Variabilité, Instabilité et Changement* (Paris: Ostrom, 1991); Ana Gordoa and Joseph Hightower, "Changes in Catchability in Bottom-trawl Fishery for Cape Hake," *Canadian Journal of Fisheries and Aquatic Sciences* (1991) 48, pp. 1887-95.

54. *Report of the Study Group on Ecosystem Effects of Fishing Activities* (Lowestoft: International Council for the Exploration of the Sea, 1991).

55. Milton Freeman, "Nature and Utility"; Douglas J. Nakashima, "Application of Native Knowledge in EIA: Inuit, Eiders and Hudson Bay Oil," Report prepared for the Canadian Environmental Assessment Research Council (1990).

56. Alison Carter, Research Notes on the Indigenous Knowledge of Fishing People in Newfoundland and Labrador. Prepared for Barbara Neis, Dept. of Sociology, Memorial University of Nfld., St. John's, (1991), p. 2.

57. Neils Daan, "The Ecological Setting of North Sea Fisheries." *Dana* 8 (1989), pp. 21-22.

58. A. C. Finlayson, *Fishing for Truth: A Sociological Analysis of Northern Cod Stock Assessments from 1977-1990* (St. John's: Institute of Social and Economic Research, 1994).

59. Anonymous, Personal Communication.

60. This was also a feature of past interviews with fishers. Ed Wade, a Flatrock fisherman interviewed in 1972, commented that draggers were "catchin' all the fish and it doesn't get a chance ... to get in." (MUNFLA, C1309 72-115). Sam Lee, a Petty Harbour fisherman interviewed for this project noted that increasing the mesh size in cod traps would simply allow cod to return to the offshore where they would be caught by trawlers.

61. Peter Sinclair, *From Traps to Draggers: Domestic Commodity Production in Northwest Newfoundland, 1850-1982* (St. John's: Institute of Social and Economic Research, 1985).

62. Leslie Harris, *Independent Review of the State of the Northern Cod Stock* (Ottawa: Department of Fisheries and Oceans, 1990); Jeff Hutchings, Barbara Neis and Paul Ripley, "The 'Nature of Cod' (*Gadus Morhua*): Perceptions of Stock Structure and Cod Behaviour by Fishermen, 'Experts' and Scientists from the Nineteenth Century to the Present," paper presented to the conference Marine Resources and Human Societies in the North Atlantic since 1500 (St. John's: Memorial Univ. Nfld., October, 1995).

63. The large "mother" fish that Petty Harbour fishers maintain were wiped out by gill nets may have included some that overwintered in deep holes off the coast in the St. John's area. Recent research in Trinity, Bonavista, and Placentia Bays suggests a similar pattern of fishing out populations of large, old "mother fish" in these areas as well. Eythorsson emphasizes the importance of local stocks to the fjord fishery in Norway. He also points out that the fjord stocks "do not seem to exist in the language of fisheries management officials." See Einar Eythorsson, "Sami Fjord Fishermen." The zone system of management adopted in the Atlantic Canadian fishery had the effect of overriding attention to local stocks. See R.G. Halliday and A.T.Pinhorn "The Delimitation of Fishing Areas in the Northwest Atlantic," *Journal of Northwest Atlantic Fishery Science* 10 (special issue, 1990) for an interesting examination of the history of zone boundary development and negotiations around stock delineations.

64. Larry Coady, Personal Communication.

65. Leslie Harris, "Presentation to the Memorial University Fisheries Forum," St. John's, September, 1992.

66. Daan, "The Ecological Setting."

67. Daan, "The Ecological Setting"; Cury and Roy, "Pecheries Ouest Africaines"; Manuel Gomes and Richard Haedrich "Predicting Community Dynamics from Food Web Structure" in *Deep-Sea Food Chains and the Global Carbon Cycle,* edited by G.T. Rowe and V. Pariente (Netherlands: Kluwer Academic Publishers, 1992), pp. 277-93.

68. Hutchings and Myers, "What can be learned."

69. James Wilson and Peter Kleban, "Practical Implications of Chaos in Fisheries: Ecologically Adapted Management," *Marine Anthropological Studies* 5(1), 1992, pp. 67-75.

70. Carter, "Research Notes," p. 2.

71. Personal Communication.

chapter eighteen

THE COMMERCIAL ANNIHILATION OF NORTHERN COD: THE FATE OF THE 1986 AND 1987 YEAR-CLASSES

Donald Harold Steele and Raoul Andersen

On 2 July 1992, the Minister of Canada's Department of Fisheries and Oceans (DFO) announced a two-year moratorium on the commercial exploitation of northern cod, the stock complex found in regions 2J3KL. Its purpose was to protect the spawning stock which, by that time, consisted almost entirely of the 1986 and 1987 year-classes (estimated initially to be larger than average in size). Scientific advice indicated that Total Allowable Catch (TAC) at $F_{0.1}$ of 50,000 metric tonnes would not be commercially viable. The initial expectation was that a fishery could resume in 1994. However, the stock continued to decline and the moratorium remains in place in 1996. Moreover, moratoria have had to be declared subsequently for other cod stocks and for other species within Canada's 200-mile jurisdictional zone, and even the traditional subsistence fishery has been halted. This study describes the fate of the 1986 and 1987 year-classes through a summary of extant theories of fish population decline, an analysis of various reports and publications documenting the decline, and conclusions about the pitfalls in scientific estimation of stock recruitment.

For some time we have known that overfishing has three phases: In the first phase, known as "growth overfishing," there is a significant decline in the abundance of large fish. Catches may remain high, if the size and age at recruitment decreases, but the landings consist mainly of small-sized fish. By itself a significant decrease in the average sizes of the fish caught is therefore an indication of overfishing. It is easily detected and observed by fishers. The change in the size structure of the stock is not usually considered important because management is based on biomass which may remain relatively high. However, there may be significant biological changes in the stock because size and age at maturity, egg production per unit weight, and spawning success are all liable to be altered when the stock consists of small fish.

In the second phase, "recruitment overfishing," the fishery becomes dependent on new recruits entering the fish stock. Since recruitment varies from year to year, the fishery in this phase is inherently unstable because there are few, if any, older fish to buffer the variable recruitment.

Finally comes the third phase, "commercial extinction." If the fish are caught at a small size before they have a chance to reproduce, recruitment is reduced to a level at which a directed commercial fishery is no longer viable. Some fish still persist and will be caught as bycatch in fisheries for other, usually less desirable, species. The effect is that the landings of the less desirable species subsidize the catches of the commercially extinct species and as a result the latter cannot recover. The final result is that eventually the fishery becomes dependent on less and less desirable species, as has happened, for example, in the fisheries in the Great Lakes, Long Island Sound, and historic international whaling.

In the late 1960s and early 1970s offshore pre-spawning aggregations of northern cod were subjected to intense exploitation and the stock declined. After adoption of the 200-mile jurisdictional limit in 1977, management of most of the northern cod stock came under Canada's control. Only an estimated 5% of the biomass occurs beyond the 200-mile limit on the so-called nose and tail of the Grand Bank, and this area is managed by NAFO (Northwest Atlantic Fisheries Organization).

The initial target fishing mortality set by Canada in 1979 was lower than the normally advised $F_{0.1}$ in order to allow the stock to rebuild until the spawning stock biomass reached 1.5 million tonnes. However, before rebuilding reached the targeted spawning stock, the fishing mortality was increased to the $F_{0.1}$ level. Unfortunately, the actual fishing mortalities were three to four times the projected values. As a result the stock declined and quickly went through the phase of growth overfishing and became dependent on recruitment.

Although the annual Atlantic Groundfish Management Plans relate recruitment to spawning stock biomass in Rule No. 1 for setting Total Allowable Catch (TAC),[1] the view of many scientists has been that recruitment and spawning stock in cod are not correlated.[2] This latter view now has been adopted by textbooks in marine biology.[3] If true, it would mean that it would not be necessary to have a large spawning stock in order to insure that recruitment is not limited by the number of spawners, and it would provide much more flexibility for management decisions.[4] In addition, the current methods of estimating the size of fish stocks—[Virtual Population Analysis (VPA) and Sequential Population Analysis (SPA)]—provide poor estimates of the number of recruits and there should be alternative confirming estimates of recruitment.[5]

Despite these problems, management has continued to rely on estimates of recruitment to manage the northern cod stock. This was crucial in 1990 when the northern cod fishery became almost entirely dependent on the 1986 and 1987 year-classes. These year-classes were originally believed to be much larger than the long-term average, and this bolstered the view that large year-classes could be produced by a small spawning stock and that it was not necessary to limit fishing on spawning grounds.

The first DFO estimates of 1986 and 1987 year-classes formed the basis for the expectation in 1990 that fishers could expect a boom.[6] They figured prominently in the conclusion that the spawning stock of northern cod was not on the verge of collapse as was argued by the Newfoundland Inshore Fisheries Association (NIFA) in its 1990 application for an injunction to limit fishing on the spawning grounds by offshore otter trawlers. Rather W.G. Doubleday predicted that "spawning biomass would increase by over 50% by 1994,"[7] and CAFSAC (Canadian Atlantic Fisheries Scientific Advisory Committee) stated that the 1986 year-class was "estimated to be about 1.5 times stronger than the average year-class size since 1977, that is about 475 million fish."[8]

Another CAFSAC Advisory Document stated: "The assessment suggests that the stock will increase faster than was forecast in last year's assessment, *largely because of the strength of the 1986 and 1987 year-classes....Therefore, CAFSAC suggests that the existing multiyear management plan be maintained*" (emphasis added).[9] It is evident that the management advice was based heavily on the optimistic estimates of the sizes of the 1986 and 1987 year-classes, and at the time there was no intention of protecting them from intensive exploitation. The implication is, in fact, that they would bear the brunt of the fishery.

However, by 1992 there was implicit recognition of recruitment overfishing and the perilous state of the spawning stock that consisted essentially of only two year-classes:

1986 and 1987. CAFSAC and NAFO reports trace the recognition of this decline. In 1991 a report stated that "[t]he numbers at age 3 in 1989 (1986 year-class) and 1990 (1987 year-class) were estimated by the accepted ADAPT formulation to be 505 and 645 million fish respectively."[10]

By early 1992 the optimism continued in spite of slight down-grading of stock abundance in a report that noted: "The results of the preliminary assessment suggest that the 1986 year-class is still amongst the highest in the past 20 years while the 1987 year-class is the highest since that of 1968, but both are now estimated to be less abundant than in the previous assessment."[11]

In July 1992, CAFSAC reported that "the entire stock is now essentially composed of the 1986 and 1987 year-classes which may be followed by four weak year-classes. Fisheries in the next few years would depend heavily on the 1986 and 1987 year-classes and would exploit them before they have made their full contribution to the spawning stock."[12]

By 1993 fisheries scientist Jake Rice stated that the 1986 and 1987 year-classes "did not recruit in the expected numbers. However, the assessment did not revise their historic size downward dramatically. The work I saw confirms that those year-classes did start out strong. The damned things just disappeared from 1990 to 1991."[13] Other reports document the downward revisions that occurred between 1992 and 1994.[14] By 1995 when they would have been fully recruited and contributing to the fishery and it should have been possible to estimate their numbers accurately, there had been a decrease of almost 90% in the estimated size of the 1987 year-class since it was first estimated by Rice in 1990. In fact, they disappeared so completely that they were no longer explicitly discussed by DFO in their Atlantic Fisheries Stock Status Report 1995.

Although great concern has been expressed about the general loss of the older, larger fish, the huge reductions in the optimistic estimates of the sizes of the 1986 and 1987 year-classes passed with little public discussion, even though they were the basis for the original moratorium. There are two possible explanations for the decrease in the estimates of recruitment over time: 1) an inability to estimate the numbers of young fish accurately. Later estimates made when the fish have been caught are much more accurate. In this case, the differences between the first and the last estimates would be expected to show a uniform or random variation. 2) The initial estimates were subject to relatively minor errors and the latest estimate differs only slightly from the first. In this case a significant decrease would indicate that the fish disappeared and presumably were caught before they were recruited to the fishery. A comparison of the decreases in the estimates of the 1986 and 1987 year-classes with 1984 indicates that the latter explanation is correct.

To examine these relationships more closely, the decreases in the estimated numbers of pre-recruits of the various year-classes were determined. The data are available in the DFO assessments back to 1978 as three-year-olds and back to 1968 as four-year-olds. The results are shown as percentage decreases, that illustrate that the decreases are not random but systematic, with peaks in the 1970s and late 1980s, both periods of declining landings. On the other hand, when catches were increasing from year to year (1968, 1978 to 1981) the decreases in the estimated numbers of pre-recruits were relatively small. The decreases in the estimates of four-year old pre-recruits correlated with decreases in the size of both Canadian and total landings. After 1977 and the extension of jurisdiction, landings became more stable. However, throughout the 1980s the reductions in the estimates of the sizes of the recruiting year-classes increased dramatically. This suggests that the landings were maintained only by harvesting small fish, and the initial estimates of recruitment were not grossly inaccurate, but the fish disappeared from the stock *before they were recruited to the stock*.

Catches made by research vessels in 2J3KL showed similar changes. Prior to the late 1980s survival of the pre-recruits (three-and four-year-olds) was over 100%, as would be expected since fish were still being recruited to the fishery. However, in later years survival was less than 100% showing that the pre-recruits had disappeared before they entered the fishery.

These observations have a number of significant implications: first, it is clear that there are serious problems in estimating recruitment at the present time, and therefore *such estimates should not form the basis for stock management.* Given the difficulties in estimating recruitment, stock assessors have usually assigned to it the long term average. However, if the stock is in the phase of recruitment overfishing, such an approach will inevitably be disastrous. Since the fishery is dependent on recruits, the inevitable occurrence of a poor year-class (below the average) will increase the exploitation of the preceding year-classes in order to maintain the fishery. On the other hand, it is folly to exploit heavily good year-classes (above the average) since they should be available in the future when smaller year-classes occur.

The advice offered by CAFSAC in 1991—to continue fishing at the same rate because of the estimated strength of the 1986 and 1987 year-classes—was disastrous. DFO failed to protect these two year-classes from exploitation. By the time the moratorium was declared in July 1992, it was too late since most of the fish had already disappeared when they were pre-recruits. As a result, the boom predicted for these year-classes never materialized. In addition, the production of eggs by these two year-classes has been very small. Thus the stock has not been able to recover quickly even with the moratorium.[15]

This situation is an almost identical replay of what occurred in the early 1970s when very high TACs were set for northern cod. According to Munro, "the argument given in 1972 for setting a large TAC for 1973 was based on the expectation that two particularly large year-classes of cod would replenish the fishery; hence it was safe to set the TAC in excess of [the Maximum Sustainable Yield] MSY."[16] However, these year-classes also failed to recruit as expected.

The second implication is that the error in overestimating recruitment from the number of pre-recruits (three- or four-year-olds) is less than 10% when landings are increasing and there are presumably no catches of pre-recruits being made. The errors in estimating recruitment are therefore not excessive.

Third, when landings are decreasing, many pre-recruits disappear before they recruit to the fishery. In the case of the 1987 year-class this amounted to almost 90%. The most likely cause for their disappearance is that they were caught and many were discarded, i.e. the catches were "high-graded." Unfortunately there are little hard data concerning this problem, but during the late 1980s there was considerable anecdotal evidence for the catching of small cod. Sinclair, et al., have provided data indicating extensive discarding of small cod in the Gulf of St. Lawrence during this period. Another report provides evidence of the processing of small-sized cod:

> Our partnership visited a number of fish plants where we toured idle equipment. There was abundant evidence of the way in which an industry adapted and continued to profit from a dying resource. A new filleting machine had been purchased, capable of processing smaller cod while simultaneously replacing eight to 10 people in the plant. The industry also adapted to smaller fish by shipping out cod that had been merely headed and gutted for later drying in Denmark or Portugal.[17]

In analyzing reasons for the collapse of cod stocks, Myers, Hutchings, and Barrowman concluded that there is a positive correlation between the fishing mortality of adults and the mortality of pre-recruits, and "that high levels of discarding and catch misreporting occur and that these levels increase with declines in population abundance and concomitant increases in fishing mortality," and "high discarding reduced the number of fish entering the fishery until the populations were reduced to the point of commercial extinction."[18]

The fourth implication is that correlations between recruitment and environmental factors or spawning stock will be present or can disappear depending on the year when the estimates are made. For example, using estimates of abundance made in 1992, Myers, et al., concluded that recruitment of Newfoundland cod stocks was correlated with salinity.[19] However, Shelton and Atkinson, using estimates of abundance made in 1993, concluded that salinity failed to predict recruitment in the same stocks,[20] and Hutchings and Myers have subsequently repudiated their earlier conclusion. It is apparent that by choosing the estimates it is possible to obtain any result that is desired. Until reliable estimates of the numbers of pre-recruits and recruits are available, correlations with environmental factors or egg production will be futile.

Finally, it is important to study and analyze the behaviour of fishers during periods with both increasing and declining catches in order to measure the sizes, amounts, and species that are discarded. Such information should form an important component of management decisions.

A review of the avalanche of material dealing with the recent decline of northern cod provides few reasons for optimism about the future, since there is no consensus as to what happened, nor any evidence of a commitment to find out. Moreover, the various interest groups, from scientists to fishers to government managers, appear to have little awareness and virtually no serious interest in each other's views or concerns. They exist in a multitude of solitudes.

Thus fisheries scientists are split into factions, each holding differing views on the causes of the collapse. There is little evidence that the members of one faction read or take into account in a serious manner the views of the other factions. It is sobering to read Smith's account of the history of fisheries biology from 1855 to 1955 and to realize that the same debates on the causes of the variations in fish landings have been going on for the past 150 years.[21] Moreover, it is the conclusion of Frank and Leggett that fisheries ecology has had relatively little interchange with ecological theory, despite the massive amounts of data that have been accumulated by fisheries scientists.[22]

Further, the fisheries scientists are generally isolated from economists and other social scientists, even though there may be a slow diffusion of concepts, such as the "Tragedy of the Commons," between them. These specialists are isolated from the fishers who operate with differing technologies (inshore and offshore), and the fishers in turn may be alienated from each other. The current "sentinel fishing project" that brings inshore fishers into a working relationship with fisheries scientists may be an exception, but it involves relatively few individuals. Even this project has created conflicts between the fishers who were selected and those who were not. Its degree of success will have to be assessed, but it is evident that there exists a huge amount of mistrust and misunderstanding that must be overcome. The fisheries managers, policy-makers, and politicians are most removed from all the above groups, but how they influence the acquisition of knowledge and arrive at their decisions remains largely unexplored.

To the extent that these interest groups fail to communicate with each other and thus remain unaware or indifferent to information outside of their respective narrow spheres of interest leads to mistrust and conflict. Part of the mandate of the Fisheries Resource Conservation Council created in 1993 is to bridge these communication gaps. However, this expectation is unrealistic, since its mandate and process are restrictive. Unless the situation is corrected so that the collected wisdom of all participants is applied, it is unlikely that the management of northern cod will ever be placed on a sound foundation.

Notes

1. "If the stock assessment provides evidence of levels of spawning stock biomass likely to endanger recruitment, fishing effort (and thus fishing mortality) in the coming year will be reduced to allow immediate growth in spawning stock biomass."

2. Kozlow, J.A. 1994. "Recruitment patterns in Northwest Atlantic fish stocks." *Can. J. Fish. Aquat. Sci.* 41: 1722-1729. See also Sinclair, A. 1994. "Recent declines in cod stocks in the northwest Atlantic." NAFO SCIENTIFIC RESEARCH DOCUMENT 94/73; and Rice, J. (Affidavit, Federal Court of Canada, Trial Division, 2717-89, April 1990).

3. Milne, D.H. 1995. *Marine life and the Sea.* Wadsworth Pub. Co., Belmont, Calif.

4. It should be noted however, that the importance of spawning biomass for recruitment has recently been reaffirmed in Myers, R.A., et al. "In search of thresholds for recruitment overfishing." *I.C.E.S. Journal of Marine Science.* 51: 191-205.

5. Pope, J.G. 1982. "Background to scientific management advice on fisheries management." Laboratory Leaflet, Lowestoft No. 54.

6. Rice, J. (Affidavit, Federal Court of Canada, Trial Division, 2717-89, April 1990) "concluded from statistical calculations based upon reliable research vessel data that estimated sizes of the 1986 and 1987 year-classes are 412 million and 659 million at age 3."

7. Doubleday, W.G. (Affidavit, Federal Court of Canada, Trial Division, 2717-89, April 1990).

8. CAFSAC ADVISORY DOCUMENT 90/5.

9. CAFSAC ADVISORY DOCUMENT 91/13 and 91/13 (Revised).

10. CAFSAC RESEARCH DOCUMENT 91/53.

11. CAFSAC ADVISORY DOCUMENT 92/2.

12. CAFSAC ADVISORY DOCUMENT 92/7 and 92/7 (Revised).

13. Jake Rice quoted in Finlayson, C. 1993. *Fishing for Truth: A Sociological Analysis of Northern Cod Assessments from 1977 to 1990.* Social and Economic Studies No. 52. Memorial University of Newfoundland, St. John's.

14. These reports document a progressive reduction in the estimates for these two year-classes:
NAFO SCIENTIFIC RESEARCH DOCUMENTS 92/18 AND 92/20—"The ADAPT calibration indicated the 1986 and 1987 year-classes to be above average at 385 and 520 million fish. The 1978-91 geometric mean recruitment for this stock since 1978 is now estimated to be about 270 million fish."
NAFO SCIENTIFIC RESEARCH DOCUMENT 93/86—"The 1986 and 1987 year-classes were previously considered to be well above average. The current analyses estimate these year-classes *to be at or above the 1978-91 geometric mean (ADAPT 225 million) or below the mean (L/S 200 MILLION)*" (emphasis added).
NAFO SCIENCE COUNCIL REPORTS—June 1993—"The 1986 and 1987 year-classes are still dominating the stock although their relative abundance may be closer to the recent average for the stock rather than the previous above average estimate."
DFO ATLANTIC FISHERIES STOCK STATUS REPORTS—July 1993—The 1986 and 1987 year-classes are still dominating the stock despite being less abundant than the average for the corresponding age.
NAFO SCIENTIFIC RESEARCH DOCUMENT 94/40 — "Research survey results....suggest that the size of the 1986 and 1987 year-classes, which were originally considered to be above average size may have been well below average, particularly that for 1987.' The 1986 and 1987 year-classes

were previously considered to be well above average. The current analyses suggest that *these year-classes may have been well below the average, particularly that for 1987.*" (emphasis added).

NAFO SCIENCE COUNCIL REPORTS—June 1994—"The 1986 and 1987 year-classes were originally estimated to be strong, but subsequent analysis suggested a downward revision of the estimates such that they now appear to have been below average."

DFO Atlantic Fisheries Stock Status Report—Sept. 1994—"The 1986 and 1987 year-classes were originally estimated to be strong but subsequent analyses result in downward revisions of the estimates such that they now appear to be below average."

15. The moratorium is still in effect at the time of publication, and it may have to be maintained for many more years.

16. Munro, G.R. 1980. "A promise of abundance: Extended fisheries jurisdiction and the Newfoundland economy." Report to the Economic Council of Canada.

17. *The Report of the Newfoundland and Labrador Round Table on the Environment and Economy* (1995).

18. In press.

19. Myers, R.A., K.F. Drinkwater, N.J. Barrowman and J. W. Baird 1992. "The influence of salinity on cod recruitment in the Newfoundland region. CAFSAC Res. Doc. 92/98. Myers, R.A., K.F. Drinkwater, N.J. Barrowman and J.W. Baird 1993. Salinity and recruitment of Atlantic Cod (*Gadus morhua*) in the Newfoundland region." Can. J. Fish. Aquat. Sci. 50: 1599-1609.

20. Shelton, R.A. and D.B. Atkinson 1994. "Failure of the Div. 2J3KL cod recruitment prediction using salinity." DFO Atlantic Fisheries Research Document 94/66.

21. Smith, T. 1994. *Scaling fisheries: The science of measuring the effects of fishing, 1855-1955.* Cambridge University Press.

22. Frank, K.T. and W.C. Leggett 1994. "Fisheries ecology in the context of ecological and evolutionary theory." *Annu. Rev. Ecol. Syst.* 215: pp. 401-22.

GLOSSARY

Acipenser oxyrhyrchus - sturgeon.

admiral - title given to captain of the first ship to arrive in a harbour for the dry cod fishery.

Alosa pseudoogarengus - gaspereau.

Alosa sapidessima - shad.

American eel - *Anguilla rostrata.*

Anadromous species - species that migrate from the sea to rivers to spawn.

Anguilla rostrata - American eel.

armateur - or armador, organized victualling in the early transatlantic fishery.

artificial dryers - used for drying salted cod.

Atlantic salmon - *Salmo salar.*

aunier - principal investor and manager of a 16th-century French transatlantic fishing vessel.

bales - cured fish prepared for shipping.

ballast - weight in the bottom of a ship to maintain stability.

banked - a term used to refer to a ship's arrival at the banks.

banker - vessel fitted for fishing on the banks.

barges - small, undecked sailboats approximately 25 ft. in length and shaped like whalers, used in the 19th-century Gaspé fishery.

Basque - Spanish ship of the 16th century.

bawns - small rocky beaches, on which salted cod are spread to dry.

beach crowd - female workers of the Newfoundland bank fishery who sun-dried fish.

beach-master - a skilled hand of the 17th-century dry cod fishery who supervised the salting and drying of fish.

beachworkers - primarily women workers who cured the fish for market in the Newfoundland bank fishery.

bel - a moveable wooden gallery, overhanging the bulwarks, installed on a French banker during the Atlantic crossing, from which men fished.

blast freezing - cold air method of freezing fish.

blaze boughs - spruce branches left over from cutting timber that were gathered and used by beachwomen to make fires on the beach to cook lunch.

boatswain - (bosun) sailor who runs the deck crew under the mate.

Bonodemequiche - Mi'kmaq name for tom cod.

bourgeois - predominant shareholder in a ship of the early French transatlantic fishery.

boss woman - worker of the bank fishery, who was employed by the merchant to assemble and supervise a beach crowd.

boyard - a stretcher on the chaloupe used to carry the cod to the salters.

B.P. - Before Present. (Present reckoned to be 1950).

brine immersion - saline method of preserving fish.

brook trout - *Salvelinus fontinalis.*

bulter - also known as dory trawling. System of fishing introduced by the French employing a longline to which short lines with baited hooks were attached at regular intervals.

bultow - variation of French *bulter.* A long line to which short lines with baited hooks were attached at regular intervals.

Canadiens - migrant seasonal workers from Quebec who lived in the Gaspé for only four or five months of the fishing season in the early-19th century.

cantle - see quintal.

caplin - a small pelagic fish rarely longer than 8 or 9 in. that swims in large schools in the north Atlantic.

carniau - chute on a French banker that took the fish from the cutting table directly to the hold.

Catadromous species - species that migrate down river to spawn.

catcher boat - small boat that supplies a mother ship with fish to process.

chaloupe - a small four or five ton fishing boat, about 10 m. in length that usually had one small sail.

Coaker Regulations - legislation introduced in 1920 by William Coaker, Minister of Marine and Fisheries of Newfoundland, to standardize the cure and regulate the prices of saltfish.

cod - *Gadus morhua.*

core fishery - a fishery made up of people with two or more valid fishing licenses.

Crassostrea virginica - oyster.

cuirier - a large leather apron treated with grease and oil worn by the lignotiers in the green cod fishery.

cull - the method by which fish were graded for market. Also refers to the actual grades into which fish were sorted.

cullage - an inferior grade or cure of dried and salted codfish.

doors - heavy weights attached to otter trawl cables, which act to keep the mouth of the trawl open.

dragger - any of various motorized trawlers, usually exceeding 100 ft. in length.

dry cod fishery - inshore fishery conducted from boats that returned to shore each day. The fish was headed, gutted, split, salted, and then dried on beaches or flakes.

dun - term used to describe fish when it received too much sun and was unsuitable for the export market.

esclipot - a case in a chaloupe used to hold cod. The *esclipot* had a sloped bottom which could slide to one side allowing the cod to slip into the boyard.

étesteur - see "header."

ethnography - writing about a people or a culture.

factory trawler - a vessel that both catches and processes fish. The factory trawler is distinguished from the "mother" ship because the "mother" ship only processes the catch from catcher boats.

faggot - term describing piles of dried fish that had been taken up from the beach or flake.

Fisheries Research Board - an independent Canadian biological research organization founded in 1937.

fish book - book in which the quintals of culled fish are recorded.

fishing room - area of shore occupied by migratory fishing station.

fishmaids - a bank fishery term used to describe women who worked at drying fish.

fish stage - a jetty extending into the water on which a shed was erected and where initial processing of fish occurred.

flakes - open, elevated wooden platforms resting on stakes that were driven into the ground for support. Layers of spruce boughs were often placed over the flakes before fish were laid on them to dry.

flounder - *Pseudopleuronectes americanus.*

fonds mort - a fund set up by 19th-century Gaspé merchants that came from any profit that a fisher obtained after a good season. It was carried over on the merchants' books as a guarantee against future advances.

freshers - a British term for ice-filled side trawlers.

Gadus morhua - cod.

gaffer - overseer or boss of a group of workers.

gantry mast - tall, two-legged mast used for heavy lifting.

Gaspé cure - dried cod produced along the Gaspé coast.

gaspereau - *Alosa pseudoogarengus.*

gill net - vertical fishing net that catches fish by the gills as they attempt to swim through the webbing.

ghost nets - lost or abandoned gill nets.

grand maison - the owner's or company agent's home, in a Gaspé fishing plant.

green cod fishery - offshore fishery on the Grand Banks. The fish was salted on board and taken back to Europe for sale. Cod were fished from the decks of ships using hook and line or jiggers.

green fish - heavily salted, undried fish of the bank fishery.

grey seal - *Halichoerus grypus.*

grommets - French term for unskilled apprentices in the dry cod fishery.

Halichoerus grypus - grey seal.

handline - hemp line with a lead sinker at the end to which two snoods and their hooks were attached for catching cod.

handline fishery - fishery using handlines.

hardtack biscuit - a hard saltless biscuit used as rations in the migratory fishery.

header - codfish handler who cut the head off the cod.

hogshead - a barrel used to hold fish.

Homarus americanus - lobster.

industrial fishing - the capture and conversion of abundant small fish to fishmeal and oil.

inshore fishery - collective name for any coastal fishery, especially for cod, utilizing small boats for catching the fish.

javelles - piles of cod in the form of haystacks.

jigger - unbaited, weighted hook(s) used with a line to catch fish by giving a sharp, upward jerk.

killick - hand-made anchor.

kingpost - cargo mast usually associated with freighters.

lapstrake - the hull of a boat built with overlapping strakes, clinker-built.

Laws of Oleron - 12th-century international laws of the sea.

leister - three-pronged fish spear used as a hunting weapon during the Ceramic Period.

lignotier - fisher of the green cod fishery who actually caught the fish.

lobster - *Homarus americanus*.

Loligo paeleii - squid.

longers - stakes driven into the ground to support wooden platforms for flakes.

longliner - decked boat 35-65 ft. long that initially used longlines and employed mechanical haulers and shooting gear. Subsequently adapted to use gill nets and lobster pots. Also name given to fishers who serve on such vessels.

longlining - fishing with a long central line to which hundreds or even thousands of hooked snoods are attached.

loury - cloudy.

making fish - drying salted cod.

Mallotus villosus - smelt.

manigots - leather mitts worn by *lignotiers.*

Maritimes - the maritime provinces of Canada: Nova Scotia, New Brunswick, and Prince Edward Island.

master - individual who represented the crew in the early transatlantic fishery.

merchantable cod - high quality cod.

Microgadus **tom cod** - tom cod.

middens - accumulations of discarded shells of mollusks including clams, quahogs, mussels, and occasionally oysters and scallops. Prehistoric shell middens are associated with the Ceramic Period.

migratory fishery - European fishery to northwest Atlantic, originating in early-16th century.

mooring - an anchor with an attached buoy used to hold fishing nets in place.

mother fish - "bottom fish," larger fish that live primarily on the bottom of the ocean that are left by handline and trap fishers to spawn and breed.

mother ship - a large processing ship that processes catches from smaller catcher boats.

mixed fishery - French method of obtaining cod in the 18th century. Ships would fish for cod on the banks for several weeks or months. The cod would be split and salted on board but would be brought to land for drying.

mizzen sail - a fore-and-aft sail set on a mizzenmast, also known as a culetin, ring-tail, crossjack, or spanker.

MSY - maximum sustainable yield; the maximum level at which a stock could be fished and still retain the ability to replenish itself.

nautier - ship's boy who detached the fish's air bladder.

NIFA - Newfoundland Inshore Fisheries Association, founded in 1986.

outport - any fishing community outside a regional centre.

oyster - *Crassostrea virginica.*

piquoir - a long wooden gaff.

pitot - a kind of large clam.

pot de vin - advance payment received by Breton fishers in the 18th century, and possibly earlier.

pounds - term used to describe wooden crates (10-12 sq. ft.) with openings along the bottom used to wash fish after it had taken the salt, and prior to spreading the fish to dry.

pre-contact fishing - the fishery undertaken by aboriginals in North America before European contact.

Pseudopleuronectes americanus - flounder.

purse seiner - a ship that uses a seine in the shape of a purse to encircle schools of fish.

quintal - weight measurement equivalent to 112 lb./50.8 kg; hundredweight; cantle.

refuse - fish damaged during curing.

regalos - a Spanish term for personal supplies.

roadstead - a nautical road, shipping lane, or entrance to a port.

Roccus saxatilis - striped bass.

round fish - fish that have not been gutted or cleaned.

rum - a space dug in salt to place the cod in; a compartment in a *chaloupe.*

sack - ship that specialized in carrying dried fish to market.

saline - salt house or shed (French).

Salmo salar - Atlantic salmon.

Salvelinus fontinalis - brook trout.

saltbulk phase - phase of fish processing in which split cod are salted and stacked prior to air-drying.

salter - fish handler who applies salt to the split cod and arranges it in layers.

saque - see "sack."

schooner - large wooden sailing vessel with fore and aft rigging.

scull - arrival of the spawning caplin when they cast their milt and roe in the surf and it is washed ashore. Also known as the "caplin rolling."

seine - a fishing net that hangs vertically in the water with floats at the top and sinkers at the bottom.

seine club - group of fishers who joined together to purchase large seine nets to fish for sand eel.

seine master - one who heads the seine club.

seiner - a boat used for seine fishing.

shad - *Alosa sapidessima.*

side trawler - a ship that lowers fishing trawl nets over the side, rather than through the stern.

skiff - a small, partly-decked boat, utilizing oars or a small sail, and used in the inshore fishery.

sloop - small sailing vessel.

smelt - *Mallotus villosus.*

snood - very short secondary line to which a hook is fastened.

splitter - fisher who would open up the cod with a cleaving knife and remove the bones and part of the spine.

splitting strop - doubling becket, bull rope, or bag becket, a noose of cable laced into the girth of the cod end used to tighten trawl nets when hauling them in to contain the cod.

sprit - a small pole or spar crossing a fore-and-aft sail diagonally.

sturgeon - *Acipenser oxyrhyrchus.*

squid - *Loligo paeleii.*

squid jigs - small lead weights painted red with 20 or more sharpened hooks for catching squid.

stage - see fish stage.

stagehead - a sort of wharf extending from the stage where fishing boats unloaded their catch.

steaming time - time expended in getting to and from distant fishing grounds.

stern trawler - a ship that uses a stern ramp or slipway to haul trawls in and let them out via the rear of the ship rather than over the side.

stower - person who stows the fish in the hold.

straddling stocks - fish stocks that straddle 200-mile limits.

strake - plank of a ship's hull.

striped bass - *Roccus saxatilis.*

sturgeon - *Acipenser oxyrhyrchus.*

sun bonnet - an ornately crafted hat that served to protect a beachwoman's face and neck from the sun and wind. Also known as sunshade, skully, scolly, slouch, thrum cap, and Dolly Varden.

sunker - rock that appears at low tide.

TAC - total allowable catch, a method of conservation that has dominated management of the fish stocks since the 200-mile limit was enacted.

tail grabber - Model 59 Schwanzlaufer, complex filleting machine used on factory trawlers that grasps the fish's tail in a vise-like grip, measures it size, and rapidly slices out the fillet with revolving knives.

tally - see fish book.

tarpaulin - old sail canvas.

throater - worker who cuts the cod's throat and slits open its belly.

tidewaiter - a customs inspector who boarded ships at dockside.

tierces - a measure of capacity equivalent to 42 wine gallons.

tom cod - *Microgadus* tom cod, small cod, Bonodemequiche.

tonguer - also known as the *élangueur*, an iron stake onto which the *lignotier* would hook the cod.

trap fishery - a fishery that uses traps instead of lines to catch fish.

trawl, trawl line - a strong fishing net that is dragged along the seabed and held open by trawl doors. Also a buoyed line with numerous short lines with baited hooks.

trawler - a large ship using trawl nets.

trawl doors - otter boards, "parotters," used for spreading trawl nets apart as they are pulled across the seabed to catch fish.

trouncing - a method of baiting used by handline and trap fishers by which iron rings were attached to a rope and dragged along the bottom of the ocean.

truck - a form of payment between fishers and merchants in which there was a cashless barter of fish for goods.

ulu - semi-lunar knife associated with the ground stone technology of the Pre-Ceramic Period.

victuals - provisions.

waterhorse - stack of split codfish piled to drain after salting.

wattle - fir posts, woven with boughs, sealed on the inside with fir rinds and roofed with rinds and turf or a sail.

winch - machine used for hauling cable or lifting.

wrapper apron - customarily worn by the beachwomen of the bank fishery, made from a potato sack material.

yaffles - small armloads of fish.

year-class - an age designation for fish spawned in one year, used for determining the quantity of biomass or stock in a fishery.

CONTRIBUTORS

DARLENE ABREU-FERREIRA completed her PhD in History at Memorial University of Newfoundland. Her dissertation was titled "The Cod Trade in Early-Modern Portugal: Deregulation, English Domination, and the Elimination of Female Cod Merchants." She has received several post-doctoral fellowships and is affiliated with the University of Toronto working on the role of women in early-modern Maritime communities.

RAOUL ANDERSEN, a social anthropologist at Memorial University, St. John's Newfoundland, has studied Canadian and other fisheries for nearly 30 years. His early works include *North Atlantic Fishermen, North Atlantic Maritime Cultures*, "The Need for human sciences in Atlantic Coast Fisheries," "Millions of Fish," and "'Chance' and 'Share.'" His future work includes a Newfoundland deep sea fisher's biography and a collaborative volume on healthways in Newfoundland's coastal communities earlier in this century.

SHEILA ANDREW teaches Acadian history and Canadian women's history at St. Thomas University in Fredericton. Her publications include articles on 19th-century Acadians and a forthcoming book, *The Development of Acadian Elites in New Brunswick, 1861-1881* (McGill-Queen's University Press).

A graduate of Memorial University of Newfoundland and the University of Western Ontario, **MELVIN BAKER** has written extensively on 19th- and 20th-century Newfoundland history. He is currently writing an official history of Memorial University of Newfoundland.

B.A. BALCOM is a historian and military curator with Parks Canada at the Fortress of Louisbourg National Historic Site. His publications include *The Cod Fishery of Isle Royale* and *History of the Lunenburg Fishing Industry*.

RAYMOND B. BLAKE, the director of the Centre for Canadian Studies at Mount Allison University, is the author of *Canadians at Last: Canada Integrates Newfoundland as a Province.*

CYNTHIA BOYD is a folklorist and lecturer at Memorial University. She has worked on the Seamen's Museum Project in Grand Bank, Newfoundland, and her publications include "'Just Like One of the Boys'" in *Undisciplined Women*.

JEAN-FRANÇOIS BRIÈRE specializes in French colonial history, and publishes extensively on the history of the French at Newfoundland. His latest book is an annotated translation of the Abbe Gregoire's "De la Litterature des Negres" (University of Massachusetts Press, 1996). He currently chairs the Department of French Studies at the State University of New York in Albany.

JAMES E. CANDOW is a project historian with Parks Canada in Halifax, N.S. His publications include *Of Men and Seals: A History of the Newfoundland Seal Hunt*.

CAROL CORBIN is assistant professor of communication at the University College of Cape Breton. Her research areas include the politics of small communities and environmental issues.

STEPHEN A. DAVIS (DPhil Oxford) is a member of the Department of Anthropology at Saint Mary's University. He has published numerous works on the early history of Atlantic Canada and its Native peoples.

D.A. MACDONALD is a social anthropologist by training. He was awarded his doctorate by Simon Fraser University in 1988 for a study of Newfoundland fish merchants Newman and Company. He is now a consultant in St. John's.

BARBARA NEISS is a member of the sociology faculty at Memorial University of Newfoundland. Her publications include "What are we managing anyway? The need for an interdisciplinary approach to managing fisheries ecosystems" that appeared in *Dalhousie Law Journal* and "Female Fish Processing Workers' Occupational Health and the Fishery Crisis in Newfoundland and Labrador" in *Chronic Diseases in Canada*.

PETER E. POPE teaches historical archaeology in the Department of Anthropology at Memorial University of Newfoundland. He is currently revising publications on the 17th-century English settlement of Newfoundland and on public memory of the voyages of John Cabot.

SHANNON RYAN, PhD, Professor of History, Memorial University of Newfoundland is author of *Fish out of Water: The Newfoundland Saltfish Trade, 1814-1914* and *The Ice Hunters: A History of Newfoundland Sealing to 1914*.

Attached to Parks Canada in Quebec City since 1975, **ROCH SAMSON** has published *Fishermen and Merchants in the 19th Century Gaspé* for which he obtained the Michel-Brunet award from the Institut d'histoire de l'Amérique française in 1985. Recently, in 1996, as scientific director attached to Quebec's Institute national de la recherche scientifique (INRS - Culture et Société), he has published *Histoire de Lévis-Lotbinière* which is part of a collection on the history of the Quebec regions. He has also published studies on the Forges du Saint-Maurice National Historic Site, and is currently editing a book on the history of the site.

DONALD HAROLD STEELE is a graduate of McGill University and is currently Professor in the Department of Biology, Memorial University, St. John's, Newfoundland. He is the author of over 100 scientific contributions.

An historian, **ROBERT SWEENY** lives with Elizabeth-Anne Malischewski and their daughter Charlotte in St John's, Newfoundland.

JUDITH TULLOCH (M.A. Dalhousie 1972) has worked as a historian for Parks Canada since 1968. Major research projects have included the restoration of both Province House National Historic Site in Charlottetown, PEI, and Cape Spear lighthouse at Cape Spear National Historic Site in St. John's, Newfoundland.

WILLIAM W. WARNER is the Pulitzer Prize-winning author of *Beautiful Swimmers: Watermen, Crabs and the Chesapeake Bay.* His *Distant Water: The Fate of the North Atlantic Fisherman* was nominated for the National Book Critics Circle.

MIRIAM CAROL WRIGHT holds a PhD in History from Memorial University of Newfoundland. She is studying the history of fisheries development policies in Newfoundland.